29th Winter Workshop on Nuclear Dynamics

(WWND2013)

Journal of Physics: Conference Series Volume 458

Squaw Valley, California, USA
3-10 February 2013

ISBN: 978-1-62993-096-1
ISSN: 1742-6588

Printed from e-media with permission by:

Curran Associates, Inc.
57 Morehouse Lane
Red Hook, NY 12571

CURRAN ASSOCIATES INC.
proceedings
.com

Some format issues inherent in the e-media version may also appear in this print version.

Printed by Curran Associates, Inc. (2013)

For permission requests, please contact the Institute of Physics
at the address below.

Institute of Physics
Dirac House, Temple Back
Bristol BS1 6BE UK

Phone: 44 1 17 929 7481
Fax: 44 1 17 920 0979

techtracking@iop.org

Additional copies of this publication are available from:

Curran Associates, Inc.
57 Morehouse Lane
Red Hook, NY 12571 USA
Phone: 845-758-0400
Fax: 845-758-2634
Email: curran@proceedings.com
Web: www.proceedings.com

TABLE OF CONTENTS

Journal of Physics: Conference Series **458** (2013) 011001 doi:10.1088/1742-6596/458/1/011001

Preface

These are the proceedings of the 29th Winter Workshop on Nuclear Dynamics, which was held in Squaw Valley, California, USA in February 2013. As in previous years, the unique character of this conference series has allowed us to bring together nuclear scientists, with very different interests, to discuss recent progress and scientific achievements. A total of 63 selected talks were presented at WWND-2013. The proceedings of 31 of these talks can be found here. The topics ranged from novel ideas for hydrodynamics modeling to the very exciting first results from the p-Pb runs at the LHC.

On the experimental side, the RHIC experiments were highlighted in contributions summarizing the beam energy scan program as well as the first results from the U-U collision program. The LHC experiments presented new results, in particular in the heavy flavor and quarkonia sector. On the theory side, the most recent continuum extrapolated results from lattice QCD were complemented with a wide variety of phenomenological approaches and their latest progress in successfully modeling relativistic heavy ion collisions. These proceedings of the 29th Winter Workshop on Nuclear Dynamics provide a snapshot of the status of the field. The articles, many of which were written by some of the most promising young scientists in the field, are documenting the excitement and achievements that are characteristic for modern day nuclear science.

Rene Bellwied
University of Houston

Journal of Physics: Conference Series **458** (2013) 011002 doi:10.1088/1742-6596/458/1/011002

Peer review statement

All papers published in this volume of *Journal of Physics: Conference Series* have been peer reviewed through processes administered by the proceedings Editors. Reviews were conducted by expert referees to the professional and scientific standards expected of a proceedings journal published by IOP Publishing.

Journal of Physics: Conference Series **458** (2013) 012001 doi:10.1088/1742-6596/458/1/012001

STAR Results from the RHIC Beam Energy Scan

Hui Wang for the STAR collaboration

RM 1-174, Building 510A, Brookhaven National Laboratory, Upton, NY, 11973

E-mail: wanghui6@bnl.gov

Abstract.

The collision of two gold nuclei at top RHIC energy ($\sqrt{s_{NN}}$=200 GeV) creates a new phase of matter, the quark-gluon plasma (QGP). The QGP exists at very high temperatures, T, and low baryo-chemical potentials, μ_B. Both lattice QCD and experimental data indicate this transition from hadronic matter to Quark Gluon Plasma is an analytical transition (cross-over). On the other hand, systems formed at larger values of the baryo-chemical potential may undergo a first-order transition.Thus there can be an end point of the first-order transition on the QCD phase diagram, the critical point.

To study the QCD phase boundary and search for the possible QCD critical point, RHIC launched a Beam Energy Scan Program. In 2010 and 2011, we collected data at $\sqrt{s_{NN}}$=7.7, 11.5, 19.6, 27, 39 and 62.4 GeV, which covers a wide range of baryon chemical potential from μ_B 420 to 40 MeV. In this article, we will report the latest results of the Beam Energy Scan Program from the STAR collaboration.

1. Introduction

Current results from RHIC and LHC indicate the existence of a deconfined Quark Gluon Plasma (QGP) phase at high energy in A+A collisions. One major challenge, however, is to understand the structure of the QCD phase diagram. If the temperature is high and μ_B is relatively small, both lattice QCD and experimental data indicate this transition from hadronic matter to Quark Gluon Plasma is an analytical transition (cross-over) [1], while some theoretical calculations predict that the transition at lower temperatures and high μ_B is a first order phase transition [2]. If a phase transition exists at higher μ_B, with a cross-over at $\mu_B = 0$, the phase transition would end in a critical point at finite μ_B. However, due to the difficulty of lattice QCD calculations at finite μ_B, accurate predictions of the critical point location are still lacking [3]. Therefore it falls to experiment to search for traces of the existence of the critical point of QCD.

To further explore the QCD phase diagram, a Beam Energy Scan (BES) proposal was made by the STAR Collaboration [4], which aims to search for the turn-off of QGP signatures, signals for first order phase transition and the critical point. The first phase of the BES program was started in 2010 with collisions recorded at $\sqrt{s_{NN}}$ = 7.7, 11.5, 39 GeV and finished in 2011 with collisions at $\sqrt{s_{NN}}$=19.6 and 27 GeV. In this paper, a few selected results from the STAR BES program will be discussed.

2. Accessing the Phase Diagram

One important measurement related to the QCD Phase Diagram is the chemical freeze-out and kinetic freeze-out parameters from different collision energies and centralities. If we assume thermodynamical equilibrium, a statistical thermal model (THERMUS)[5] can be used to determine the chemical freeze-out temperature $T_{\rm ch}$ and baryon chemical potential μ_B. At STAR, the THERMUS model was used to fit mid-rapidity particle ratios including yields of π, K, p, $K^0{}_S$, Λ and Ξ measured in Au+Au collisions at $\sqrt{s_{NN}}$= 7.7, 11.5, 39 and 200 GeV [6]. Figure 1 shows centrality and energy dependence of the extracted chemical freeze-out parameters using the Grand-Canonical Ensemble (GCE) approach of THERMUS, while the curves represent different parametrizations [7, 8]. A large μ_B range of 20 to 400 MeV is covered with the BES energies.

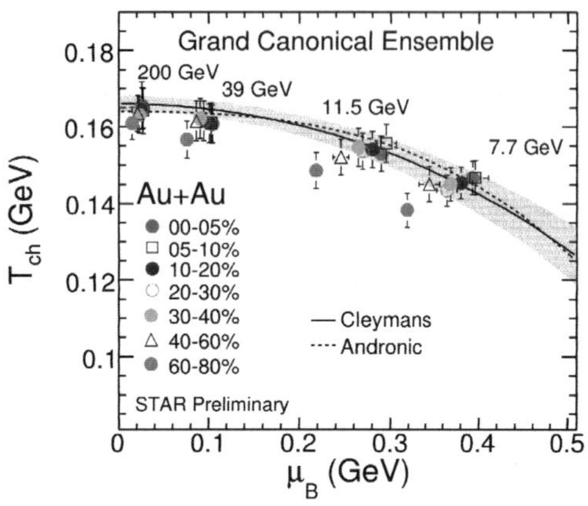

Figure 1. Centrality and energy dependence of the extracted chemical freeze-out parameters.

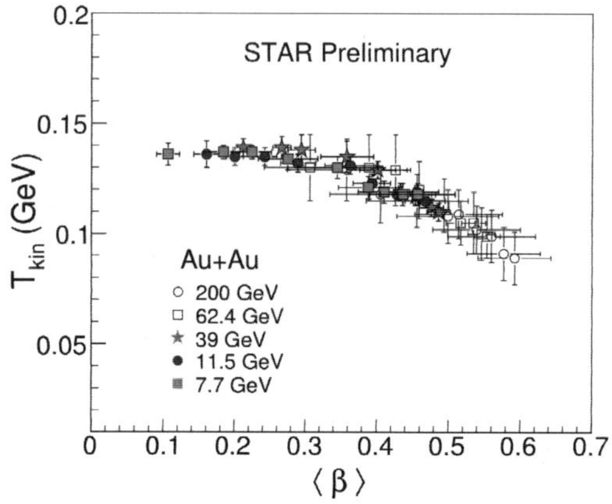

Figure 2. Centrality and energy dependence of the extracted kinetic freeze-out parameters.

The kinetic freeze-out is the point where all elastic collisions among the particles stop. By doing the simultaneous fits of π, K, and p transverse momentum spectra to a blast-wave model, one can determine the kinetic freeze-out temperature T_{kin} and average flow velocity $\langle\beta\rangle$. Figure 2 shows the variation of $\rm T_{kin}$ as a function of $\langle\beta\rangle$ at $\sqrt{s_{NN}}$= 7.7, 11.5, 39, 62.4 and 200 GeV. The $T_{\rm kin}$ decreases with increasing collision centralities and energy, while the $\langle\beta\rangle$ increases with increasing collision centralities and energy.

3. Search for Turn-off of QGP Signatures

3.1. the Balance Function

The balance functions, which measure the correlation between the opposite sign charge pairs, are sensitive to the mechanisms of charge formation and the subsequent relative diffusion of the balancing charges [9]. Due to conservation laws like electric charge conservation, particles and their anti-particles are pair produced and correlated initially in coordinate space, if a delayed hadronization occurs, the lower temperature and less expansion and diffusion will result in a narrower charge balance function. It has been reported that the balance function for $\Delta\eta$ narrows at top RHIC energies [10]. Thus the balance function could be used to probe the evolution of the system hadronization time vs. energy and search for possible turn-off of QGP at lower energies.

Figure 3 shows the balance function in terms of $\Delta\eta$ for all charged particles. The most central events (0-5%) are shown for seven incident energies. The data in the figure are the balance function results from real data corrected by subtracting the balance function calculated using mixed events. We can see that, for all the energies shown here, the balance functions from data are narrower than the ones from shuffled events. To quantify the narrowing of balance function,

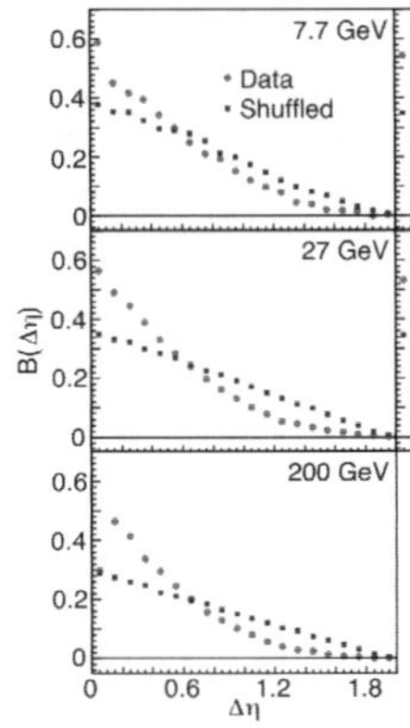

Figure 3. The balance function in terms of $\Delta\eta$ for all charged particles. Central events (0-5%) are shown here with $\sqrt{s_{\rm NN}}$ from 7.7 to 200 GeV.

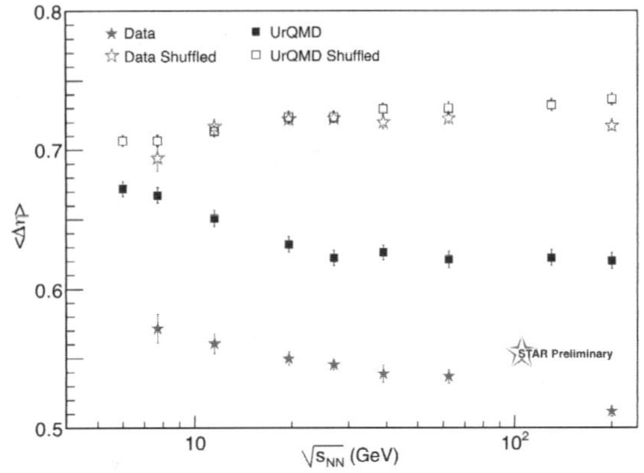

Figure 4. Energy dependence of the balance function width $\langle\Delta\eta\rangle$ for central Au+Au collisions (0-5%) compared with shuffled events. Both data and UrQMD calculations are shown here.

figure 4 shows the energy dependence of the balance function width for central Au+Au collisions. The data show a smooth decrease of $\langle\Delta\eta\rangle$ with increasing energy. UrQMD calculations predict a similar trend but over predict the observed results. Since the balance function is sensitive to the hadronization time and relative diffusion after hadronization, this decrease in balance function width could mean a longer lived QGP phase at higher energies. The UrQMD model is a hadronic model that does not have a deconfined phase transition and has little flow. This early hadronization time combined with strong interaction between final particles leads to a wider balance function in UrQMD. In the same figure, the shuffled events from both data and UrQMD show a wider balance function that slightly increases with increasing energy. Since the shuffled events represent the widest balance function within STAR's acceptance, the change of the balance function calculated using shuffled events is due to the slight changes in STAR's η acceptance with energy.

3.2. Elliptic Flow

Elliptic flow is the second harmonic coefficient of the Fourier expansion

$$\frac{dN}{d\phi} \propto 1 + 2\sum_{n\geq 1} v_n \cos\left[n(\phi - \Psi)\right] \tag{1}$$

where ϕ is the azimuthal angle of the particles and Ψ is the reconstructed event plane azimuthal angle. Elliptic flow is generated by the initial pressure gradient created by non-central heavy ion collisions. One major evidence that a deconfined quark gluon plasma is produced in Au+Au collisions at $\sqrt{s_{NN}} = 200$ GeV is the number-of-constituent quark (NCQ) scaling of v_2 versus transverse momentum p_T for hadrons at intermediate p_T (2 to 5 GeV/c) [11, 12]. Deviations from such a scaling at lower beam energies could be an indication for the absence of the deconfined phase [13].

Figure 5 shows the differences in v_2 between particles X (p, Λ, Ξ^-, π^+, K^+) and corresponding anti-particles $\overline{X}$ ($\bar{p}$, $\overline{\Lambda}$, $\overline{\Xi}^+$, π^-, K^-) with $\sqrt{s_{NN}}$. Larger v_2 values are found for particles than for antiparticles, except for pions for which the opposite ordering is observed. The difference increases with decreasing beam energy and is larger for baryons compared to mesons [14].

As discussed previously, the universal NCQ scaling of v_2 at $\sqrt{s_{NN}} = 200$ GeV suggests strongly interacting partonic matter is produced. The observed difference in v_2 at lower beam energies demonstrates that this common NCQ scaling of particles and anti-particles splits. Such a breaking of the NCQ scaling could indicate increased contributions from hadronic interactions in the system evolution with decreasing beam energy, or could be related to the larger values of μ_B.

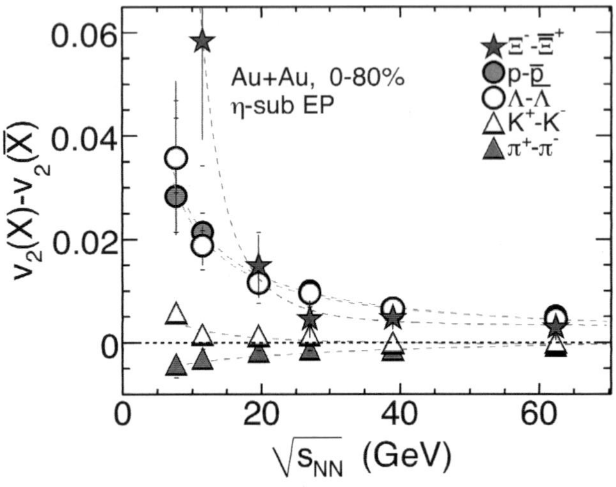

Figure 5. The difference in v_2 between particles (X) and their corresponding anti-particles ($\overline{\mathrm{X}}$) (see legend) as a function of $\sqrt{s_{NN}}$ for 0–80% central Au+Au collisions. The dashed lines in the plot are fits with a power-law function [14].

4. Search for Critical Point

4.1. Particle Ratio Fluctuations

The energy dependence of particle-ratio fluctuations is also an interesting topic. Enhanced fluctuations are one of the possible signatures of a phase transition near a critical point. The observable ν_{dyn} for kaons and pions can be written as

$$\nu_{\mathrm{dyn},K\pi} = \frac{< K(K-1) >}{< K >^2} + \frac{< \pi(\pi-1) >}{< \pi >^2} - \frac{2 < K\pi >}{< K >< \pi >} \tag{2}$$

Figure 6 shows the $\nu_{\mathrm{dyn},K\pi}$ results for 7.7 - 200 GeV [15, 16, 17]. STAR results are approximately independent of collision energy. This disagrees with NA49's results, which show a strong increase with decreasing incident energy. The same figure also shows model calculations. The points labeled STAR UrQMD represent UrQMD calculations with STAR acceptance cuts, which show little energy dependence and over predict the magnitude of the data. The HSD model predicts increased fluctuations at low energies and agrees with the NA49 measurements at the lowest energies but over predict the data at higher energies. None of the models presented here can fully describe the incident energy dependence of the data.

Unlike the results for K/π fluctuations, the results for p/π fluctuations are affected by resonance correlations (e.g. Δ, Λ, Σ all decay to p, π). These correlations increase the cross-correlation terms of ν_{dyn} and produce a negative ν_{dyn} value. Figure 7 shows the incident energy

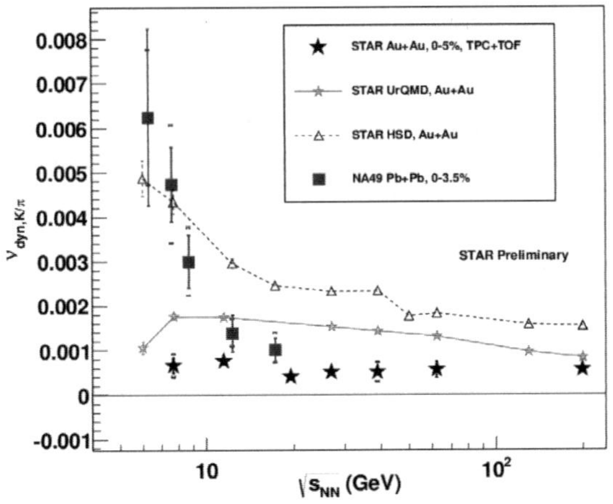

Figure 6. Energy dependence of K/π fluctuations. Only central events are shown here (0-5% for STAR Au+Au collisions,0-3.5% for NA49 Pb+Pb collisions). UrQMD and HSD calculations are also shown.

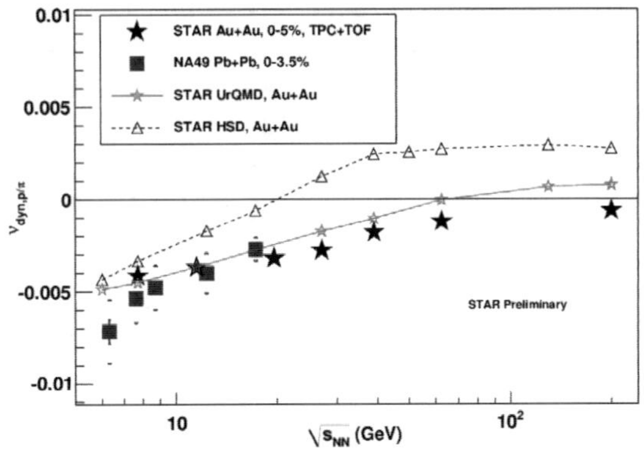

Figure 7. Energy dependence of p/π fluctuations expressed as $\nu_{\mathrm{dyn,p}/\pi}$. Only central events are shown here (0-5% for STAR Au+Au collisions,0-3.5% for NA49 Pb+Pb collisions). UrQMD and HSD calculations are also shown.

dependence of $\nu_{\mathrm{dyn},p\pi}$. The STAR and NA49 results for p/π fluctuations show good agreement. They are both negative and increase with increasing collision energy. The UrQMD model describes the data well at SPS energies, which supports the resonance correlations interpretation because UrQMD is a hadronic transport model. However, UrQMD becomes positive and over predicts the data at higher energies.

p/K fluctuations, which are related to baryon-strangeness correlations, can be used as a tool to study the deconfinement phase transition. Figure 8 shows the incident energy dependence of $\nu_{\mathrm{dyn},Kp}$ results. The STAR data show a smooth decrease with decreasing collision energy and disagree with NA49 data at 7.7 GeV. Further study is still needed to understand the differences between the two experiments. A UrQMD calculation with the STAR acceptance filter is also shown in the same figure. UrQMD always over predicts fluctuations and becomes positive at high collision energies. The HSD model always predicts positive ν_{dyn} results.

4.2. p_t Fluctuations

The p_t fluctuations could also serve as a signal for the QCD critical point or the occurrence of thermalization and collectivity [18, 19]. One observable for the event-by-event two particle momentum correlation is defined [20] as

$$\langle \Delta p_{t,i} \Delta p_{t,j} \rangle = \frac{1}{N_{\mathrm{event}}} \sum_{k=1}^{N_{\mathrm{event}}} \frac{C_k}{N_k(N_k-1)}, \tag{3}$$

where

$$C_k = \sum_{i=1}^{N_k} \sum_{j=1, i\neq j}^{N_k} \left(p_{t,i} - \langle\langle p_t \rangle\rangle\right)\left(p_{t,j} - \langle\langle p_t \rangle\rangle\right). \tag{4}$$

Figure 9 shows the incident energy dependence of p_t correlations. The STAR data shows a rapid increase from 7.7 to 62.4 GeV and then little energy dependence up to 2.76 TeV. UrQMD

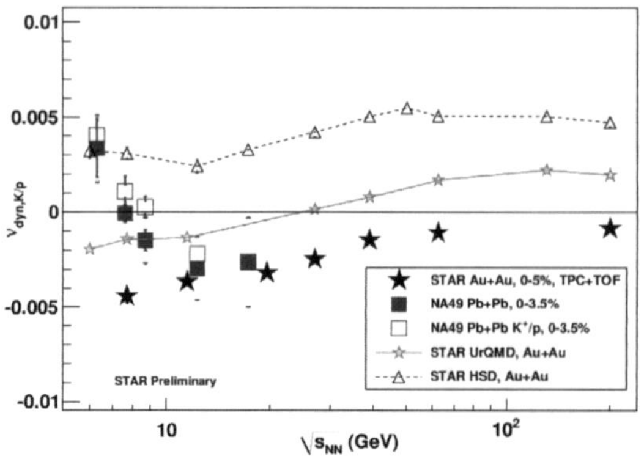

Figure 8. Energy dependence of p/K fluctuations. Only central events are shown here (0-5% for STAR Au+Au collisions,0-3.5% for NA49 Pb+Pb collisions). UrQMD and HSD calculations are also shown.

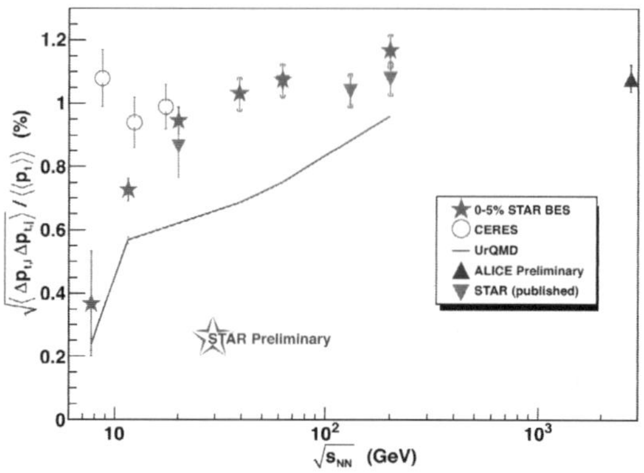

Figure 9. $\sqrt{\langle \Delta p_{t,i} \Delta p_{t,j} \rangle} / \langle\langle p_t \rangle\rangle$ as a function of incident energy. Only central events from STAR, ALICE and CERES are shown here. UrQMD calculations are also shown in the same figure.

shows a similar increasing trend but under predicts the measured correlations. The CERES data deviates from STAR at lower energy. Effects due to different experimental acceptances are still under investigation.

5. Search for First Order Phase Transition

One important goal of the STAR BES program is to search for the evidence of a first order phase transition. The HBT technique can be used to determine the freeze-out eccentricity ε_F. A non-monotonic behavior of ε_F as a function of energy could indicate a soft point in the equation of state. Figure 10 shows the excitation function of the freeze-out eccentricity. The combined E895 and STAR data shows a smooth decrease of ε_F with energy. Also, the UrQMD model reproduces both E895 and STAR data. Overall, no non-monotonic behavior is observed in ε_F.

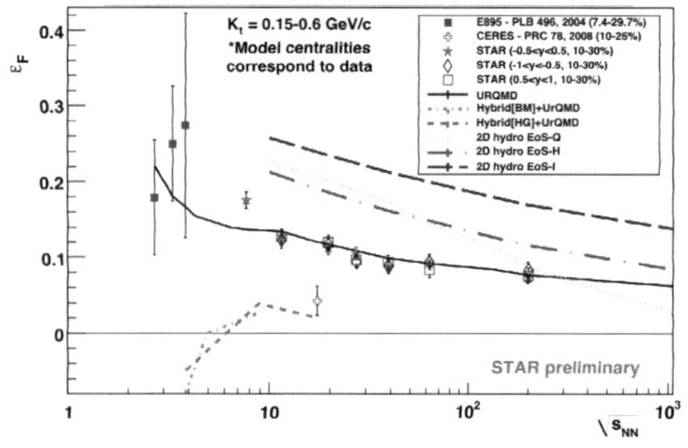

Figure 10. Freeze-out eccentricity, ϵ_F , as a function of $\sqrt{s_{NN}}$ for data and models [22].

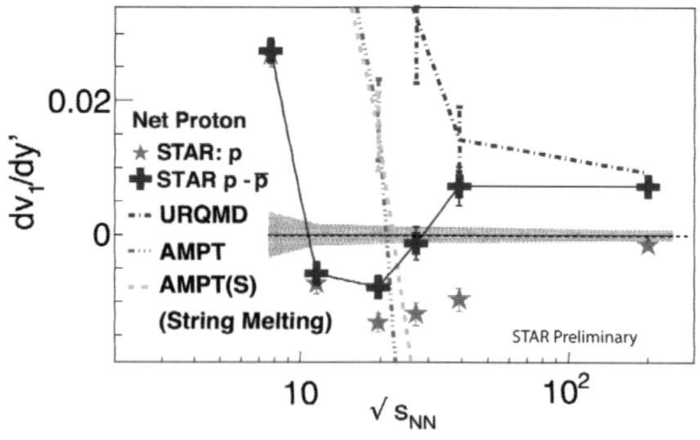

Figure 11. Directed flow slope (dv_1/dy') near mid-rapidity as a function of beam energy for mid-central(10-40%) Au+Au collisions, where the primed quantity y' refers to normalized rapidity $y/y_{\rm beam}$ [23].

Directed flow, which measures the "side-splash" motion of the collision products, is sensitive

Collider mode $\sqrt{s_{NN}}$ (GeV)	Fixed-target mode $\sqrt{s_{NN}}$ (GeV)	μ_B (MeV)
19.6	4.5	585
15	4.0	625
11.5	3.5	670
7.7	3.0	720
5	2.5	775

Table 1. Summary of collider mode $\sqrt{s_{NN}}$, fixed target mode $\sqrt{s_{NN}}$, and corresponding μ_B values [24].

to the equation of state (EOS) and hence can be considered a first order phase transition signal [21]. Figure 11 shows the energy dependence of directed flow slope (dv_1/dy') near mid-rapidity for 10-40% central Au+Au collisions. The v_1 slope for net protons is calculated via the relation $F = rF_{\bar{p}} + (1-r)F_{\mathrm{transp}}$, where r is the observed ratio of antiprotons to protons among the analyzed tracks. The net proton v_1 slope changes sign twice and shows a minimum at $\sqrt{s_{NN}}$ =10 to 20 GeV. This result is qualitatively different from UrQMD and AMPT transport models, which both predict a monotonic trend throughout $\sqrt{s_{NN}}$ =7.7 to 200 GeV [23]. Further studies are needed to understand the current results and their implications for the Equation of State.

6. BES Phase II

As discussed above, The BES Phase I program from STAR has some interesting results. However, there are several other observables such as higher moments of net-protons distributions and ϕ-meson v_2, which show some hints but are not statistically proficient enough to draw quantitative conclusions. To confirm these results, we need more statistics at lower energies. Another energy point at around 15 GeV is also necessary to fill the large gap in μ_B between 11.5 and 19.6 GeV. To address these problems, STAR has proposed a BES phase II program, which requests more data at these energies. With the help of electron cooling and longer bunches, we are expecting a luminosity increase of a factor of about 3-5 at 7.7 GeV and about 10 around 19.6 GeV. STAR is also planning to upgrade the inner sectors of the existing Time Projection Chamber (iTPC). This will increase the acceptance at low p_t and extend the pseudo-rapidity coverage by approximately half a unit. The increased acceptance combined with higher statistics will improve the current measurement significantly and make possible measurements of rare probes such as dilepton and hypertriton production.

In order to reach higher μ_B, STAR has also proposed to run in "fixed-target mode" by putting a gold target inside the beam pipe [25] . This will allow STAR to run at energies lower than 7.7GeV and allow the μ_B to be extended from 400MeV to 800MeV without affecting normal RHIC operation. A previous study from NA49 has reported signatures of the onset of deconfinement near $\sqrt{s_{NN}}$ =7.7 GeV [26]. To confirm this, it is important to take data both below and above the transition energy. Table 1 shows a summary of collider mode, fixed target mode $\sqrt{s_{NN}}$, and corresponding μ_B values.

7. Summary

We have presented the latest results from the STAR BES Phase I program. Most results show a smooth change vs. incident energy. We do see significant differences in particle and anti-particle v_2, which indicates the breaking of the NCQ scaling. A possible minimum at $\sqrt{s_{NN}}$ =10–20 GeV is also observed for the net proton v_1 slope. More statistics are needed to confirm a few other interesting observables such as higher moments of net-protons distributions and ϕ-meson v_2. We are looking forward to the BES Phase II program with STAR iTPC upgrade and fixed-target mode.

References

[1] Y. Aoki *et al.* Nature **443**, 675-678 (2006)
[2] Shinji Ejiri, Phys. Rev. D **78** 074507 (2008)
[3] M. A. Stephanov PoS LAT2006:024,2006
[4] M.M.Aggarwal *et al.* [STAR Collaboration], arXiv:1007.2613v1
[5] J. Cleymans et al., Computer Physics Communications,180, 84 (2009).
[6] Sabita Das, [STAR Collaboration], arXiv:1210.6099
[7] A. Andronic et al., Nucl. Phys. A 834, 237 (2010).
[8] J. Cleymans et al., Phys. Rev. C 73, 034905 (2006)
[9] S.A. Bass, P. Danielewicz, and S. Pratt, Phys. Rev. Lett. **85**, 2689 (2000).
[10] B. I. Abelev *et al.* [STAR Collaboration], Phys. Rev. C **82**, 024905 (2010).
[11] J. Adams *et al.* [STAR Collaboration], Phys. Rev. Lett. **95**, 122301 (2005).
[12] B. I. Abelev *et al.* [STAR Collaboration], Phys. Rev. C **75**, 054906 (2007).
[13] M. M. Aggarwal *et al.* [STAR Collaboration], arXiv:1007.2613 [nucl-ex] (2010).
[14] L. Adamczyk *et al.* [STAR Collaboration], Phys. Rev. Lett. **110**, 142301 (2013).
[15] C. Alt et al., Phys. Rev. C **79**, 044910 (2009).
[16] J. Adams *et al.* [STAR Collaboration], Phys. Rev. Lett. **103**, 092301 (2009)
[17] Terence J Tarnowsky [STAR Collaboration], arXiv:1110.2222 [nucl-ex].
[18] M.A. Stephanov, K. Rajagopal and E.V. Shuryak Phys. Rev. Lett. **81** 4816-9 (1998).
[19] S. Gavin Phys. Rev. Lett. **92** 162301 (2004).
[20] J. Adams *et al.* [STAR Collaboration], Phys. Rev. C **72**, 044902 (2005).
[21] R. Snellings *et al.*, Phys. Rev. Lett. **84**, 2803 (2000); J. Brachmann *et al.*, Phys. Rev. C **61**, 024909 (2000); L. P. Csernai and D. Rohrich, Phys. Lett. B **458**, 454 (1999); H. Stoecker, Nucl. Phys. A **750**, 121 (2005).
[22] N. Shah [STAR Collaboration], arXiv:1210.5436 [nucl-ex].
[23] Yadav Pandit [STAR Collaboration], arXiv:1210.5315[nucl-ex] (2012).
[24] J. Cleymans *et al.*, Phys. Rev. C **73**, 034905 (2006).
[25] RHIC Beam Use Request For Runs 13 and 14
[26] NA49 Collaboration, Phys. Rev. C 77, 024903 (2008).

Journal of Physics: Conference Series **458** (2013) 012002 doi:10.1088/1742-6596/458/1/012002

Beam energy scan results from PHENIX

Abhisek Sen for the PHENIX Collaboration

University of Tennessee

E-mail: `asen1@utk.edu`

Abstract. Probes like J/ψ and π^0 production are very important for studying the properties of the strongly interacting partonic medium created in heavy ion collisions. Their production is strongly suppressed in $\sqrt{s}$=200 GeV $Au + Au$ collisions, in comparison to the expectation from binary collision scaled $p + p$ collisions. The recent low-energy scan at RHIC provided the PHENIX collaboration with an opportunity to study the evolution of the suppression at $\sqrt{s}$=39 and 62.4 GeV center of mass energies in order to disentangle multiple contributing mechanisms. The suppression of J/ψ observed is similar to those previously measured at 200 GeV. In contrast neutral pion suppression shows a distinct energy dependence at moderate p_T region of the central collisions.

1. Introduction

The theory of Quantum Chromodynamics (QCD) predicts a phase transition from hadronic matter to a deconfined Quark Gluon Plasma (QGP) at high temperature and energy density. The Relativistic Heavy Ion Collider (RHIC) was build to achieve these conditions by colliding heavy nuclei at very high energies in order to test the prediction of QCD and understand the properties of the medium. One of the powerful probes is a hard parton scattered in the early stage of the collisions. The scattered parton then pass through the created matter and fragments. If a dense, colored medium is formed in Au+Au collisions, these hard-scattered partons may lose energy while traversing some of it. In case of quarkonia ($q\bar{q}$) bound states like $J/\psi(1S)$, the suppression mechanism is thought to be very different as there will be other additional effects like color screening in the medium which may hinder its production. Therefore, the observed hadron yield will be lower than that expected from binary collision scaling. This suppression is quantified in terms of the nuclear modification factor R_{AA}:

$$R_{AA} = \frac{1}{\langle T_{AA} \rangle} \frac{dN^{AA}/dy}{d\sigma^{pp}/dy} \tag{1}$$

where dN^{AA}/dy is the invariant yield in $Au + Au$ collisions, $d\sigma^{pp}/dy$ is the $p + p$ cross-section and $\langle T_{AA} \rangle$ is the nuclear overlap function.

The PHENIX experiment previously measured R_{AA} for J/ψ and π^0 production at 200 GeV Au+Au collision and found a strong suppression in the most central collisions [1]. The measured J/ψ suppression was similar to that measured at CERN-SPS energies in Pb+Pb collisions at

$\sqrt{s_{NN}}$ = 17.2 GeV. This is in contradiction with the color screening interpretation that the suppression is expected to increase at higher temperature. So additional effects are need to be investigated. On the other hand the π^0 production is suppressed in $Au + Au$ collisions while it is observed to be enhanced in the lighter system Cu+Cu at $\sqrt{s_{NN}}$ =22.4 GeV. In order to study the transition of enhancement to suppression of π^0 R_{AA} and shine light on the J/ψ puzzle, RHIC started the beam energy scan program in 2010 by varying the beam energy and studying the effect on R_{AA}. This is an unique characteristic of RHIC to be able to collide different combination of species in a wide range of energies. Measurement of R_{AA} in a wide range of system energies is an important diagnostic means that will help quantifying the medium properties. We present new measurements of π^0 and J/ψ R_{AA} at lower energies at $\sqrt{s_{NN}} = 39$ and 62.4 GeV $Au + Au$ collisions.

2. π^0 Measurement at 39 and 62.4 GeV

Neutral pion productions were measured at midrapidity ($|y| < 0.35$) in several energies of $Au+Au$ collisions. Our earlier π^0 measurements in $\sqrt{s_{NN}}$=130 and 200 GeV of $Au+Au$ collisions revealed a strong suppression at high p_T, which was interpreted very well by the partonic energy loss model in the medium [1]. Additionally data from $\sqrt{s_{NN}}$ = 200 GeV $d + Au$ collisions showed no suppression or enhancement indicating that the hadron suppression is a final state effect.

Moreover, the PHENIX experiment also studied π^0 production in a lighter collision system, $Cu + Cu$, at three energies ($\sqrt{s_{NN}}$=22.4, 62.4 and 200 GeV). An enhancement was observed at 22.4 GeV but their production was suppressed at higher energies. This was interpreted as the interplay of the multiple soft scattering (Cronin effects) in the medium at various energies. During the beam scan program in 2010, PHENIX collected data at $\sqrt{s}$ =39 and 62.4 GeV $Au+Au$ collisions. These energies were specifically chosen to study this transition from enhancement to suppression as a function of collision energy in order to constrain the energy loss models.

Neutral pions are reconstructed via their $\pi^0 \to \gamma\gamma$ decay with the electromagnetc caloriemeter (EMCal). Yields were extracted on statistical means by subtracting the photon pairs from the combinatorial background estimated by event-mixing in each p_T and centrality bin [2]. In order to compute R_{AA}, baseline $p + p$ measurements are needed. Corresponding $p + p$ reference data at 62.4 GeV were collected earlier in 2006 by PHENIX but RHIC was never run at 39 GeV $p+p$ collisions. Therefore, $p+p$ data from Fermilab E706 were used as a baseline for 39 GeV. Figure 1 shows the nuclear modification factor in most central $0 - 10\%$ and mid-peripheral $40 - 60\%$ of $Au + Au$ collisions at $\sqrt{s_{NN}}$ = 39, 62.4 and 200 GeV. There is a significant suppression in the most central collisions at all three energies. In the mid-peripheral bin (40-60%), 39 GeV points are consistent with unity above $p_T > 3$ GeV/c but there is a suppression in all p_T region at both 62.4 and 200 GeV. The R_{AA} match well at all three energies at low or high p_T but it gets weaker in the intermediate-p_T region for the lower energies. pQCD calculations are also shown (solid lines) for the most central $(0 - 10\%)$ collisions which were used to successfully describe the 200 GeV $Au + Au$ data but failed to describe the new measurements at 39 and 62.4 GeV. Only a qualitative agreement of turnover point of the R_{AA} curves moves to higher p_T at lower collision energies is noticeable. The bands are the newly calculated results with a similar framework but with reduced Cronin-effect which are in better agreement with data [2].

Figure. 2 is the p_T averaged R_{AA} in the region above $p_T > 6$ GeV as a function of the number of participants. A similar suppression is observed at 62.4 and 200 GeV but it gets

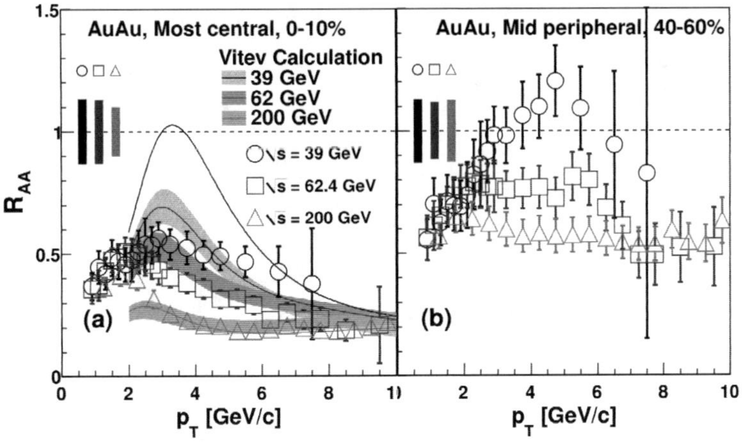

Figure 1. π^0 Nuclear modification factor in $Au + Au$ collisions for the most central 0-10% and mid-peripheral 40-60% bin. Also shown for central collisions are two pQCD calculations with Cronin-effect (solid lines) and with the Cronin-effect reduced (bands) for all three energies.

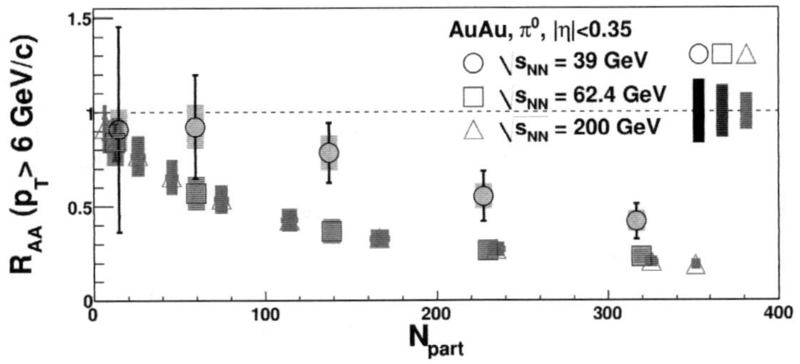

Figure 2. π^0 Nuclear modification factor for $p_T > 6$ GeV.

weaker the lower energy. The new measurements of π^0 suppression over wide energy range will help constrain the energy-loss models.

3. J/ψ Measurement at 39 and 62.4 GeV

Quarkonia bound states are expected to be suppressed in the Quark Gluon Plasma (QGP) due to color screening [3]. PHENIX measured a strong J/ψ suppression (a factor of ~ 5 for the most central collisions) at both mid and forward rapidities in $Au + Au$ collisions at 200 GeV. This suppression is very similar to that measured at the CERN-SPS at $\sqrt{s_{NN}} = 17.2$ GeV $Pb + Pb$ collisions [4]. This contradicts the color screening interpretation that the dissociation of the quarkonia states will increase with energy density. It was clear that there are additional classes of effects like "cold nuclear matter" effects that are not due to the QGP and additional QGP mechanisms (e.g. coalescence, energy loss) are in play. PHENIX extended the J/ψ measurements to the lower energies $\sqrt{s_{NN}} = 39$ and 62.4 GeV [5] in order to disentangle different competing physics processes which might contribute. A broad measurement over $\sqrt{s_{NN}}$ will not only vary the temperature and density of the medium but also the $c\bar{c}$ production and cold nuclear effects (CNM).

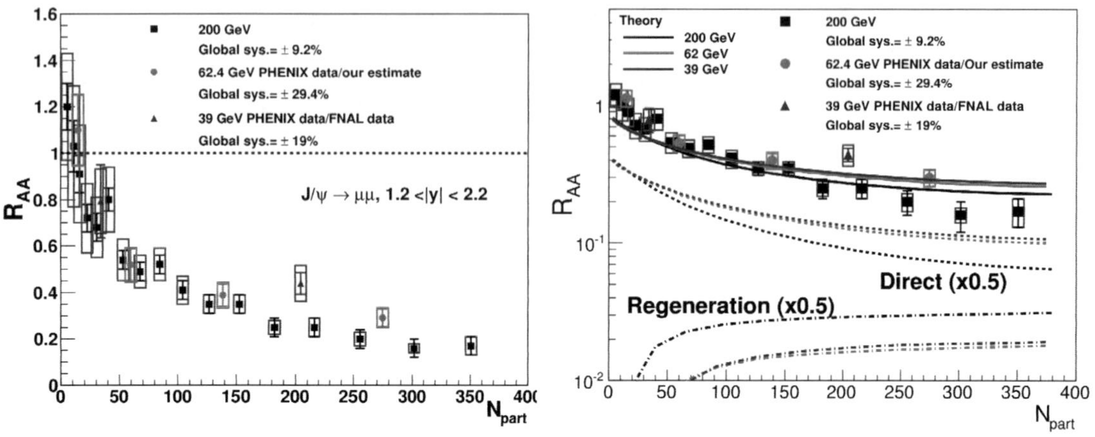

Figure 3. J/ψ R_{AA} and model comparison.

Two forward spectrometers composed of the Muon Tracker and Muon Identifiers are used to reconstruct the muon tracks. The J/ψ candidates were reconstructed via the dimuon channel in the rapidity range $1.2 < |\eta| < 2.2$ and full azimuth. There are no PHENIX measurements for J/ψ reference data in $p+p$ collisions at these energies since there is no run in 39 GeV $p+p$ and only a limited dataset was previously collected at 62.4 GeV $p+p$ collisions, which could not provide a reasonable baseline for J/ψ. Hence a compilation of the measurements from other experiments in addition to a theoretical estimate (Color Evaporation Model by R. Vogt) were used to make a best estimate for the baseline cross-section. The details can be found in [5].

Figure 3 (right) shows the J/ψ suppression at forward rapidities at all three energies as a function of N_{part}. At lower energies J/ψ suppression is similar to that at 200 GeV at forward rapidity. Although there is a modest decrease of suppression in central collisions at 39 GeV, overall they agree within systematic errors. The similarity between different energies are the cumulative effect of the different competing physics processes. However the strength of these processes remain unclear without detailed understanding of the production of J/ψ in nuclear target, termed cold nuclear matter effects. These effects are expected to be different in at each collision energy. It is therefore important to measure these effects experimentally in d(p)+Au collisions at the same energies as $Au + Au$ collisions.

A new theory calculation which includes cold nuclear matter effects, regeneration and QGP suppression for the J/ψ suppression at forward rapidity is compared to the data in figure 3 (left). The contribution of direct J/ψ and regeneration are shown separately. It appears that the QGP suppression decreases while going from 200 GeV to 39 GeV as seen in the direct component but there will be a strong regeneration effect at higher energies due to higher number of $c\bar{c}$ pairs. The regeneration contribution is expected to the increase with collision energy due to the increase in the total number of charm pairs produced. The inclusion of the regeneration component resulted in a similar suppression at all the energies and PHENIX measurements are consistent with the theoretical calculations within the global systematic uncertainties.

4. Summary

The PHENIX experiment has measured both π^0 and J/ψ production at $\sqrt{s_{NN}}$ =39 and 62.4 GeV $Au + Au$ collisions. Using suitable $p + p$ reference estimates from other experiments and theory calculations when our own measurements are not available, R_{AA} was calculated and compared with our earlier 200 GeV measurements. The observed π^0 production is strongly suppressed in most central collisions at all energies but the suppression gets weaker at lower energies in the moderate p_T region. The J/ψ R_{AA} results are similar to that measured in 200 GeV with slightly less suppression at these lower energies. The J/ψ results are consistent with the theoretical calculation which shows a balance in effects of more QGP suppression as well as more regeneration at higher energy collisions.

References

[1] A. Adare *et al.*, "Neutral pion production with respect to centrality and reaction plane in Au+Au collisions at $\sqrt{s_{NN}} = 200$ GeV," *Phys. Rev. C*, vol. 87, p. 034911, Mar 2013.

[2] A. Adare *et al.*, "Evolution of π^0 Suppression in Au + Au Collisions from $\sqrt{s_{\rm NN}} = 39$ to 200 GeV," *Phys. Rev. Lett.*, vol. 109, p. 152301, Oct 2012.

[3] T. Matsui and H. Satz, "J/ψ suppression by quark-gluon plasma formation," *Phys. Lett. B*, vol. 178, pp. 416–422, Jul 1986.

[4] A. Adare *et al.*, "J/ψ suppression at forward rapidity in Au + Au collisions at $\sqrt{s_{NN}} = 200$ GeV," *Phys. Rev. C*, vol. 84, p. 054912, Nov 2011.

[5] A. Adare *et al.*, "J/ψ suppression at forward rapidity in Au + Au collisions at $\sqrt{s_{NN}} = 39$ and 62.4 GeV," *Phys. Rev. C*, vol. 86, p. 064901, Dec 2012.

Journal of Physics: Conference Series **458** (2013) 012003 doi:10.1088/1742-6596/458/1/012003

Azimuthal Anisotropy in U+U Collisions at $\sqrt{s_{NN}} = 193$ GeV with STAR Detector at RHIC

Yadav Pandit (for the STAR Collaboration)
Department of Physics, UIC, Chicago, USA
E-mail: ypandit@uic.edu

Abstract. We report the measurement of first ($n = 1$) and higher order(n = 2-5) harmonic coefficients (v_n) of the azimuthal anisotropy in the distribution of the particles produced in U+U collisions at $\sqrt{s_{NN}}$ =193 GeV, recorded with the STAR detector at RHIC. The differential measurement of (v_n) is presented as a function of transverse momentum (p_T) and centrality. We also present v_n measurement in the ultra-central collisions. These data may provide strong constraints on the theoretical models of the initial condition in heavy ion collisions and the transport properties of the produced medium.

1. Introduction

The study of azimuthal anisotropy, based on Fourier coefficients, is widely recognized as an important tool to probe the hot, dense matter created in heavy ion collisions [1]. These measurements are sensitive to early stage evolution of the system. In a picture of hydrodynamic expansion of the system formed in the collisions, these final state momentum anisotropies are expected to rise due to the initial state anisotropy driven pressure gradients and subsequent interactions of the constituents. Specially the differential measurement of azimuthal anisotropy have been found to be sensitive to initial condition, thermalization, the equation of state and transport coefficients of the medium in the heavy ion collisions. Azimuthal anisotropy quantified by the Fourier coefficient of the particle distribution can be written as

$$E\frac{d^3N}{d^3p} = \frac{1}{2\pi}\frac{d^2N}{p_T dp_T dy}(1 + \sum_{n=1}^{\infty} 2v_n \cos n(\phi - \psi_R). \tag{1}$$

where p_T is the transverse momentum, y is rapidity and ϕ is the azimuthal angle of each particle and ψ_R is the true reaction plane angle, defined by beam direction and impact parameter vector between two colliding nuclei. The sine terms in the Fourier expansion vanish due to the reflection symmetry with respect to the reaction plane. The reaction plane ψ_R is not measurable directly a priori, so the Fourier coefficients are determined with respect to the estimated participant event planes [1],

$$v_n = \langle \cos n(\phi - \Psi_n) \rangle. \tag{2}$$

where Ψ_n are the generalized participant event planes at all orders for each event. The estimated event plane from detected particles is corrected for event plane resolution. The measured v_n with respect to Ψ_n for reasonable event-plane resolutions are closer to the root-mean-square values [2] than the mean values. The first harmonic coefficient v_1, called directed flow, has two components rapidity odd directed flow generated due to bounce off motion of the nuclei and rapidity even directed flow due to dipole asymmetry in the initial geometry. The directed flow measurement should suppress the the contribution from global momentum conservation especially at peripheral collisions along with other non-flow contributions. The rapidity odd $v_1(y)$ component at 200 GeV Au+Au collision is found very small at mid-rapidity and becomes significant at forward rapidity [3]. The rapidity even component $v_1(y)$ recently reported at 200 GeV Au+Au collisions [4] form STAR collaboration is found to have weak rapidity dependence as predicted by some models [5]. The transverse momentum dependence of rapidity event $v_1(p_T)$ is of interest in the present study. The second harmonic coefficient v_2, popularly known as elliptic flow have been extensively studied both experimentally and theoretically. Recently higher order harmonics have also gained attention both from theory and experimental community. These higher order harmonics can provide valuable information about the initial state of the colliding system [6, 7, 8] and also provide an natural explanation to the ridge phenomena in heavy ion collisions [9]. Measurement of these harmonics in different systems and collisions energy helps to further understand the heavy ion physics in general.

In uranium - uranium (U+U) collisions, there is the potential to produce more extreme conditions of excited matter at higher density and/or greater volume than is possible using spherical nuclei like gold or lead at the same incident energy [10]. Uranium has quadrupole deformed shape. So U+U collisions may offer an opportunity to explore wider range of initial eccentricities. The collisions of special interest are the "ideal tip-tip" orientation in which the long axes of both deformed nuclei are aligned with the beam axis at zero impact parameter, and the "ideal body-body" orientation in which the long axes are both perpendicular to the beam axis and parallel to each other at zero impact parameter. The "ideal tip-tip" and "ideal body-body" collision events allow to test the prediction of hydrodynamical model(s) by varying the transverse particle density at spatial eccentricity similar to central Au+Au collisions.

We report here the first v_n measurements in U+U collisions from the STAR experiment at the Relativistic Heavy Ion Collider (RHIC) at $\sqrt{s_{NN}} = 193$ GeV. These data were collected during RHIC run-XII in the year 2012. These measurements are presented as a function of transverse momentum (p_T) and centrality.

2. The STAR experiment and Analysis Details

Data reported in this proceedings were collected in U+U collisions at $\sqrt{s_{NN}} = 193$ GeV in the year 2012 with a minimum bias trigger and the ultra central events were taken with a dedicated central trigger. Main detector used in the present measurements is the Time Projection Chamber (TPC) [11], the primary tracking device at STAR. TPC has full azimuthal coverage and uniform acceptance in ± 1.0 units of pseudorapidity. The charged particle momenta are measured by reconstructing their trajectories through the TPC. The events with the primary collision vertex position along the beam direction (V_z) within 30 cm of the center of the detector are selected for this analysis. Event vertex is further required to be in the transverse direction within 2.0 cm from the center of the beam pipe. Centrality classes in U+U collisions at $\sqrt{s_{NN}} = 193$ GeV are defined using the number of charged particle tracks reconstructed in the Time Projection Chamber(TPC) within pseudorapidity $|\eta| < 0.5$ and passing within 3 cm of interaction vertex. The uncorrected, i.e. not corrected for acceptance and reconstruction efficiency, multiplicity ($N_{\rm ch}$) distribution for events with a reconstructed primary vertex is shown in Fig. 1.

This analysis was carried out on tracks that had transverse momenta $p_T > 0.15$ GeV/c, passed within 3 cm of the primary vertex, had at least 15 space points in the main TPC acceptance ($|\eta| < 1.0$) and the ratio of the number of actual space points to the maximum possible number of space points for that track's trajectory was greater than 0.52.

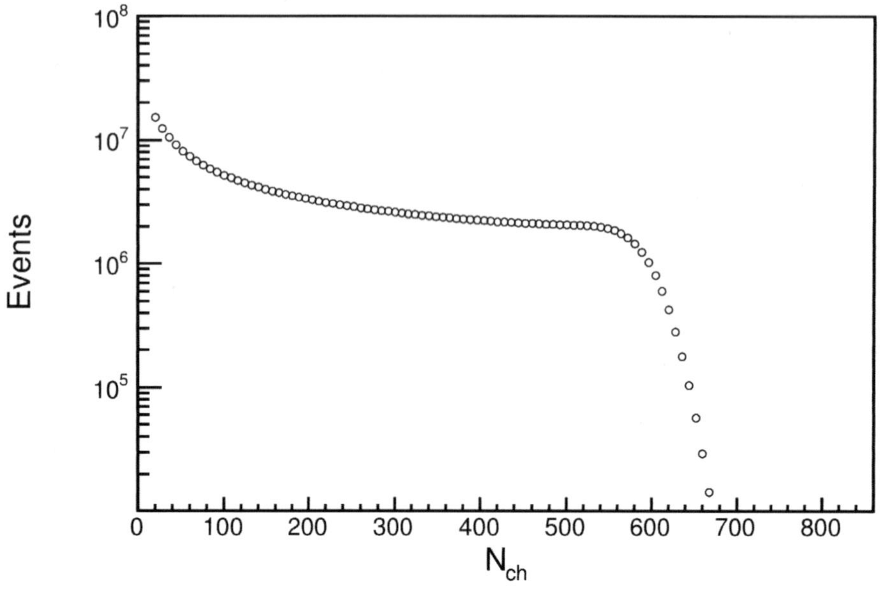

Figure 1. Uncorrected multiplicity distribution with $|\eta| < 0.5$ in U+U collisions at $\sqrt{s_{NN}} =$ 193 GeV. Events with $N_{ch} > 10$ are selected for the present analysis.

For any Fourier harmonic n, the event flow vector (Q_n) and the event plane angle (Ψ_n) are defined by [1],

$$Q_n \cos(n\Psi_n) = Q_{nx} = \sum_i w_i \cos(n\phi_i), \tag{3}$$

$$Q_n \sin(n\Psi_n) = Q_{ny} = \sum_i w_i \sin(n\phi_i), \tag{4}$$

$$\Psi_n = \left(\tan^{-1}\frac{Q_{ny}}{Q_{nx}}\right)/n, \tag{5}$$

where sums extend over all particles i used in the event plane calculation, and ϕ_i and w_i are the laboratory azimuthal angle and the weight for the i-th particle, respectively. The event plane vector is reconstructed from tracks with transverse momentum (p_T) up to 2 GeV/c and within $|\eta| < 1.0$. For first harmonics(n=1), the weight is taken as,

$$w_i = p_T - \frac{< p_T^2 >}{< p_T >}, \tag{6}$$

and for higher harmonics ($n > 1$) the weight is equal to the transverse momentum

$$w_i = p_T. \tag{7}$$

The $< p_T >$ and $< p_T^2 >$ represent event averaged quantities. The choice of this weight corrects the effect of the momentum conservation and also it cancels out the conventional pseudorapidity asymmetric directed flow [5, 12].

We used scalar product method as well as the event plane method to measure the signal. In the scalar product method, the particles with $0.5 < \eta < 1.0$ were assigned to one subevent and

Journal of Physics: Conference Series **458** (2013) 012003 doi:10.1088/1742-6596/458/1/012003

particles with $-1.0 < \eta < -0.5$ to the other subevent, separated by η gap of 1.0 units between two subevents and at least 0.5 unit with the particle of interest and the subevent to which it is correlated. In order to remove the acceptance effects we applied recentering correction [1] to the flow vectors. With unit vector defined as $u_{ni} = e^{in\phi}$, the first harmonic coefficient v_n is evaluated as,

$$v_n(\eta > 0) = \frac{\langle Q_{na}(\eta < 0)u^*_{ni}\rangle}{\sqrt{\langle Q_{na}Q_{nb}\rangle}} \quad (8)$$

$$v_n(\eta < 0) = \frac{\langle Q_{nb}(\eta > 0)u^*_{ni})\rangle}{\sqrt{\langle Q_{na}Q_{nb}\rangle}} \quad (9)$$

In the event plane method, the particles with $0.1 < \eta < 1.0$ were assigned to one subevent and particles with $-1.0 < \eta < -0.1$ to the other subevent, so that subevents were separated by η gap of 0.2 units. Larger pseudorapidity separation was desired but we are limited by the event plane resolution. In this method, we first evaluate the event plane angle Ψ_n from the event plane vector (Q_n) and the event plane angle was flattened using a shifting correction method to correct for the detector acceptance effects. The particle correlation with event plane angle Ψ_n is given by,

$$v_n(\eta > 0) = \frac{\langle \cos n(\phi - \Psi_{na}(\eta < 0))\rangle}{\sqrt{\langle \cos n(\Psi_{na}(\eta < 0) - \Psi_{nb}(\eta > 0))\rangle}}. \quad (10)$$

$$v_n(\eta < 0) = \frac{\langle \cos n(\phi - \Psi_{nb}(\eta > 0))\rangle}{\sqrt{\langle \cos n(\Psi_{na}(\eta < 0) - \Psi_{nb}(\eta > 0))\rangle}}. \quad (11)$$

3. Result and discussion

Results are presented with only statistical errors. In these studies, contribution from short range correlation such as Bose-Einstein correlations, coulomb interactions are studied using a pseudorapidity gap of at least 0.5 units in pseudorapidity between the event vector and the particle of interest using scalar product method. Results from both methods, scalar product with larger psuedorapidity gap and event plane method with smaller pseudorapidity gap are consistent with each other, which suggests that non-flow contribution from short range correlations are small. To reduce the effects from high p_T particles in the estimation of the event plane, we used p_T weight only up to 2 GeV/c. The non-flow contribution from the jets/minijets is not known and might be a significant contributor to the systematic uncertainties especially at peripheral collisions.

Figure 2 (a) presents the v_n integrated in transverse momentum p_T ($0.15 < p_T < 2GeV/c$) and pseudorapidity η ($|\eta| < 1.0$) as a function of centrality. Except for v_2, we observe very weak centrality dependence. The initial overlap geometry, which makes the dominant contribution to the elliptic flow, changes from central to peripheral collisions, giving rise to the strong centrality dependence of v_2. For v_1, centrality dependence is observed only at more peripheral collisions (40-80%) where non flow contribution may be significant. The non flow contribution from the momentum conservation effect [12] may hot have perfectly suppressed in peripheral collisions in the measurement of v_1. The centrality dependence of the geometry fluctuations are not yet understood. The weak centrality dependence of the flow coefficients which are believed to originate from geometry fluctuations indicates that geometry fluctuations weakly depend on centrality.

Figure 2 (b) presents the elliptic flow v_2 as a function of centrality compared with a model prediction based on glauber based [13] prediction. Model curve is for 200 GeV U+ U collisions from reference [14]. For central collisions, experimental data are lower than the prediction.

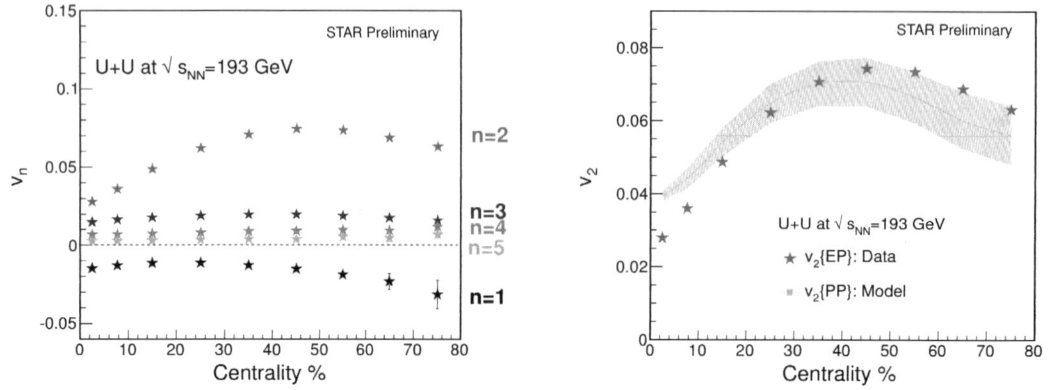

(a) v_n integrated in $p_T < 2$ GeV/c and $|\eta| < 1.0$ as a function of centrality

(b) v_2 as a function of centrality compared to the prediction based on MC glauber model [14]

Figure 2. v_n as a function of centrality for U+U Collisions

The v_n of all charged hadrons, for $n = 1, 2, 3, 4$ and 5 as a function of p_T at various centralities, as shown in Fig. 3. The data for 0–5% centrality are shown in the upper panels, and for intermediate centrality (30–40%) in the lower panels. All v_n measurements show an increasing trend as a function of p_T. Except for v_2, we observe very weak centrality dependence. Also shown are the same measurement from Au+Au collisions at 200 GeV for the comparisons. We observe the difference in v_n for n=1 and n=2 at 0-5% central between U+U and Au+Au collisions. This may hint to initial overlap geometry difference in the central collisions between two systems Au+Au and U+U collisions. The difference diminishes in higher harmonics and more peripheral collisions.

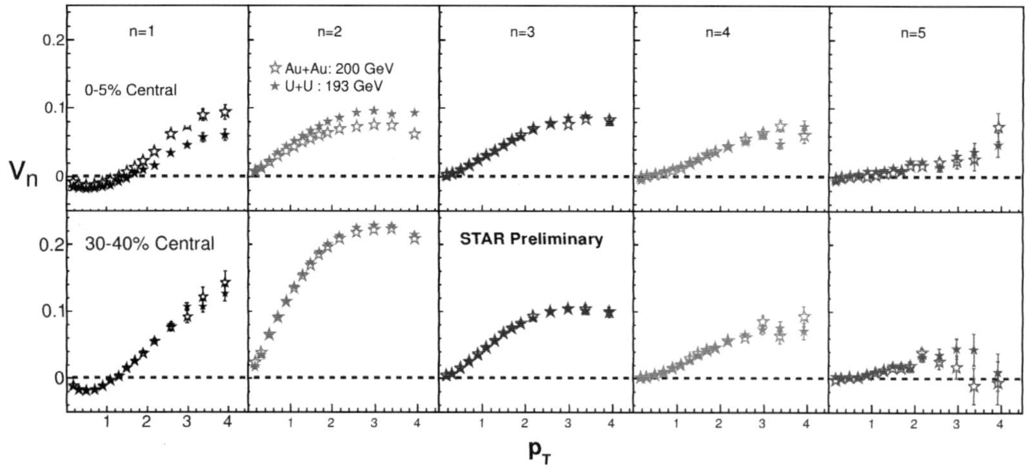

Figure 3. v_n measurement at 0-5% central in upper panels and mid central (30-40%) collisions in the lower panels as a function of p_T for U+U collisions at 193 GeV compared with v_n measurement for Au+Au collisions at 200 GeV [15]

We also report measurement of v_n for ultra central U+U collisions. The centrality selection is based on reference multiplicity. We subdivide 0-5% central bin into 10 smaller centrality bins upto most central 0-0.5% centrality. The most central 0-0.5% collisions should have have dominant contribution from tip-tip collisions. In Fig. 4 (a), v_n as a function of centrality for ultra central collisions is shown. Other than second harmonic coefficient, the v_n do not change in this centrality range. We observe small change for v_2 since it still has some contribution from initial overlap geometry. This observation suggests that there is still a small contribution from body-body collisions.

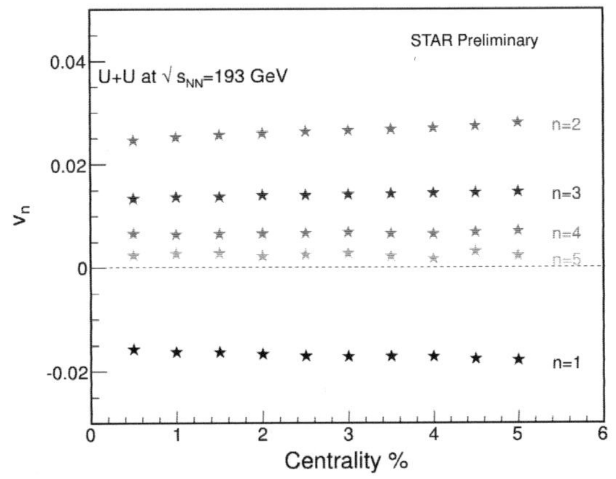

(a) v_n as a function of centrality at 0-5% central collisions

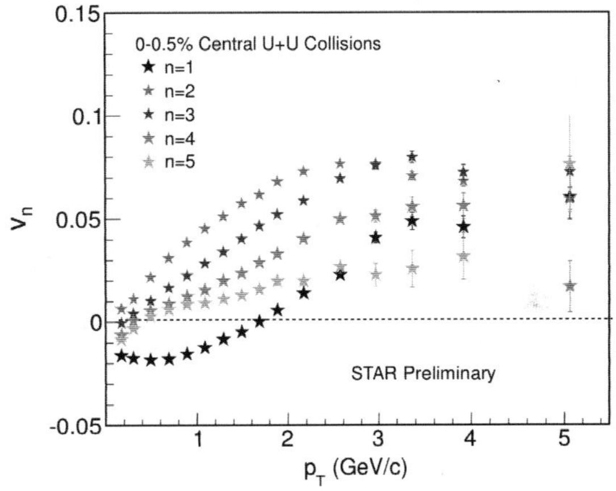

(b) v_n as a function of p_T at ultra central(0-0.5%) collisions

Figure 4. v_n as a function of centrality and transverse momentum for ultra central collisions

In Fig. 4 (b), the v_1 signal along with v_2, v_3, v_4 and v_5 a function of transverse momentum up to $p_T \sim 5$ GeV/c is shown for ultra central (0-0.5%) collisions. In these most central events we find that higher harmonics are also significant in magnitude compared with second harmonics in the intermediate transverse momentum. Contributions from higher harmonics should not be overlooked interpreting dihadrom correlation data at intermediate transverse momentum. These higher harmonics may offer natural explanation to the novel ridge phenomena in heavy ion collisions [6]. Model comparisons with these new data may help us to better understand the medium properties.

4. Summary

We report the first measurement of azimuthal anisotropy v_n for n = 1-5 as a function of transverse momentum p_T and centrality in U+U collisions at $\sqrt{s_{NN}}$ = 193 GeV, recorded with the STAR detector at RHIC. Centrality dependence is week for harmonics other than second harmonics. For higher harmonics and mid central collisions, v_n(U+U) is similar to v_n(Au+Au), the difference appears at central collisions for v_1 and v_2. At intermediate p_T ranges 3-5GeV/c, v_n are comparable to the v_2 signal in ultra central collisions. Model calculation specially at ultra central collisions may be useful to constrain the initial condition and transport coefficient of the medium produced in heavy ion collisions.

References

[1] A. M. Poskanzer and S. A. Voloshin, Phys. Rev. C **58**, 1671 (1998).
[2] J. -Y. Ollitrault, A. M. Poskanzer and S. A. Voloshin, Phys. Rev. C 80, 014904 (2009).
[3] Y. Pandit for the STAR Collaboration, 2011 J. Phys.: Conf. Ser. 316 012001.

[4] Y. Pandit for the STAR Collaboration, arXiv:1211.7162 [nucl-ex].
[5] Matthew Luzum and J. Y. Ollitrault Phys. Rev. Lett. **106**, 102301 (2011).
[6] P. Sorensen, B. Bolliet, A. Mocsy, Y. Pandit and N. Pruthi, Phys. Lett. B **705** 71 (2011).
[7] B. Alver and G. Roland, Phys. Rev. C **81**, 054905 (2010).
[8] B. Schenke, S. Y. Jeon, and C. Gale, Phys. Rev. Lett. **106**, 042301 (2011).
[9] D. Teaney and L. Yan Phys. Rev. C**83**, 064904 (2011).
[10] C. Nepali et al . Phys. Rev. C, 73, 2006.
[11] M. Anderson *et al.*, Nucl. Instr. Meth. A **499**, 659 (2003).
[12] E Retinskaya et al Phys. Rev. Lett. **108**, 252302 (2012)
[13] M. L. Miller, K. Reygers, S. J. Sanders and P. Steinberg, Ann. Rev. Nucl. Part. Sci. **57**, 205 (2007); B. Alver et al., Phys. Rev. C **77** , 014906 (2008).
[14] H. Masui et. al, Phys. Lett. B 679, 440(2009)
[15] Y. Pandit for the STAR Collaboration, arXiv:1210.5315 [nucl-ex].

Journal of Physics: Conference Series **458** (2013) 012004 doi:10.1088/1742-6596/458/1/012004

U+U and Cu+Au results from PHENIX

Aneta Iordanova, for the PHENIX Collaboration
University of California, Riverside
E-mail: anetai@ucr.edu

Abstract. The flexibility of RHIC to collide different nuclei provides experiments with a rich set of data to systematically test models and scaling behaviors in various collision systems. The latest RHIC run collided U+U and Cu+Au nuclei. These collisions promise an array of unique initial geometrical configurations. For example, in U+U collisions the slightly elongated nuclei overlap in a variety of different ways such that, even at zero impact parameter, distinct configurations exist. In central Cu+Au collisions the Cu nucleus is completely embedded within the Au. Such geometries present an opportunity to measure the wide range of initial energy densities of these systems. They also allow the study of some unique features arising from these configurations. In particular, the odd harmonics from the Cu+Au system offer sensitivity to v_3 generated from the collision geometry as opposed to fluctuations in a symmetric system.

In these proceedings the analysis status of the recently taken U+U and Cu+Au data in PHENIX is presented. The results from the global particle production and the challenges in analyzing these asymmetric systems is discussed.

1. Introduction

The versatility of RHIC to provide heavy ion collisions at different center of mass energies and for different ion species has been further demonstrated during run 12. The latest improvements in the ion source and the stochastic beam cooling lead to two new unique collision environments: first U+U collisions at 193 GeV and Cu+Au at 200 GeV. Both systems serve as an important test of the initial geometry.

As the U nucleus is deformed, the initial collision geometry will change event-by-event. The $tip - tip$ orientation, where the two oval-like nuclei collide along their semi-major (long) axis, is of a special interest. Early theoretical works [1] predicted a much higher energy density (up to 55%) for this configuration in central collisions than for the $body - body$ U+U or Au+Au collisions at the same centrality. Thus it was suggested that U+U collisions would provide an excellent opportunity to further study the initial geometry in heavy ion collisions and its relation to the resulting elliptical flow, parton energy loss, and potentially other observables. The key point for such promising measurements is to be able to experimentally separate such distinct initial configurations.

Cu+Au collisions stirred interest due to the initial asymmetry of the colliding system. The $core - corona$ description of the initial collision overlap [2], shows this asymmetric density profile in the core, and also in the corona, where the corona is larger on the Au-side of the collision area, see Fig. 1. Such a substantial initial asymmetry could lead to naturally arising odd harmonics, from both the core and/or the corona. The case where the Cu nucleus is completely swallowed by the Au, for central collisions, could also be potentially very interesting.

Journal of Physics: Conference Series **458** (2013) 012004 doi:10.1088/1742-6596/458/1/012004

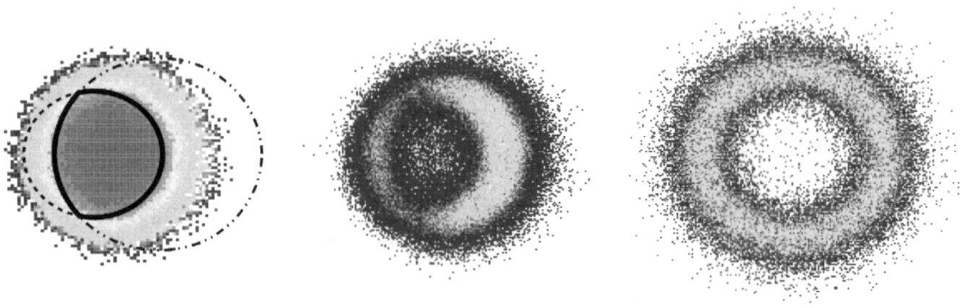

Figure 1. Illustration of the initial geometry of Cu+Au collisions, from a Glauber model simulation [3]. The left panel shows the *core*, or the participant density, for mid-central collisions. The middle panel shows the singly-interacting nucleons, *corona*, at the same centrality. The right panel shows the spectator nucleons for the special case of very central collisions, where the Cu nucleus is completely swallowed by the Au.

2. Global particle production

The charged particle multiplicity, $dN/d\eta$, was measured in U+U collisions and compared to other symmetric systems measured by PHENIX (Fig. 2). $dN/d\eta$ increases with increasing collision energy, and an increase for more central collisions compared to peripheral for all energies is observed. U+U and Au+Au collisions are similar and there is no significant excess of particle production in very central U+U collisions compared to Au+Au at the same energy. This is contrary to the expectation that *tip−tip* U+U collisions should yield a much higher multiplicity. Note that the currently employed centrality selection cannot separate this configuration from the rest of the events and instead the most central 1% class is shown (the rightmost point), which is believed to contain an enhanced number of this type of collisions.

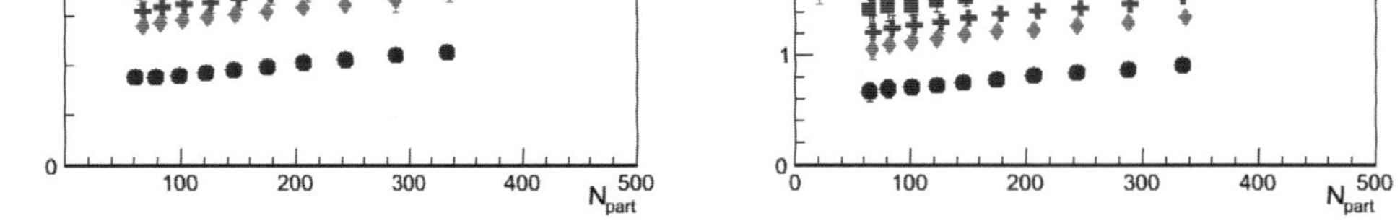

Figure 2. $dN/d\eta$ (left) and $dE_T/d\eta$ (right panel) as a function of centrality in U+U collisions at 193 GeV compared to various collision energies in Au+Au. Centrality selection for the U+U system is defined in 5% bins. The rightmost point shows the 1% most central data.

The measured transverse energy density in U+U, $dE_T/d\eta$, is also very similar to that in Au+Au at 200 GeV (see the centrality distribution in Fig. 2, right panel). A small increase is observed for the most central U+U data, which translates to a new record of energy density

Journal of Physics: Conference Series **458** (2013) 012004 doi:10.1088/1742-6596/458/1/012004

measured at RHIC. The value of the Bjorken energy density, ϵ_B, in most central U+U collisions is 6.15 GeV/fm^2/c. Although the increase from the most central Au+Au collisions to the most central U+U is only moderate, 20%, this represents an increase in energy density by a factor of 3.8 compared to the lowest collision energy of 7.7 GeV.

From this new U+U data, the centrality dependence of the transverse energy production is seen to be independent of the collision system. This was also observed in the comparison of Au+Au and Cu+Cu at 200 and 62.4 GeV, shown in Fig. 3. Surprisingly, the new Cu+Au results at 200 GeV exhibit a similar centrality trend, but an overall higher E_T. This was not expected, and may reflect the larger core in the Cu+Au system at the same N_{part} value compared to the symmetric systems.

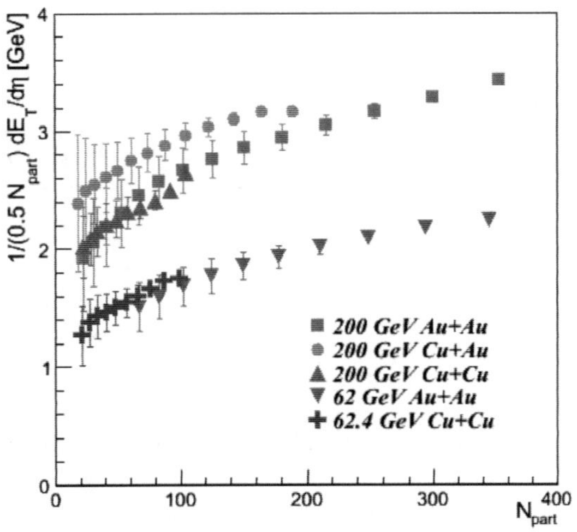

Figure 3. $dE_T/d\eta$ as a function of centrality in Cu+Au collisions at 200 GeV compared to Au+Au and Cu+Cu at 200 and 62.4 GeV.

3. Mid-rapidity identified flow

Measuring the elliptic flow, v_2, is one of the main initial geometry tests for the U+U collision system. Comparison of v_2 of pions and protons in Au+Au and U+U shows mass ordering at low-p_T, with similar shapes across systems and centrality, see Fig. 4. There is a slight difference in the slope of the proton v_2 in the 0-10% central U+U collisions. This effect of flattening at low-p_T was further examined by splitting the central data into smaller percentile bins; it was observed that this flattening is most pronounced in the very central 0-2% class and goes away for the 6-10% class. More studies are needed to examine if this is caused only by the moderate increase (20%) in the energy density from central Au+Au to most central U+U, or if the proton radial flow plays a major role.

Previous PHENIX measurements [4] have shown that the identified charged hadron v_2 obeys number of the valance quark scaling rule. For identified hadrons in Au+Au at 200 GeV this scaling holds for the most central data (0-10%) and starts to break in the next (10-20%) centrality class. For the U+U system the number of valence quark scaling is studied for pions and protons (Fig. 5). Here, the scaling holds for all centrality bins. Note that future measurements, utilizing the full statistics of the run, will further extend the measured range in p_T.

The expectations for the v_n measurements in Cu+Au take into consideration the asymmetric nature of the colliding system. The initial asymmetric density profile (Fig. 6 left) will lead to

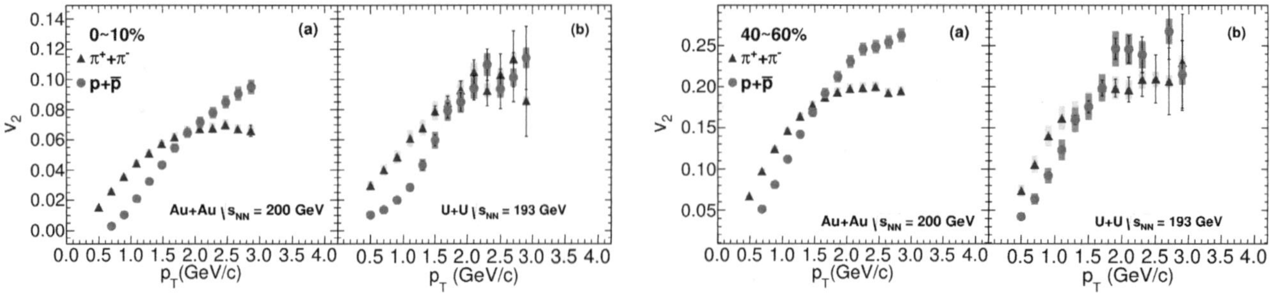

Figure 4. Comparison of elliptic flow for identified hadrons in Au+Au and U+U for 0-10% central data (left pair) and 40-60% mid central data (right pair).

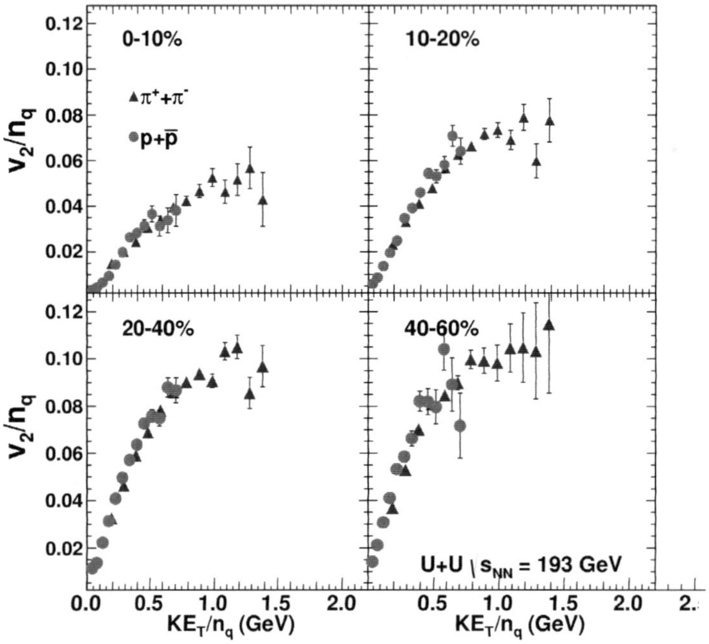

Figure 5. Number of valence quark scaling for v_2 of identified charged hadrons in U+U collisions.

an asymmetric pressure gradient, possibly resulting in an odd harmonic flow. The direction of the true (spectator) plane is defined by the Au spectators, which produce more yield, as shown by simulations (Fig. 6, right). The corresponding spectator reaction plane in data, $\psi_1 smd$, is measured by the shower max detector in the ZDCs (neutrons).

As expected, and asymmetric $dN/d\Delta\phi$ is observed at mid-rapidity, $|\eta|<0.35$ (Fig. 7 left), with a maximum along the reaction plane direction; more particles are emitted from the Au-side than from the Cu-side of the colliding system. This angular particle distribution is fitted with a Fourier function and the v_1, v_2 and v_3 coefficients are extracted. Hadrons at mid-rapidity have large v_1 component in Cu+Au, which is not observed in the Au+Au collision system. This could be naively explained by the asymmetric *corona*. There is also a large v_2 component but surprisingly, the extracted v_3 is consistent with 0. (Note that this measurement of v_3 – with respect to the v_1-reaction plane – does not exclude the possibility of a significant v_3 signal with respect to the v_3-reaction plane, arising from participant nucleons.) The results are compared with model predictions (Fig. 7 right). The trend of v_2 as observed in data is expected by the models. The predicted v_1 has wrong sign compared to that in data.

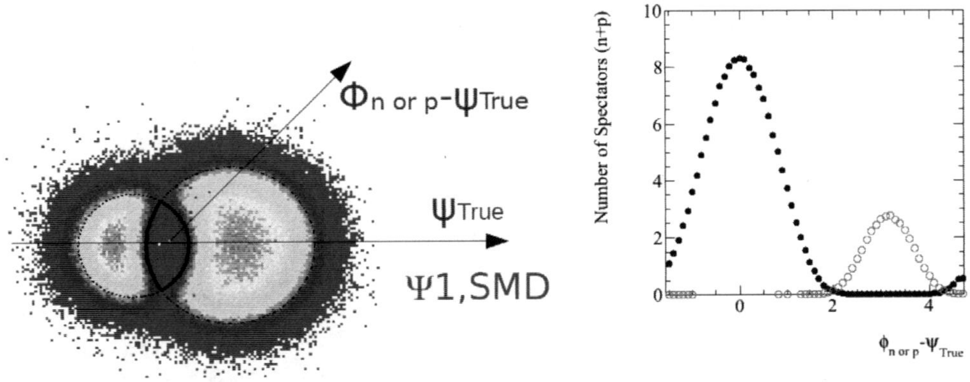

Figure 6. Glauber simulation of the Cu+Au spectator collision profile in 200 GeV (left panel). The true reaction plane, ψ_{True}, is oriented along the Au-side of the collision. Simulation of the angular distribution of the number of spectators from the Au nucleus (black) and from the Cu nucleus (red) (right panel).

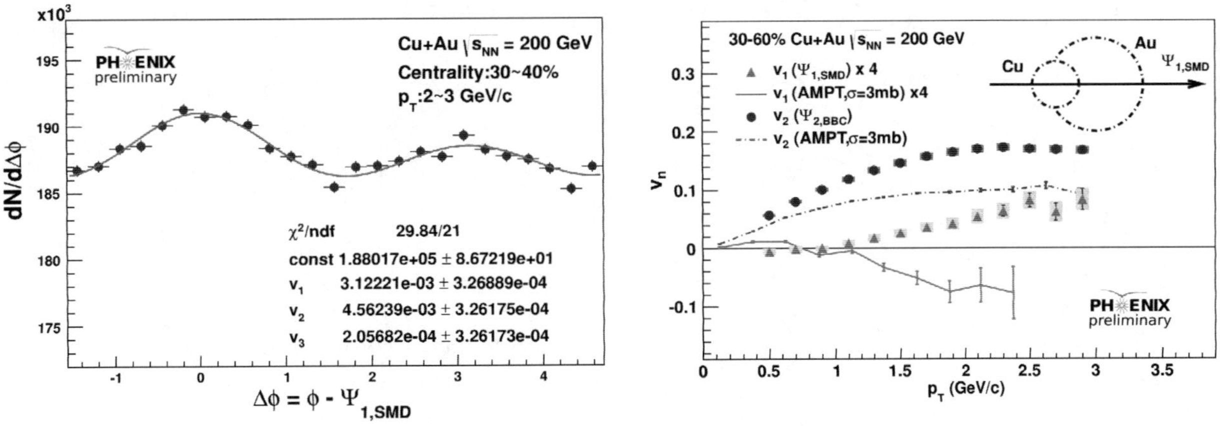

Figure 7. The measured angular distribution of the charged particles at mid rapidity with respect to the $\psi_1 smd$ in Cu+Au at 200 GeV (left). The line represents a fit with Fourier components v_1,v_2,v_3. AMPT model predictions [5] (lines) for the measured v_1 and v_2 in Cu+Au data (points), (right panel).

Fig. 8 shows the measured mid-rapidity pion v_1 in Cu+Au at 200 GeV for a few centrality bins. Sizable positive v_1 is observed for $p_T > 1$ GeV/c at mid-rapidity. The signal increases with p_T in all centrality bins shown. This signal may be due to the initial asymmetric density profile of the system. The results for protons require the use of full data statistics and are not shown. The v_2 for pions and protons (Fig. 9) are measured as a function of p_T and centrality.

The v_2 mass ordering is observed at low p_T for all centrality bins. The number of valence quark scaling also holds in Cu+Au collisions over the measured p_T range and centrality. For the data presented here, the v_2 signal was measured with respect to the second order reaction plane from the BBCs, $\psi_2 bbc$. As a cross check, $\psi_1 smd$ was also used and the measure v_2 for hadrons was found to be consistent with that from $\psi_2 bbc$.

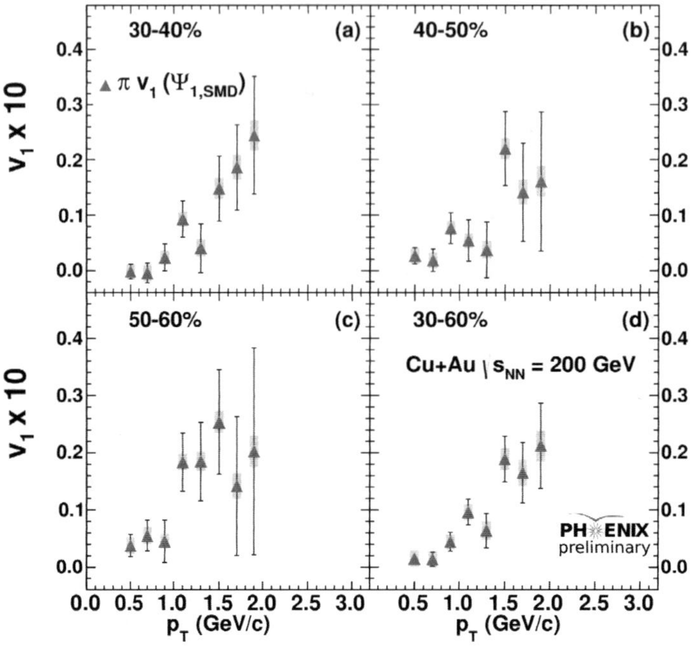

Figure 8. Pion v_1 at mid-rapidity measured in Cu+Au collisions at 200 GeV for a few centrality bins.

4. Identified charged particle spectra

The identified charged particle spectra at mid-rapidity in Cu+Au collisions at 200 GeV were measured for pions and protons. The pion nuclear modification factors, R_{AB}, were formed and then compared to the ones observed in Au+Au collisions at the same center of mass energy, and for an equivalent N_{coll} number, see Fig. 10. The pion nuclear modification factors are very similar in both asymmetric (Cu+Au) and symmetric (Au+Au) systems. Then the baryon-to-meson ratio is examined by forming the spectra ratios between pions and protons, separately for the positive and negative particles, see Fig. 11. Significant baryon enhancement is observed in central collisions, with similar magnitude for the positive and negative ratios. This is quite different than the previous observations in Au+Au and Cu+Cu at the same collision energy, [6]. The pbar/p ratio at mid-rapidity in symmetric systems depends on the collision energy only, with a small centrality and p_T dependence. It is approximately 0.8 in 200 GeV collisions and 0.5 in 62.4 GeV [7]. In Cu+Au collisions the pbar/p ratio is ≈ 0.9. There seems to be less protons produced and the question is why? We have measured a higher ϵ_B for an equivalent N_{part} in this asymmetric system than in the symmetric Cu+Cu and Au+Au systems, possibly due to a larger *core*. Further measurements are needed to find the answer.

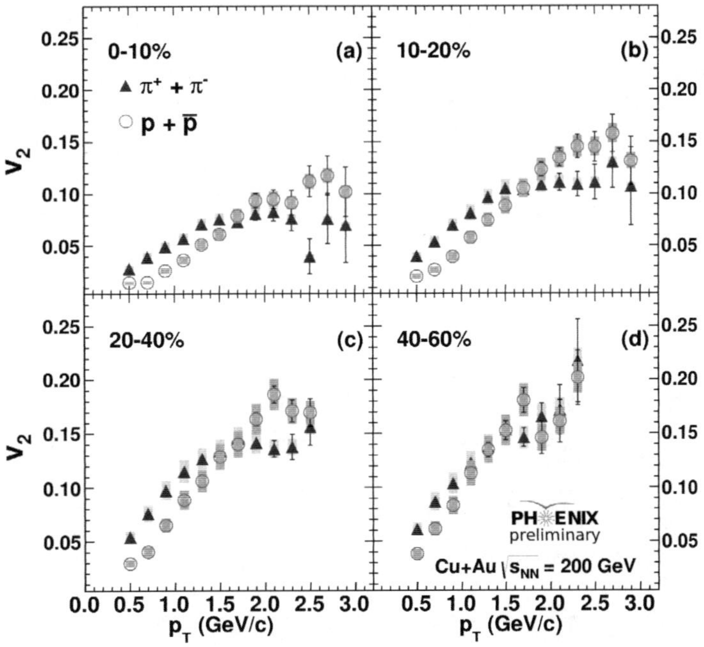

Figure 9. Pion and proton v_2 at mid-rapidity measured in Cu+Au collisions at 200 GeV for a few centrality bins.

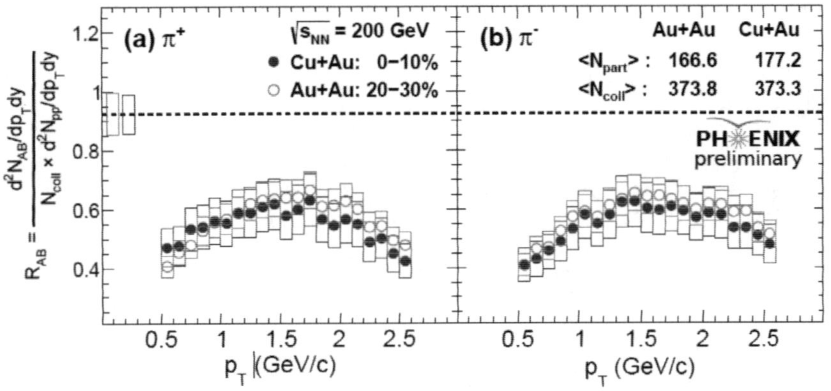

Figure 10. Pion nuclear modification factor, R_{AB}, in Cu+Au collisions at 200 GeV (closed circles) compared to that from Au+Au (open circles) at an equivalent N_{coll}. The results for π^+ are in panel a), for π^- in panel b).

5. Summary

PHENIX has shown first measurements at RHIC from two very unique systems: U+U at 193 GeV and Cu+Au at 200 GeV. Data from both systems are attempting to shed more light on the role of the initial collision configuration and the final data observables. The U+U system holds the highest energy density so far measured at RHIC. The initial results on global particle production in these collisions follow the observed trends in other systems, but more experimental

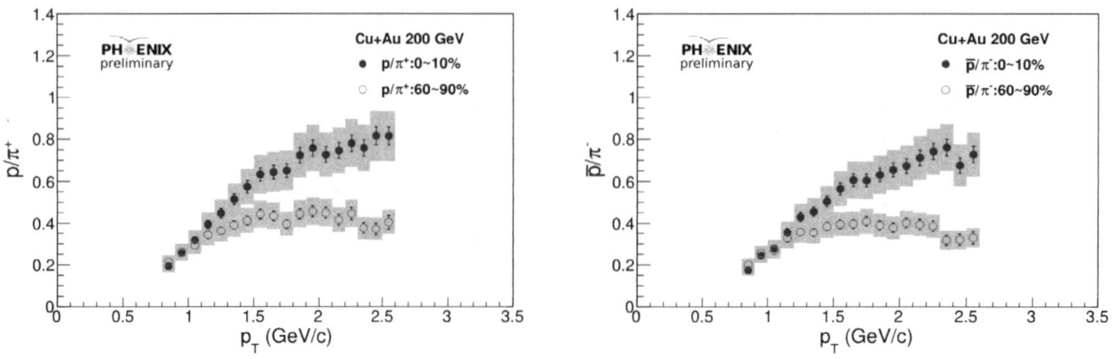

Figure 11. Ratios of proton to pion spectra for two centrality bins measured in Cu+Au collisions at 200 GeV. Ratios from positive particles are shown in the left panel, and for the negative particles in the right panel.

work is needed to further explore the initial collision configurations and to separate $tip - tip$ from $body - body$ collisions.

The asymmetric Cu+Au system has shown surprising results for the measured v_n harmonics at mid-rapidity. Non zero v_1 is observed, possibly due to the nature of this system's initial asymmetry. On the other hand, no v_3 component was observed. The higher energy density in Cu+Au compared to that in other symmetric systems at the same energy and N_{part} is also an interesting observation. We have also seen a different pbar/p ratio compared to that in symmetric systems at the same center of mass energy. PHENIX plans in future to provide further systematic measurements from both U+U and Cu+Au collisions.

References

[1] U. Heinz and A. Kuhlman, Phys. Rev. Lett. **94** (2005) 132301.
[2] R.S. Hollis, Nucl. Phys **A904** (2013) 507c
[3] B. Alver *et al.*, arXiv:0805.4411v1.
[4] A. Adare *at al.* Phys. Rev. C **85** (2012) 064914
[5] H. Ruiz and J. Nagle, AMPT (V1.21) calculation with string melting cross-section σ=3 mb, arXiv:1011.4534
[6] A. Adler *et al.* Phys. Rev. **C74** (2006) 024904
[7] M. Aggarwal *et al.*, Phys. Rev. **C83** (2011) 034910

Journal of Physics: Conference Series **458** (2013) 012005 doi:10.1088/1742-6596/458/1/012005

Rapidity Dependence of Transverse Momentum Correlations from Fluctuating Hydrodynamics

Rajendra Pokharel[a], Sean Gavin[a] and George Moschelli[b]

a)Wayne State University, 666 W Hancock, Detroit MI 48084, USA, b) Frankfurt Institute for Advanced Studies, Johann Wolfgang Goethe University 60438 Frankfurt am Main, Germany.

E-mail: rajpol@wayne.edu

Abstract. Interest in the development of the theory of fluctuating hydrodynamics is growing [1]. Early efforts suggested that viscous diffusion broadens the rapidity dependence of two-particle transverse momentum correlations [2]. That work stimulated an experimental analysis by STAR [3]. In this work we present results of correlation observables computed using the second order Israel-Stewart hydrodynamics with stochastic noise and latest theoretical equations of state and transport coefficients. We also compute these observables using the first Navier-Stokes theory. We find that the second order theory better describes the data and also provides some novel features forced by causal constraints of the second order theory.

1. Introduction

Experiments carried out at Relativistic Heavy Ion Collider (RHIC) provided some clear signatures of formation of a deconfined state of quark and gluons, or quark-gluon plasma (QGP) [4, 5]. It was concluded that QGP behaves like a nearly perfect liquid. The excellent agreement of predictions of relativistic ideal hydrodynamic models [6, 7] with experimental data [8, 9] led to RHIC to conclude that the QGP behaves like a nearly perfect fluid. The quantity often used to indicate the perfectness, or lack thereof, in relativistic hydrodynamics is the ratio of viscosity to entropy density η/s. For a perfect fluid, its value approaches toward zero. Current consensus, based on viscous hydrodynamic models and experimental data, on the values of this ratio for a quark-gluon plasma is in the range between the conjectured lower bound (the "KSS" bound) of $1/4\pi$ and ~ 0.3 [10]. Note that we are using the units in which $\hbar = k_B = c = 1$. These values of η/s are much smaller than, for example, its value the superfluid liquid helium ($= 0.7$). Anisotropic flow data of produced particles have been traditionally used to estimate η/s. In Ref. [2], Gavin and Abdel-Aziz proposed a different method for estimating this ratio. They argued that viscous diffusion results in broadening the rapidity distribution of transverse momentum correlations and by measuring the change in width between peripheral and central collisions, one can estimate the value of η/s. STAR carried out experimental analysis [3] on this suggestion found that $0.06 < \eta/s < 0.21$. Note that these values are consistent with the values found from flow data.

Our current theory is based on the second order Israel-Stewart viscous hydrodynamics. Using this hydrodynamics and techniques of stochastic methods, we develop a second oder deterministic diffusion equation for two-particle transverse momentum correlation. The diffusion coefficient of the equation encapsulates the information on viscosity, more specifically information on η/s. We use a general temperature dependent η/s based on kinetic theory and latest development in equations of state to construct a general temperature dependent diffusion coefficient. Clearly, our current goal is not to estimate η/s. The goal is to employ the current knowledge of η/s and our theory to compute how a fluctuating initial correlation responds to viscous hydrodynamic expansion. Evolution of correlation observables proposed in Ref. [2] gives a measure of the response. These observables are the same observables measured by STAR [3]. These observables give a measure of contribution of viscous expansion on short range correlations as well as the relevant part of the long range correlations called the ridge.

We organize this paper as follows. We start with a brief discussion of diffusion of transverse momentum fluctuation in Section 2. We then apply the concepts to the two-particle p_t correlations in Section 3 and briefly mention the way we obtain our equations. We also discuss the main observables in that section. Finally, in Section 5, we present our results and compare them with experimental data.

2. Diffusion of Transverse Flow Fluctuations

Equations of dissipative relativistic hydrodynamics are derived from the basic principles of conservations of charges and of energy-momentum. Depending on the order of gradients of flow included in the dissipative part of the energy momentum tensor, relativistic hydrodynamics is said to be of first or second order. The relativistic version of the well known Navier-Stokes hydrodynamics is a first order relativistic hydrodynamics. The second order relativistic hydrodynamics is the also known Israel-Stewart theory. If we take small fluctuations on quantities (like the flow u^μ) over equilibrium values, we can see that transverse modes decouple from the longitudinal modes. The longitudinal modes are propagating sound modes while the transverse modes are solely the diffusing shear modes. As we are more interested in shear viscosity, we focus on the transverse modes. Note that shear viscosity is the most important mode of dissipation in QGP. Starting with relativistic hydrodynamic equations and constitutive relation for shear tensor (see Ref. [11], for example, for such equations and relations) and then linearizing, one arrives at diffusion equations for a transverse momentum current. If δu^y is the fluctuation in transverse flow from the average value, we have $T_{0y} \approx (\epsilon_0 + p_0)\delta u^y$. Here, ε_0 and p_0 are the local equilibrium values of energy density and pressure. The first order hydrodynamics leads to

$$\frac{\partial \delta T_{0y}}{\partial t} = \nu \nabla_z^2 \delta T_{0y} \tag{1}$$

Where ν is given by

$$\nu = \frac{\eta}{\varepsilon_0 + p_0} = \frac{\eta}{Ts} \tag{2}$$

Note that Eqn. (1) is a diffusion equation and that the diffusion coefficient contains the important ratio η/s. In our work, we refer this regular diffusion equation also as a first order diffusion equation. The diffusion coefficient ν is temperature dependent not just because of the explicit presence of T in Eqn. (2), but also because, as will be elaborated shortly, η/s itself is temperature dependent.

This diffusion equation, Eqn. (1), has a well known problem. Its solutions violate causality and is therefore not suitable for describing relativistic physics of quark-gluon plasma. It is a little ironic that the equation obtained from relativistic equation gives acausal solutions. It turns out that the first order gradient corrections contained in the relativistic Navier-Stokes equations are not accurate enough to preserve causality. The situation is resolved by resorting to second order hydrodynamics. For details on causality and diffusion see [11], [12], and the references therein. The second order hydrodynamics, on the other hand, gives us

$$\tau_\pi \frac{\partial^2 \delta T_{0y}}{\partial t^2} + \frac{\partial \delta T_{0y}}{\partial t} = \nu \nabla_z^2 \delta T_{0y} \tag{3}$$

We call this equation a second order diffusion equation. It is often referred to as causal diffusion equation, since it maintains causality. Note that the second order theory introduces the relaxation time τ_π and it is crucial in forcing the causality constraint in solutions to the equation. Note that in the limit $\tau_\pi \to 0$, we retrieve the first order diffusion equation. Note also that the second order equation has both diffusion and wave parts. The relative strength of the relaxation coefficient compared to the diffusion coefficient determines whether wave or diffusion behavior predominates.

3. Two-particle transverse momentum correlations

Correlation function of two-particle transverse momentum current can be defined as

$$r = \langle T_1^{0y} T_2^{0y} \rangle - \langle T_1^{0y} \rangle \langle T_2^{0y} \rangle \tag{4}$$

where $T_1^{0y} \equiv T^{0y}(\mathbf{x}_1)$ and $T_2^{0y} \equiv T^{0y}(\mathbf{x}_2)$ are momentum currents of particle pairs at points $\mathbf{x}_1$ and $\mathbf{x}_2$, respectively. Considering small fluctuations about the average, $T^{0y} = \langle T^{0y} \rangle + \delta T^{0y}$ and treating them as stochastic quantities, we can write $r = \langle \delta T_1^{0y} \bar{T}_2^{0y} \rangle + \langle \bar{T}_1^{0y} \delta T_2^{0y} \rangle + \langle \delta T_1^{0y} \delta T_2^{0y} \rangle$. The bars over the letters represent the same average - the average over the ensembles of events. Note the inclusion of the last term here. This term becomes important in stochastic methods, and is not dismissed as a second order term. It is a stochastic noise term and vanishes in the absence of noise. Using Eqn. (3) and the expression for r and manipulating the terms we can subtract away the noise term to obtain [13]

$$\tau_\pi \frac{\partial^2 \Delta r}{\partial t^2} + \frac{\partial \Delta r}{\partial t} = \nu (\nabla_{z1}^2 + \nabla_{z2}^2) \Delta r. \tag{5}$$

Here, $\Delta r = r - r_{eql}$, with r_{eql} as the equilibrium value of Eqn. (4). Note that, although diffusion equation for the correlation r contains noise term, the second order diffusion equation for Δr is a deterministic equation.

If one starts with the first order equation for single particle case, Eqn. (1), we obtain a regular deterministic diffusion equation for Δr:

$$\frac{\partial \Delta r}{\partial t} = \nu (\nabla_{z1}^2 + \nabla_{z2}^2) \Delta r. \tag{6}$$

The last equation, Eqn. (5), was first obtained in Ref. [2] and is the basis of the alternative method of estimating η/s of quark-gluon plasma that we mentioned in Section 1. The idea proposed in that reference is to measure the broadening of the width of Δr (strictly, the width of the correlation observable $\mathcal{C}$ in relative rapidity, discussed below).

We have developed the diffusion equations in such a way as to focus on the longitudinal expansion with Bjorken boost invariance. This is, clearly, a simplified part of a more general three dimensional problem. The RHIC and LHC experiments have provided us complex two-particle correlation profiles consisting of ridges, valleys and peaks. The goal of the future extended version of our current theory is to understand the contribution of viscous hydrodynamic expansion in those correlation profiles. Currently, we limit ourselves to the correlation profiles only in rapidity window, not the correlations in azimuthal angle. However, as we will point out below that, as far as our integral correlation observables are concerned, we can leave out transverse expansion and consider only the longitudinal expansion. It turns out that the transverse expansion integrates out and becomes redundant.

Bjorken longitudinal boost invariance can be applied in the simplest way by writing equations in terms of rapidity and proper time. Bjorken boost invariant velocity $v_z = z/t$ is like one diminutional Hubble expansion along z direction. The suitable coordinate is the Milne coordinates $x^\mu = (\tau, x, y, \eta)$. Here τ and η are the longitudinal proper time and spacetime rapidity respectively: $\tau = \sqrt{t^2 - z^2}$ and $\eta = (1/2)\ \ln((t+z)/(t-z))$. [1] For longitudinal Bjorken expansion, fluid rapidity reduces to spacetime rapidity. In Minkowski coordinates $x^\mu = (t, x, y, z)$, Bjorken boost invariant flow is given by $\bar{u}^\mu = \gamma(1, 0, 0, z/t) = (t/\tau, 0, 0, z/\tau) = (\cosh\eta, 0, 0, \sinh\eta)$. In Milne coordinates, it becomes $u^\mu = (1, 0, 0, 0)$. Note that expression of Bjorken flow cannot get any simpler. A fluid particle expanding longitudinally with Bjorken flow is at rest in Milne coordinates. The affine connections (Christoffel symbols), however, are not all zero. The non-zero elements are $\Gamma^\eta_{\eta\tau} = 1/\tau$ and $\Gamma^\tau_{\eta\eta} = \tau$. Now we can use this to find the expansion rate $\nabla_{,\mu}u^\mu \rightarrow \nabla_{;\mu}u^\mu = \partial_{;\mu}u^\mu = \partial_\mu u^\mu + \Gamma^\mu_{\mu\nu}u^\nu = \Gamma^\eta_{\eta\tau}u^\tau = 1/\tau$, for example. We can use this coordinate system to derive our diffusion equations in terms of proper time and spacetime rapidity. The second order equation, Eqn. (5), takes the form

$$\tau_\pi \frac{\partial^2 \Delta r}{\partial \tau^2} + \frac{\partial \Delta r}{\partial \tau} = \frac{\nu}{\tau^2}(\nabla^2_{\eta 1} + \nabla^2_{\eta 2})\Delta r. \tag{7}$$

Clearly, η_1 and η_2 are respective spacetime rapidities for particle 1 and particle 2. Recall that we are not including the transverse expansion. Similarly, the first order equation takes the form

$$\frac{\partial \Delta r}{\partial \tau} = \frac{\nu}{\tau^2}(\nabla^2_{\eta 1} + \nabla^2_{\eta 2})\Delta r \tag{8}$$

The last two equations, Eqns. (8) and (7), are solved numerically and the solutions Δr's are used to compute the observables. The main observable is the correlation profile $\mathcal{C}$ given by (see Ref. [2] for detail):

$$\mathcal{C} = \langle N \rangle^{-2} \int \Delta r(\mathbf{x_1}, \mathbf{x_2}) d^3x_1 d^3x_2 \tag{9}$$

Again, following the Ref. [2], this observable is related to experimentally measurable p_t covariance:

$$\mathcal{C} = \langle N \rangle^{-2} \langle \sum_{i \neq j} p_{ti} p_{tj} \rangle - \langle p_t \rangle^2 \tag{10}$$

The last two relations give the connection between our theoretically computed Δr to experimentally measurable two-particle transverse momentum correlations.

The other observable is the width of $\mathcal{C}$. STAR has reported $\mathcal{C}$ and its widths for various centralities [3]. In Section 5, we present the results of our computation and compare them with the STAR data. Before that, we would like to elaborate on the temperature dependent η/s in the following section.

4. Viscosity and entropy density

Viscosity and relaxation coefficients are determined from kinetic theory. In Ref. [14], Hirano and Gyulassy have combined the latest information on η: results from perturbative QCD, kinetic theory, and $\mathcal{N} = 4$ Super Yang-Mills (SYM) at infinite coupling strengths, in the respectively applicable temperature ranges. They also propose that low values of η/s for quark-gluon plasma is not because viscosity suddenly drops in QGP phase, but because the entropy density increases when the system makes a transition to QGP from hadronic phase. We use the expressions for shear viscosity given in that reference. As for the entropy density, we make use of the values calculated using lattice QCD. Specifically, values given in s95p-v1 from Ref. [15, 16] have been used in this work. We refer to this equation of state as EOS I here. Another equation of state we use here is the standard equation of state based on the bag model, also summarized in [14].

[1] Note that the same letter η is used for shear viscosity and for the longitudinal spacetime rapidity here. The meaning should be clear from the context of its use.

With the information on $\eta(T)$ and $s(T)$, we construct the ratio of viscosity to entropy density $\eta/s(T)$. This gives us the temperature dependent diffusion coefficient $\nu(T)$. In order to get ν as a function of proper time, necessary to make our correlation Δr evolve, one needs entropy density as a function of time. We employ the entropy production equations for this purpose. It should be noted that entropy production equations are obtained from the same hydrodynamic equations and, therefore, can be of first or second order. The general equation, with Bjorken longitudinal boost invariance, can be written as [11]:

$$\frac{ds}{d\tau} + \frac{s}{\tau} = \frac{\pi}{T\tau}, \tag{11}$$

where π is the entropy production current, which comes from shear viscosity. In Milne coordinates: $\pi = \pi^{\eta}_{\eta}$ (or $\pi = \pi^0_0 - \pi^z_z$ in Minkowski coordinates). For an ideal fluid, $\pi = 0$, and we obtain the well known Bjorken relation $s(\tau) = s_0\tau_0/\tau$. For Navier-Stokes or first order theory, $\pi = 4\eta/3\tau$ and for the second order theory π is given by the differential equation

$$\tau_\pi \frac{d\pi}{d\tau} + \left(1 + \frac{\tau_\pi}{2\tau} + \frac{1}{2}\eta T \frac{d}{d\tau}(\frac{\tau_\pi}{\eta T})\right)\pi = \frac{4\eta}{3\tau}. \tag{12}$$

Since the second order diffusion and entropy production equations are obtained from the same second order Israel-Stewart hydrodynamics, we are speaking of the same relaxation time.

Like viscosity, relaxation time is obtained from kinetic theory. The latest and comprehensive treatment on kinetic theory estimate of relaxation times are in Ref. [17]. This reference gives $\tau_\pi = \beta\frac{\eta}{Ts}$ with $\beta = 5 - 6$. The value of β suggested in Ref. [18] is 6.32. These values of β are significantly higher than those which are used in hydrodynamic calculations by some other groups (for example, $\beta = 3$ in Ref. [19]). We, however, adopt the kinetic theory values. In our work here, we have used $\beta = 6$.

Before we present our results in Section 5, we would like to discuss the redundancy of transverse expansion for our observables. Here is a simple, but general, argument based on conservation of energy-momentum, $\partial T^{\mu\nu} = 0$. Integrating it over $\mathbf{r}_\perp$, which is, effectively, the same as integrating over the azimuthal angle ϕ, we obtain

$$\int d\mathbf{r}_\perp \partial_0 T^{0\nu} + \int d\mathbf{r}_\perp \partial_i T^{i\nu} = 0 \tag{13}$$

Let us take y as a transverse component, as before($\nu = y$). We get

$$\begin{aligned} 0 &= \int \mathrm{d}x\,\mathrm{d}y\,\partial_0 T^{0y} + \int \mathrm{d}x\,\mathrm{d}y\,\partial_i T^{iy} \\ &= \int \mathrm{d}x\,\mathrm{d}y\,\frac{\partial T^{0y}}{\partial t} + \int \mathrm{d}x\,\mathrm{d}y\,(\partial_x T^{xy} + \partial_y T^{yy}) + \int \mathrm{d}x\,\mathrm{d}y\,\frac{\partial T^{zy}}{\partial z} \end{aligned} \tag{14}$$

The integral of the middle term in the second line is a surface integral of a gradient. If we take the surface to "infinity", i.e., to a large $r_\perp$, this term vanishes. Note that this is the very term that contains the transverse expansion. The last one is the longitudinal expansion term. Upon linearization and using $T_{zy} = -\eta\partial_z v^y$, we obtain a diffusion equation. Thus, a three dimensional diffusion equation when averaged over $\mathbf{r}_\perp$ gives our longitudinal diffusion equations.

Thus, even if we do not include the transverse expansion, our results on the observable $\mathcal{C}$ and its widths are general in nature, as long as the conditions for the vanishing surface terms are valid.

Journal of Physics: Conference Series **458** (2013) 012005 doi:10.1088/1742-6596/458/1/012005

5. Results and discussion

We now present the results of our computation. Before that let us mention some of the parameters used in the theory. It is already mentioned that we have used $\beta = 6$. The initial or thermalization time is taken to be $\tau_0 = 1$ fm/c. Initial Δr is taken to be a gaussian in relative rapidity $\Delta\eta$, with initial width $\sigma_0 = 0.54$, making it consistent with experimental data for the most peripheral collisions. We stop the hydrodynamic evolution at a constant temperature of 150 MeV. Glauber model has been used to connect impact parameters and multiplicity of the collisions. The freezeout proper time is assumed as $\tau_F - \tau_0 \propto (R - R_0)^2$, with R as the rms participant radius. Freezeout time for most central collisions is taken to be 9 fm/c. The computed σ and $\mathcal{C}$ are compared with the values measured recently by STAR [3].

Figure 1 shows widths versus centralities computed using both first and second order theory with two different equations of state - EOS I and EOS II (equations of state used in our computations are discussed in Section 4). We observe that second order calculations agree very well with the experimental data. We also note that the choice of EOS makes a very small difference. However, there is significant difference between the first and second order diffusion, except for a few most central cases. The deviation of the first order results from data relates to how the relaxation time compares with the evolution time [13]. This clearly shows that one needs causally constrained hydrodynamic evolutions to better explain the experimental data. Figure 1 also compares results with NEXSPHERIO calculations [20]. NEXSPHERIO uses inviscid hydrodynamics in evolution of initial correlations. It reproduces most qualitative observed features of the two-particle correlations. It, however, does not reproduce the broadening of the rapidity width with increasing centrality, as one can see in Fig. 1. We attribute that to the absence of viscosity. Other explanations involving collective initial state behavior are also possible. One important reason for comparing with the NEXSPHERIO results is to look for any broadening effect of resonances, which is included in that computation. Since we do not observe any broadening in NEXSPHERIO results, we infer that resonances, a source of short range correlations, do not have important, if any, effect in our observables.

Figure 2 displays the observable $\mathcal{C}$ vs $\Delta\eta$ for various centralities from our numerical calculations and the values measured by STAR [3]. The STAR measured values have error bars in the offsets (ridge), which we have not shown here. This figure shows a good agreement of the numerical results with the experiments. The single point at $\Delta\eta = 0$ in the most central case has been attributed to the track merging and is just a detector artifact. One important observation we make here is the flattening of the peak (and to some extent double humps) in $\mathcal{C}$. The reason for the double hump (or rather the flattened peak) is a second order diffusion effect. We note that the second order diffusion equation has propagating wave part as well. Thus there are competing wave and diffusion effects, depending on the size of τ_π and ν. Wavefronts propagate in opposite directions and diffusion fills in the space in between. This can be seen more effectively in coordinate space (not shown here). In rapidity space, width saturates since the effective diffusion coefficient ν/τ^2 in Eqn. (7) decreases rapidly with time. The first order diffusion obviously does not show such humps and flattening - there are no propagating waves. Figure 3 compares $\mathcal{C}$ obtained from the solutions of the first and second order theory. It can be observed that STAR data for other centralities [21] show the humps exactly at the same centralities. Shapes show fair agreement with the calculations. Here we have restricted ourselves with the published data only for the three centralities shown in Fig 2. Finally, we just want to mention that the order of entropy production equations does not make any discernible difference in our results. Note that we have relatively large β in $\tau_\pi = \beta\eta/Ts$.

It would be interesting to measure the observable $\mathcal{C}$ from the p-A [22] and Pb-Pb [23] collisions data from LHC. In the pA case, it would be interesting to see whether there is a broadening with respect to pp. Experiments have identified a ridge in pA [22]. If hydrodynamics is applicable in the pA system, then viscous diffusion would broaden the rapidity width of $\mathcal{C}$.

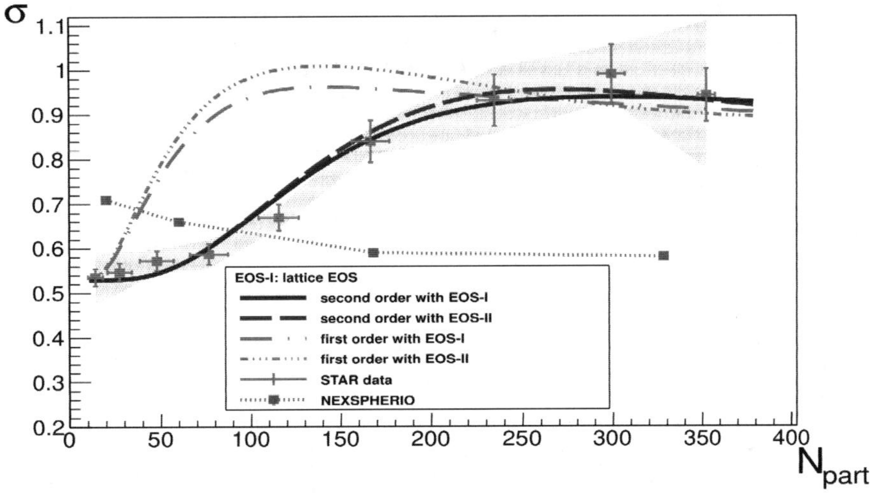

Figure 1. Second order, first order, with both EOS [13] and NeXSPHeRIO [20] calculations of width vs STAR [3] measurements.

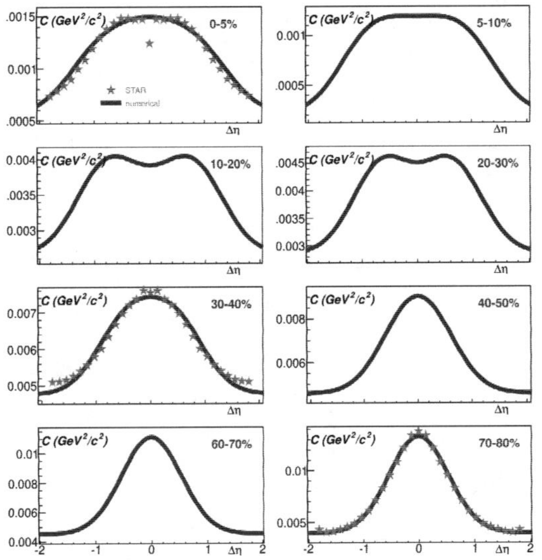

Figure 2. $\mathcal{C}$ vs $\Delta\eta$. Comparison with STAR data at different centralities. [3].

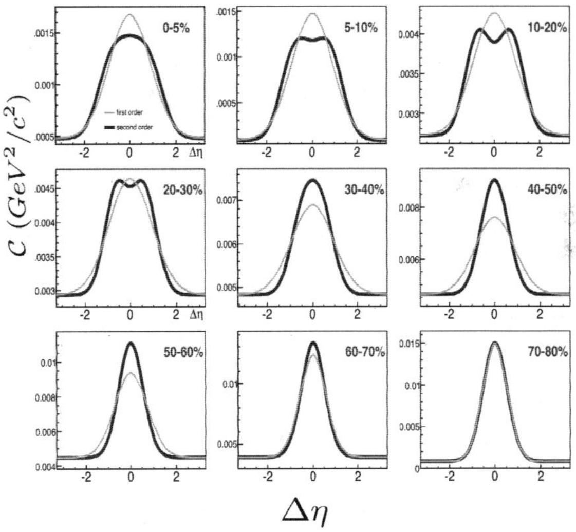

Figure 3. Profiles of correlation observable $\mathcal{C}$ obtained from the solutions of the first and second order diffusion equations.

6. Acknowledgement

We would like to thank C. Pruneau and M. Sharma for a great deal of communication and discussions of the experimental results and for generously providing STAR data. RP would like to thank P. Huovinen for his helpful responses to questions on lattice data. This work was supported by U.S. NSF grant PHY-1207687.

References

[1] Kapusta J, Muller B and Stephanov M 2012 *Phys.Rev.* **C85** 054906 (*Preprint* 1112.6405)
[2] Gavin S and Abdel-Aziz M 2006 *Phys. Rev. Lett.* **97** 162302 (*Preprint* nucl-th/0606061)
[3] Agakishiev H *et al.* (STAR Collaboration) 2011 *Phys. Lett.* **B704** 467–473 (*Preprint* 1106.4334)
[4] Adams J *et al.* (STAR Collaboration) 2005 *Nucl.Phys.* **A757** 102–183 (*Preprint* nucl-ex/0501009)
[5] Adcox K *et al.* (PHENIX Collaboration) 2005 *Nucl.Phys.* **A757** 184–283 (*Preprint* nucl-ex/0410003)
[6] Huovinen P, Kolb P, Heinz U W, Ruuskanen P and Voloshin S 2001 *Phys.Lett.* **B503** 58–64 (*Preprint* hep-ph/0101136)
[7] Teaney D, Lauret J and Shuryak E 2001 (*Preprint* nucl-th/0110037)
[8] Adler S *et al.* (PHENIX Collaboration) 2003 *Phys.Rev.Lett.* **91** 182301 (*Preprint* nucl-ex/0305013)
[9] Adams J *et al.* (STAR Collaboration) 2005 *Phys.Rev.* **C72** 014904 (*Preprint* nucl-ex/0409033)
[10] Teaney D A 2009 (*Preprint* 0905.2433)
[11] Muronga A 2004 *Phys. Rev.* **C69** 034903 (*Preprint* nucl-th/0309055)
[12] Aziz M A and Gavin S 2004 *Phys. Rev.* **C70** 034905 (*Preprint* nucl-th/0404058)
[13] Pokharel R, Gavin S and Moschelli G in preparation
[14] Hirano T and Gyulassy M 2006 *Nucl. Phys.* **A769** 71–94 (*Preprint* nucl-th/0506049)
[15] Huovinen P and Petreczky P 2010 *Nucl. Phys.* **A837** 26–53 (*Preprint* 0912.2541)
[16] https://wiki.bnl.gov/hhic/index.php/Lattice_calculatons_of_Equation_of_State
[17] York M A and Moore G D 2009 *Phys. Rev.* **D79** 054011 (*Preprint* 0811.0729)
[18] Hong J, Teaney D and Chesler P M 2011 (*Preprint* 1110.5292)
[19] Song H and Heinz U W 2009 *J.Phys.G* **G36** 064033 (*Preprint* 0812.4274)
[20] Sharma M, Pruneau C, Gavin S, Takahashi J, de Souza R D *et al.* 2011 *Phys.Rev.* **C84** 054915 (*Preprint* 1107.3587)
[21] Sharma M and Pruneau C private communications
[22] D. Velicanu's presentation, Hot Quarks 2012
[23] Aamodt K *et al.* (ALICE Collaboration) 2010 *Phys. Rev. Lett.* **105** 252302 (*Preprint* 1011.3914)

29th Winter Workshop on Nuclear Dynamics (WWND2013) IOP Publishing
Journal of Physics: Conference Series **458** (2013) 012006 doi:10.1088/1742-6596/458/1/012006

Very High-p_T Triggered Dihadron Correlations in PbPb Collisions at 2.76 TeV with CMS

Rylan Conway (for the CMS Collaboration)
University of California, Davis
E-mail: Rylan.Towne.Conway@cern.ch

Abstract. Measurements of dihadron correlations triggered by very high-p_T particles in 2.76 TeV PbPb collisions are shown. The data set, which corresponds to an integrated luminosity of 150 μb^{-1}, collected by the CMS detector is utilized. Long-range correlations are measured up to $p_T \sim 50$ GeV/c. These correlations are driven by single-particle azimuthal anisotropies and characterized by the Fourier harmonics, v_n. Once the $v_2 - v_4$ harmonic components are subtracted the associated particle yields on the near and away side of the jet-like correlations are studied. These measurements are done over a wide kinematic range in the associated and trigger particle p_T and as a function of centrality. The data are compared to pp collisions at the same energy. A suppression of about 50% from the pp reference in the away-side associated yield is observed for $p_T^{assoc} > 3$ GeV/c. For lower momentums, $p_T^{assoc} \sim 0.5$ GeV/c, the yield is found to be enhanced significantly by up to a factor of 3-4 on the away-side. On the near side, there is some evidence of a moderate enhancement.

1. Introduction

Opaqueness to high-energy quarks and gluons, known as jet quenching, is one of the key signatures of the Quark Gluon Plasma (QGP). This effect can be measured using dihadron correlations by comparing the integrated-yields in pp to PbPb collision for the near and away-side regions. The contributions from flow and anisotropies in the interaction region must be subtracted in order to isolate the underlying jet-like correlations. The I_{AA} modification factor is used to quantify the amount of suppression/enhancement with respect to a pp reference and is described in more detail in the next section. These measurements provide important constraints on energy-loss models of a QGP medium.

2. Methods

The measurements described in this section were all done using the CMS detector, which is described in detail in Ref. [1]. Two-particle associated yield distributions are constructed by dividing the signal distribution by the background distribution and integrating across the range $|\Delta\eta| < 1$ and from $0 < \Delta\phi < 1$ for the near-side and $1 < \Delta\phi < \pi$ for the away-side. The procedure is described in more detail in Refs. [2, 3]. The contributions from single-particle anisotropies in the measured correlation function need to be subtracted. These single-particle anisotropies are given by the Fourier harmonics, v_n, and were measured out to very high-p_T using the event plane method in Refs. [3, 4]. The I_{AA} modification factor, which is the ratio of the integrated associated yields in PbPb to that in pp collisions as a function of p_T^{trig} and p_T^{assoc}, is used to quantify the amount of suppression observed in the near and away sides, as in Ref. [5]. A cross check was done by calculating the near-side I_{AA} values using a long-range $\Delta\eta$ subtraction method, which is described in more detail in Ref. [3]. In this method the correlations from single-particle azimuthal anisotropies are estimated from the two-particle correlation functions averaged over "long-range" region, $1 < |\Delta\eta| < 4$, and then subtracted from the full correlation function, $|\Delta\eta| < 4$. The single-particle anisotropies are expected to extend to large values of $\Delta\eta$ and the jet-correlations on the near-side do not. This is a good alternative approach to estimating and then subtracting the background correlations. Since the jet-correlations extend to large $\Delta\eta$ value on the away-side, the long-range $\Delta\eta$ subtraction method can only be applied to the near side.

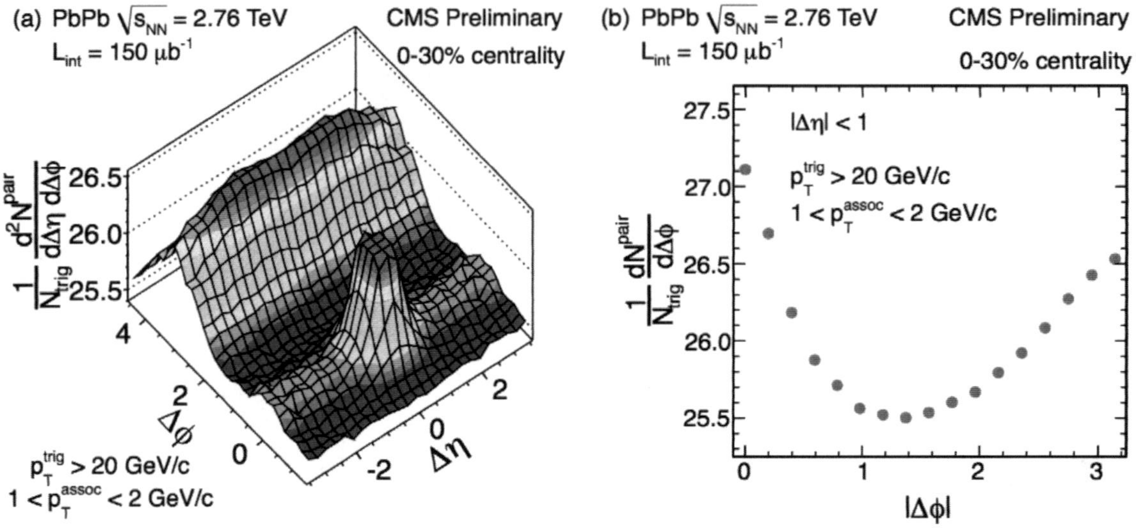

Figure 1. The per-trigger-particle associated yield of charged particles in (a) two-dimensions (2-D) as a function of $\Delta\eta$ and $\Delta\phi$ and (b) one-dimension (1-D) as a function of $|\Delta\phi|$ averaged over $0 < |\Delta\eta| < 1$ for $p_T^{trig} > 20$ GeV/c and $1 < p_T^{assoc} < 3$ GeV/c from the 0-30% centrality range of PbPb collisions at $\sqrt{s_{NN}} = 2.76$ TeV. The near-side peak in (a) is truncated to better display the surrounding structure.

3. Results

The near-side I_{AA} results are shown in Fig. 2 as a function of p_T^{assoc} for various p_T^{trig} bins and with $|\Delta\eta| < 1$. The red points on the top row are for the 0–10% most central collisions while the blue points on the bottom row are for the 50–60% central collisions. The black data points are from the long-range $\Delta\eta$ subtraction method while the colored points are calculated from the v_n subtraction method. At low p_T^{assoc} there is a noticeable enhancement of up to a factor of 2 for both methods even though they disagree slightly. This disagreement is possibly due to the fact that v_1 is not included in the v_n subtraction method but inherently included in the long-range $\Delta\eta$ subtraction method. The away-side I_{AA} are also shown as a function of p_T^{assoc} in Fig. 3 with $|\Delta\eta| < 1$ and in four different p_T^{trig} bins. There is a significant enhancement up to a factor of 3-4 at low p_T^{assoc} and a suppression of 50% at high p_T^{assoc}. The near-side I_{AA} values are shown as a function of N_{part} in Fig. 4 for four different p_T^{assoc} bins with $19.2 < p_T^{trig} < 24$ GeV/c. The blue points show the results derived from the v_n-subtraction method at $0 < |\Delta\eta| < 1$ while the black points show the results derived from the long-range $\Delta\eta$ subtraction method. There is a clear correlation with N_{part} at low p_T^{assoc} and becomes less significant at higher p_T^{assoc}. The away-side I_{AA} values are shown as a function of N_{part} in Fig. 5 for four different p_T^{assoc} bins with $19.2 < p_T^{trig} < 24$ GeV/c and $0 < |\Delta\eta| < 1$. Again, there is a very strong correlation with N_{part} which shows a large enhancement at low p_T^{assoc} and high N_{part}. This correlation changes to a suppression at higher p_T^{assoc} and high N_{part} values.

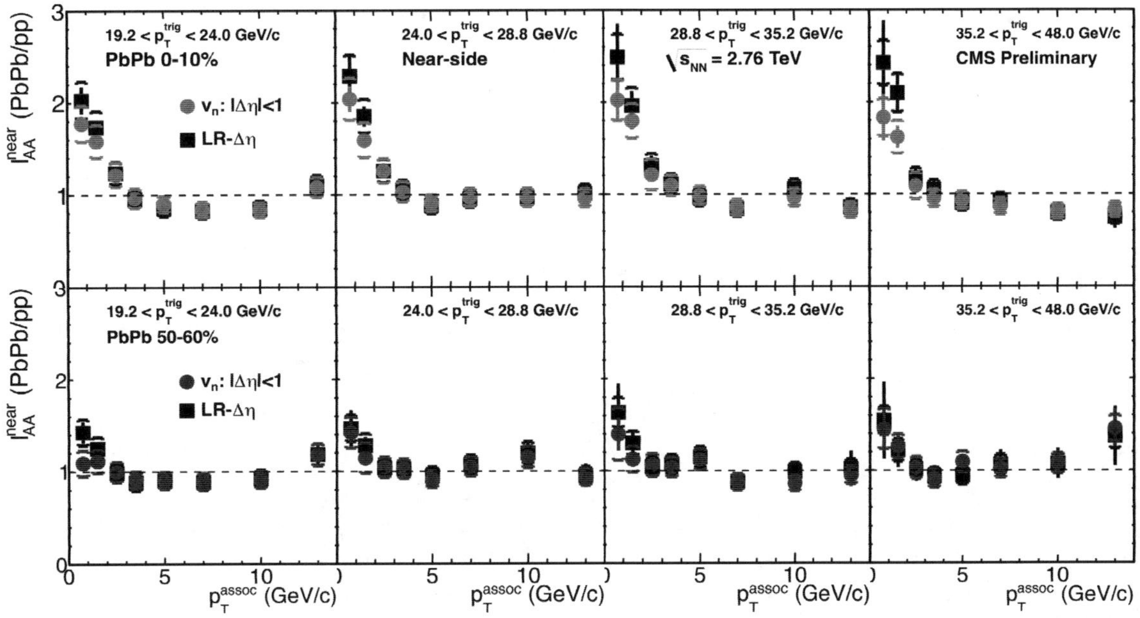

Figure 2. Near-side I_{AA} values derived from the v_n-subtraction method at $|\Delta\eta| < 1$ shown for four different p_T^{trig} ranges as a function of p_T^{trig} for PbPb collisions at $\sqrt{s_{NN}} = 2.76$ TeV, the top for 0–10% centrality and bottom row for 50–60% centrality. The black squares represent the values calculated from the long-range-$\Delta\eta$ subtraction method. The error bars represent statistical uncertainties while the brackets represent the systematic uncertainties.

4. Conclusion

The near- and away-side I_{AA} values as a function of p_T^{trig} and p_T^{assoc} for different centrality ranges were measured in PbPb collisions at $\sqrt{s_{NN}} = 2.76$ TeV. Long-range, hydrodynamic flow-like behavior is accounted for by a separate measurement of harmonic v_n coefficients using the event-plane method. There is a large enhancement of up to a factor of 3-4 in the away-side I_{AA} for 0-10% central PbPb at low p_T^{assoc} and a suppression of about 50% at high p_T^{assoc} for all p_T^{trig} ranges studied in the analysis. This is consistent with the jet-quenching picture, namely that the medium-induced energy loss from high-p_T patrons is transferred to low-p_T particles, especially below $p_T = 2$ GeV/c. There is a less significant enhancement, about a factor of 2, observed in the near-side associated particle yield at low-p_T^{assoc} in the most central collisions. The near-side I_{AA} results at intermediate and higher p_T^{assoc} ranges show a significantly smaller difference between PbPb and pp collisions since high-p_T^{trig} particles on the near-side are much more likely to come from the surface of the medium. These measurements can help provide quantitative constraints on parton energy loss mechanisms in a QGP medium.

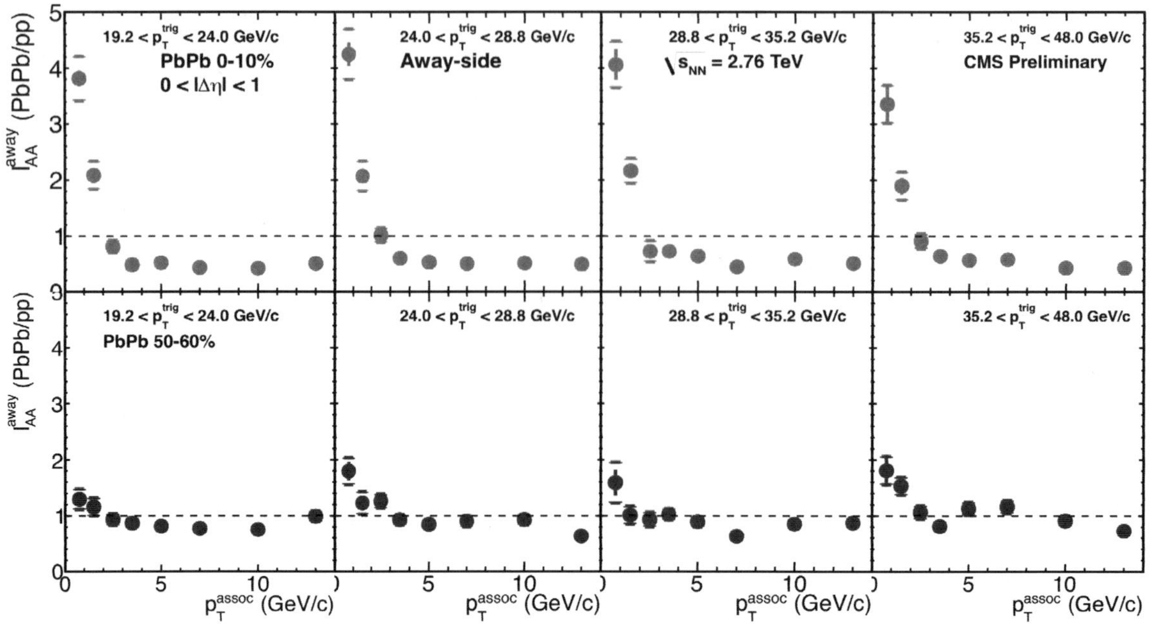

Figure 3. Away-side I_{AA} values derived from the v_n-subtraction method at $|\Delta\eta| < 1$ shown for four different p_T^{trig} ranges as a function of p_T^{trig} for PbPb collisions at $\sqrt{s_{NN}} = 2.76$ TeV, the top for 0–10% centrality and bottom row for 50–60% centrality. The error bars represent statistical uncertainties while the brackets represent the systematic uncertainties.

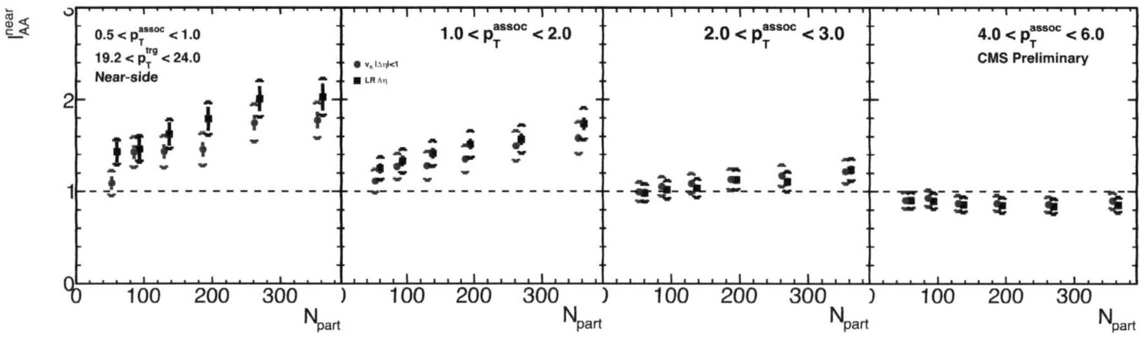

Figure 4. Near-side I_{AA} shown for four different p_T^{assoc} ranges as a function of N_{part} for $19.2 < p_T^{trig} < 24$ GeV/c in PbPb collisions at $\sqrt{s_{NN}} = 2.76$ TeV. The solid blue circles show the results derived from the v_n-subtraction method at $|\Delta\eta| < 1$, while the black squares show the results derived from the long-range $\Delta\eta$ subtraction method. The error bars represent statistical uncertainties while the brackets represent the systematic uncertainties.

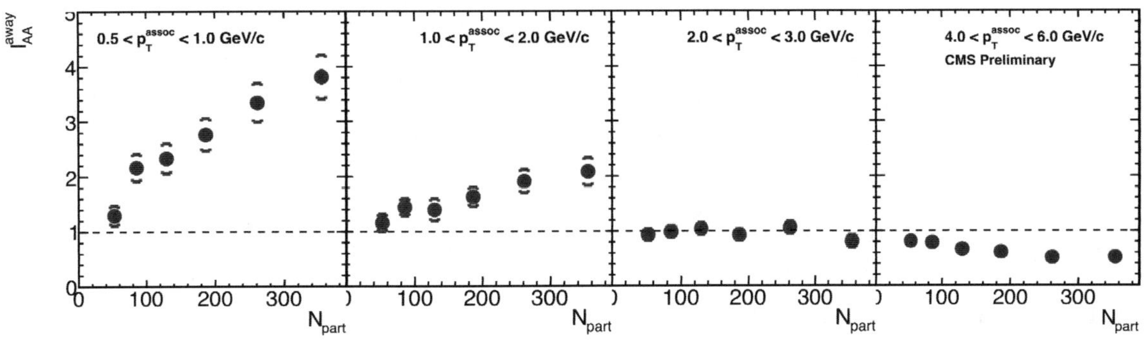

Figure 5. Away-side I_{AA} values derived from the v_n-subtraction method shown for four different p_T^{assoc} ranges as a function of N_{part} for $19.2 < p_T^{trig} < 24$ GeV/c and $|\Delta\eta| < 1$ in PbPb collisions at $\sqrt{s_{NN}} = 2.76$ TeV. The error bars represent statistical uncertainties while the brackets represent the systematic uncertainties.

5. References

[1] The CMS Collaboration, JINST 3, S08004, (2008).

[2] The CMS Collaboration, "Centrality dependence of dihadron correlations and azimuthal anisotropy harmonics in PbPb collisions at $\sqrt{s_{NN}} = 2.76$ TeV", Eur. Phys. J. C 72 (2012) 2012. doi:10.1140/epjc/s10052-012-2012-3, arXiv:1201.3158

[3] The CMS Collaboration, "Very high-p_T dihadron correlations in PbPb collisions at $\sqrt{s_{NN}} = 2.76$ TeV", CMS-PAS-HIN-12-010

[4] The CMS Collaboration, "Azimuthal anisotropy of charged particles at high transverse momenta in PbPb collisions at $\sqrt{s_{NN}} = 2.76$ TeV", arXiv:1204.1850v1 [nucl-ex] 9 Apr 2012

[5] The ALICE Collaboration, Phys. Rev. Lett. 108, 092301 (2012)

Journal of Physics: Conference Series **458** (2013) 012007 doi:10.1088/1742-6596/458/1/012007

Forward/backward $J/\psi\rightarrow\mu\mu$ production in Cu+Au collisions at PHENIX

Richard S. Hollis, for the PHENIX Collaboration
University of California, Riverside
E-mail: rhollis@ucr.edu

Abstract. RHIC is a versatile machine which has the unique capability to collide different nuclei. One new system recently explored is that of Cu on Au at 200 GeV. One of the first PHENIX measurements made in these collisions was that of forward and backward J/ψ production, where the production mechanism for J/ψ is believed to be dominated by gluon-fusion interactions. Resultant yields from such interactions can be modified in both cold and hot nuclear matter, where the asymmetric nature of the initial state of this collision system may give rise to a systematic difference in forward and backward particle production. In these proceedings, we present our latest studies and compare the resulting distributions to those from earlier heavy-ion collision systems.

1. Introduction

During the 2012 RHIC run, the flexibility of the RHIC accelerator was enacted to collide ions of asymmetric size (Cu on Au), allowing the exploration of the unique phase space created in such collisions. For very central collisions, the smaller radius Cu is entirely swallowed by the larger gold nucleus. For off-center (mid peripheral) collisions, the differing radii form naturally odd harmonics along with the expected even harmoincs (v_2) which occur in symmetric systems. In *core* and *corona* type models, these Cu+Au collisions lead to a large left/right asymmetry (largest corona on the Au-side). Aside from the transverse asymmetries, a longitudinal asymmetry exists in terms of the forward/backward momenta. The larger Au nucleus, for a given collision, contributes more participants than the Cu nucleus, which may play a role in forward (Cu-going) and backward (Au-going) particle production. The previous asymmetric system collided (d+Au) probed the role of cold nuclear matter, particle production in the absence of the hot, dense Quark-Gluon Plasma (QGP). For this new system, QGP effects are very much at the fore in terms of particle production and modification.

For this new type of collision system, PHENIX earmarked several analyses for fast-track evaluation of the collision dynamics. The soft sector focused on the azimuthal dependence on particle production, see [1], whereas the hard sector was studied via the measurement of forward/backward J/ψ yields. The production of J/ψ is a hard collision product and is dominated by gluon fusion. It was theorized [2] that this loosely bound charmonium state may not survive the hot, dense matter created in the collision of large nuclei, resulting in a suppression relative to the expected production from p+p collisions. Observance of this suppression was heralded as an initial signature for QGP formation. However, in d+Au collisions, it is observed that, even in the absence of the hot, dense QGP, the yields of J/ψ are significantly modified [3].

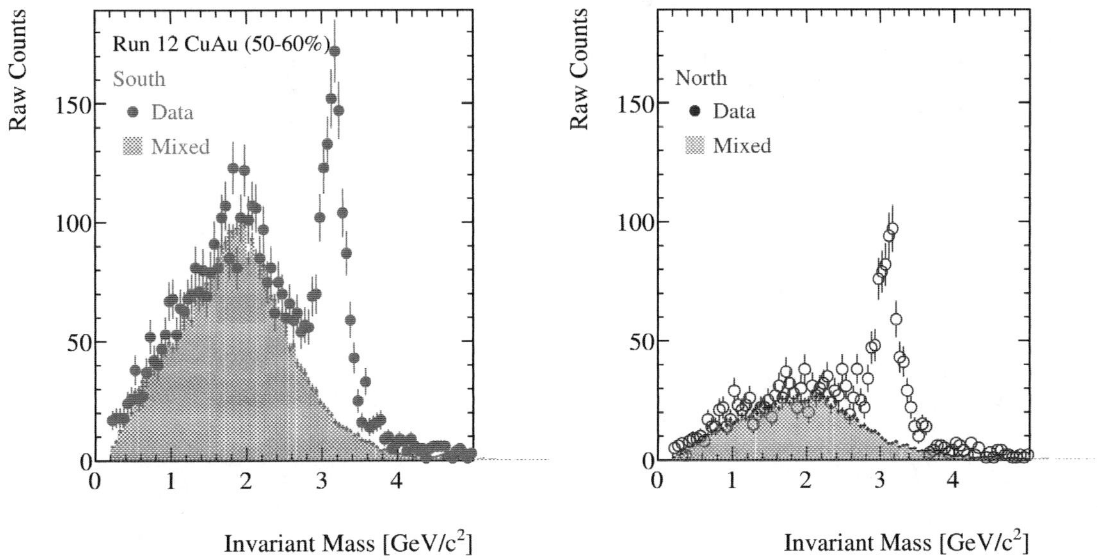

Figure 1. Raw di-μ spectrum from mid-peripheral (50-60% central) Cu+Au collisions. The left panel shows raw signal (including combinatorial background) data from the South-arm (Au-going) muon-candidate pairs. The left panel shows pairs from the North (Cu-going) spectrometer. The shaded area represents the (combinatorial) background, determined from event-mixing (see text).

It is then difficult to disentangle the complex interplay between cold (d+Au) and hot (A+A) collisional modification simply with two systems. Measurements of the modification in a smaller system, Cu+Cu [4], are found to be commensurate with the Au+Au system, leading to the conclusion that further variables are needed to disentangle all the effects. The Cu+Au system becomes a favorable source owing to its natural asymmetries in both the forward/backward and side/side regions.

There are an array of effects in both hot and cold nuclear matter which can contribute to the modification of the spectra. In cold nuclear matter (often referred to as CNM), nuclear modification to the parton distribution functions lead to a suppression of low-x gluons (known as the "shadowing region") and a compensatory enhancement of large-x gluons ("anti-shadowing region"). Further effects also measured in light-meson and baryon production include the Cronin effect, due to multiple partonic scatterings and energy loss/break-up in the nucleus. Once the initial-state (CNM) modification has occurred, the hot nuclear matter (or final state effects) further change the measured yields of J/ψs, predominantly due to color screening.

The current interpretation of the results from RHIC, and the LHC, require both cold- and hot-nuclear matter effects, although the true relative contribution is difficult to disentangle. Colliding different systems, at the same energy, provides additional insight into the relative importance of such mechanisms.

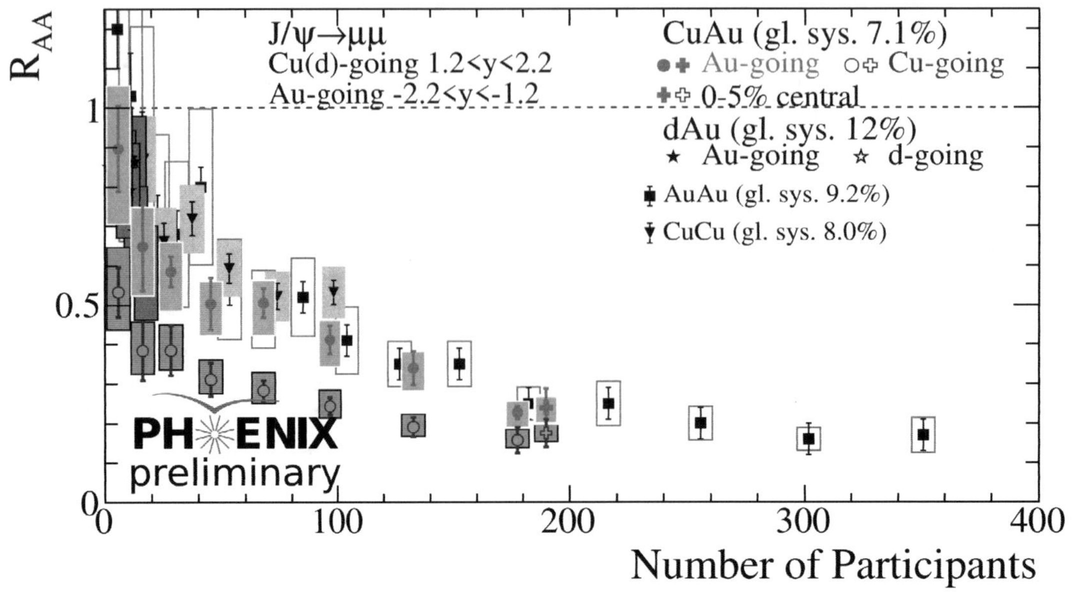

Figure 2. Comparison of the nuclear modification factor found in Cu+Au collisions (red and blue) to Au+Au (square symbols), Cu+Cu (triangles), and d+Au (stars) data. The red depict the R_{AA} in the Au-going direction, the blue show the Cu-going data. The cross-symbol shows the top 5% of collision cross-section in which the Cu-nucleus is completed swallowed by the Au-nucleus.

2. J/ψ measurement

The PHENIX experiment has three spectrometers for measuring charmonia: the central arm spectrometer with tracking and electromagnetic calorimetery, not discussed in these proceedings, and forward (north) and backward (south) muon spectrometers. These muon spectrometers are similarly constructed with a tracker inside of a magnetic field, followed by a muon identifier. The muon identifier is several layers of iron, inter-spaced with Iorocci tubes for read-out. A particle is identified as a muon if it completely penetrates the muon identifier. To reduce the number of hadrons into the muon-tracker, additional hadronic absorber material was added after RHIC Run 9. Hadrons which are identified as muons are strongly suppressed and contribute to the combinatoric background.

Further reduction of the non-muon backgrounds is made through several track-quality residual variables. The reconstruction of tracks in the muon system is broken into two broad pieces, a momentum-determining track in the tracker and a muon-identifying road in the muon-identifier. From the tracker, the track-χ^2 measures the closeness of the found track to that expected in the spectrometer assuming no multiple-scattering. Roads formed in the muon-identifier can be up to four planes "deep"; hadronic candidates are more likely to fail to penetrate through all planes and therefore requiring "deep" roads signals a likely muon candidate. Matching the road and track portions also reduces hadronic backgrounds as the larger multiple scattering for hadrons causes a mis-match in joining the road, both in actual spatial position and the slope between the two pieces.

For the first time in heavy-ion collisions, the data-rate from RHIC exceeded the PHENIX ability to collect all the data. All muon-arm triggers were retained, as well as other rare triggers

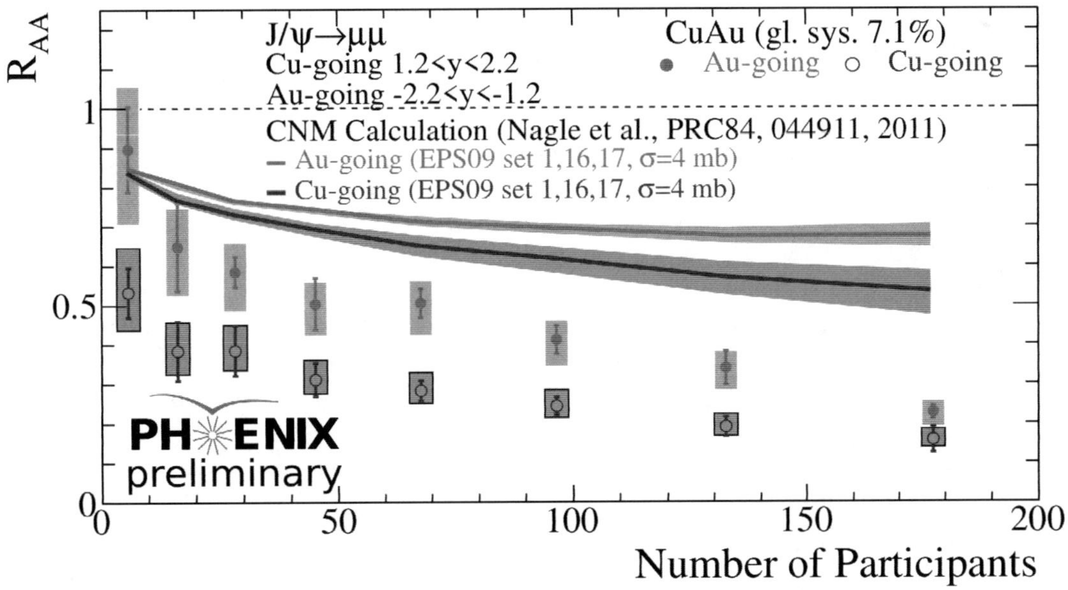

Figure 3. Comparison between the nuclear modification factor found in Cu+Au collisions (red and blue symbols) and a model (lines) which describes cold-nuclear effects, see text for a description. The red (blue) lines represent this models cold nuclear matter component of the Au- (Cu-)going data.

for other analyses, however only a fraction of minimum bias data were collected. The dataset analyzed comprises of about 12 billion minimum bias events sampled, from which over 25 K J/ψs were found.

The data were divided into ten centrality classes, based on the energy signals in the Beam-Beam-Counters (BBCs). In addition to the requirements of track quality, the event vertex was required to be within 10 cm of the nominal interaction position in PHENIX.

Figure 1 shows the measured dimuon-pairs, after the subtraction of the non-correlated background. The background subtraction was formed via a mixing of candidate tracks from two separate events, which necessarily breaks all correlations. The signal peak is fit using a double-Gaussian form (which is found to best describe the lower-background p+p data) to extract the final yields. The raw yields are corrected for detector inefficiencies via the embedding of simulated tracks into real minimum bias events.

3. Results

To show the relative suppression of the A+B data, the nuclear modification factor, R_{AB} was formed as in Eqn. 1. This ratio compares the yield in A+B collisions to that expected from p+p collisions, where the expected scaling factor between the two systems is the number of binary collisions (N_{coll}).

$$R_{AB} = \frac{1}{N_{coll}} \frac{Yield\ in\ A + B}{Yield\ in\ p + p} \tag{1}$$

The measured R_{AA} in Cu+Au collisions are shown in Fig. 2, and compared to d+Au, Cu+Cu,

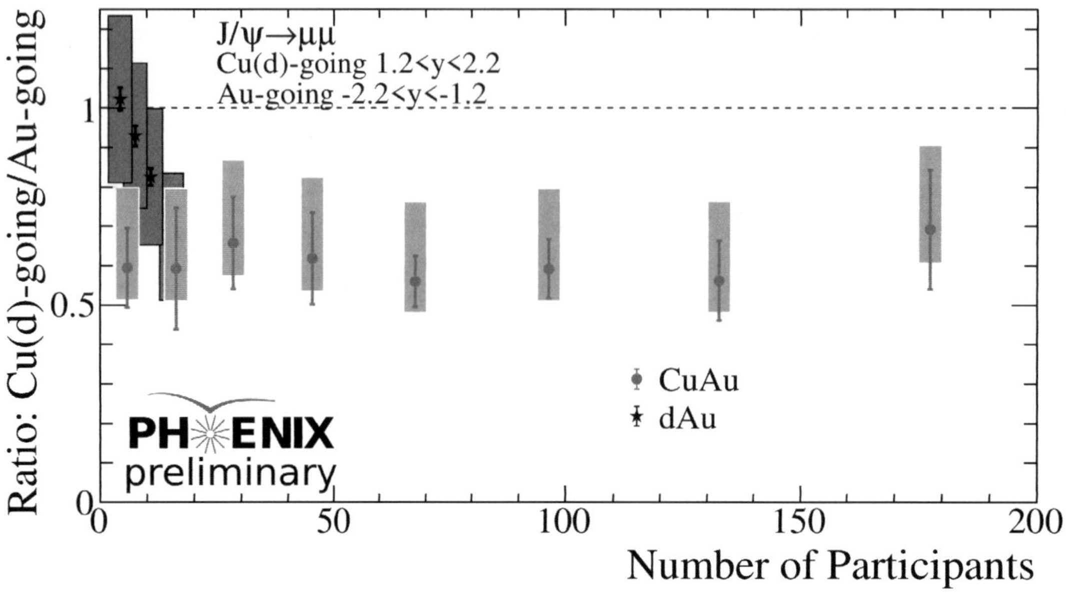

Figure 4. Ratio of the J/ψ yield from forward (Cu-going) to backward (Au-going) rapidities. The grey stars represent the same ratio in d+Au collisions.

and Au+Au collision data. Here, the red symbols depict the Au-going (backward) data, where the low-x gluon comes from the Cu-nucleus, and the blue show the Cu-going (forward) data. Cu+Au data are presented in 10% centrality slices (circles), except for the most central point, which shows a 0-5% central bin and represents events where the Cu nucleus is wholly swallowed by the Au nucleus. As is evident, the backward nuclear modification is found to be the same as for Cu+Cu and Au+Au data, whereas a stronger modification of the forward-going J/ψs is seen.

The role of cold-nuclear matter was explored using a model of the data where the initial parton distributions are modified using the EPS09 data sets [6], coupled with a nuclear break-up cross-section of the J/ψ. This model [7] was tuned to the d+Au data where a break-up cross-section of 4 mb was found. Assuming that this model can be used in A+B collisions as an estimator of the initial state modification, the same break-up cross-section was applied to the Cu+Au system. The result is shown in Fig. 3. A difference between forward and backward J/ψ is expected already due to CNM. The additional suppression, assuming this model represents the full underlying suppression, must be due to final-state effects of the QGP.

To further study this, Fig. 4, shows the ratio of J/ψ yields in the forward (Cu-going) region to the backward (Au-going) region. This ratio is surprisingly found to be independent of centrality and provides a challenge to theories trying to model the data.

4. Summary

In summary, we have measured the number of J/ψs in the Cu+Au collision system and compared the forward (Cu-going) and backward (Au-going) yields. The production relative to that expected from p+p collisions is found to be suppressed, similarly to Au+Au and Cu+Cu data. The suppression found in the forward direction is larger than that in the backward direction.

Journal of Physics: Conference Series **458** (2013) 012007 doi:10.1088/1742-6596/458/1/012007

Using a model of cold nuclear matter effects, which described well the d+Au data, shows an expected difference forward and backward, although this does not account for all the suppression observed; the remainder due, presumably, to QGP-effects.

References

[1] A. Iordanova, *these proceedings.*
[2] T. Matsui and H. Satz, Phys. Lett. **B178** (1986) 416
[3] A. Adare *et al.*, Phys. Rev. Lett **107** (2011) 142301
[4] A. Adare *et al.*, Phys. Rev. Lett **101** (2008) 122301
[5] A. Adare *et al.*, Phys. Rev. **C84** (2011) 054912
[6] K.J. Eskola *et al.*, JHEP **04** (2009) 065
[7] J. Nagle *et al.*, Phys. Rev. **C84** (2011) 044911

Journal of Physics: Conference Series **458** (2013) 012008 doi:10.1088/1742-6596/458/1/012008

Quarkonia RdAu and FVTX c/b measurement at the PHENIX

Kwangbok Lee for the PHENIX Collaboration

Physics Division, Los Alamos National Laboratory, Los Alamos, NM, 87545, U.S.A

E-mail: kwangboklee1@gmail.com

Abstract. Since quarkonia are produced dominantly by gluon-gluon fusion process, it is a good channel to explore the gluon distribution of the nucleon and its modification in nuclei. Contrasting with heavy ion collisions, d+A collisions are not expected to create a hot medium of nuclear matter. Therefore, we can study the quarkonia production in d + A collision and separate the cold nuclear matter effects from the pattern seen in the collisions of large nuclei. The PHENIX collaboration has measured interesting results for J/ψ, ψ' and χ_c at mid rapidity, $|\eta| < 0.35$, and J/ψ and Υ at forward rapidity, $1.2 < \eta < 2.2$ for $\sqrt{s_{NN}} = 200$ GeV d+Au collisions [1-4]. Recently the Forward Silicon Vertex Detector (FVTX) has been installed to the PHENIX detector and much more precise open heavy flavor and vector meson measurements will be possible in the forward rapidity region of the PHENIX Muon Arms. In this talk we will show the aforementioned quarkonia measurements from PHENIX and report on the status of the FVTX detector and physics analyses that will use the FVTX detector in conjunction with the Muon Arms.

1. Introduction

Cold nuclear matter, which is created by $p(d)$+A collisions, is interesting by itself and also important to understand hot nuclear matter by A+A collisions. In hot nuclear matter there should be contributions from cold nuclear matter effects as well as those from the hot nuclear matter effects. Thus, to separate these two types of contributions, it is necessary to measure productions of $p(d)$+A collisions. Several channels can be used to study cold nuclear matter effects and quarkonia production is one of the good probes to study initial gluon modification of nuclei comparing to a proton since they are mostly produced by gluon fusion process. But the produced quarkonia also get affected by breakup effect in the medium. The PHENIX experiment has measured considerable J/ψ data at forward-, backward- and mid rapidities with high statistics and given valuable constraints to theory calculation on the initial and final state effects [1,2]. Besides, other quarkonia like ψ', χ_c and $\Upsilon(1S + 2S + 3S)$ [3,4] have been also measured or being measured in a good shape. Using those measurements, we can test different x (momentum fraction) region and any dependence of break up effect to its binding energy and nuclear size.

Another good channel to study cold nuclear matter is decay leptons from open heavy flavors. They are ideal to test initial state effects of Cronin effect and energy loss as they don't experience any final state breakup effect. Moreover, we expect to get more precise understanding for the initial state effects with direct measurement of charm and bottom. For that need, the PHENIX has developed and implemented silicon vertex detectors of VTX and FVTX to separate charm

Journal of Physics: Conference Series **458** (2013) 012008 doi:10.1088/1742-6596/458/1/012008

and bottom, and archive other improvements. In this proceedings, we can review the PHENIX quarkonia results and see the status and analysis with the vertex detectors upgrade.

2. Quarkonia measurement

2.1. Charmonia measurement

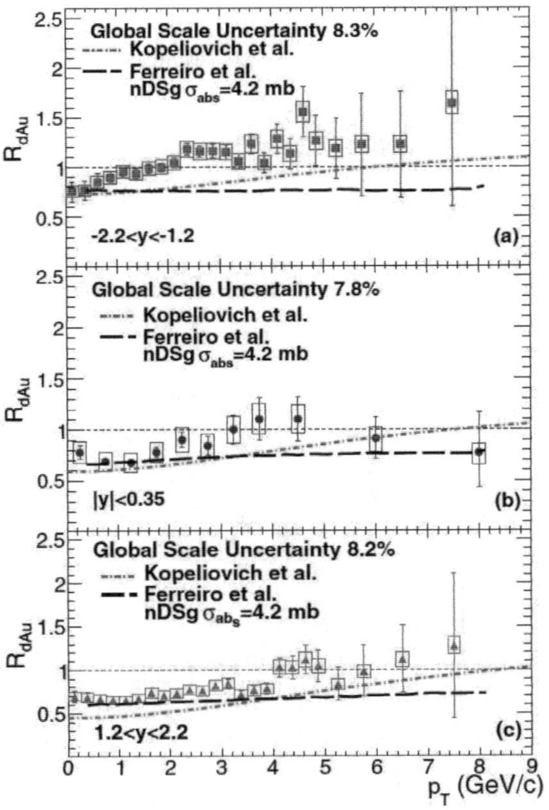

Figure 1. (Color online) J/ψ nuclear modification factor, R_{dAu}, as a function of p_T for (a) backward rapidity, (b) midrapidity, and (c) forward rapidity 0-100% centrality-integrated d+Au collisions.

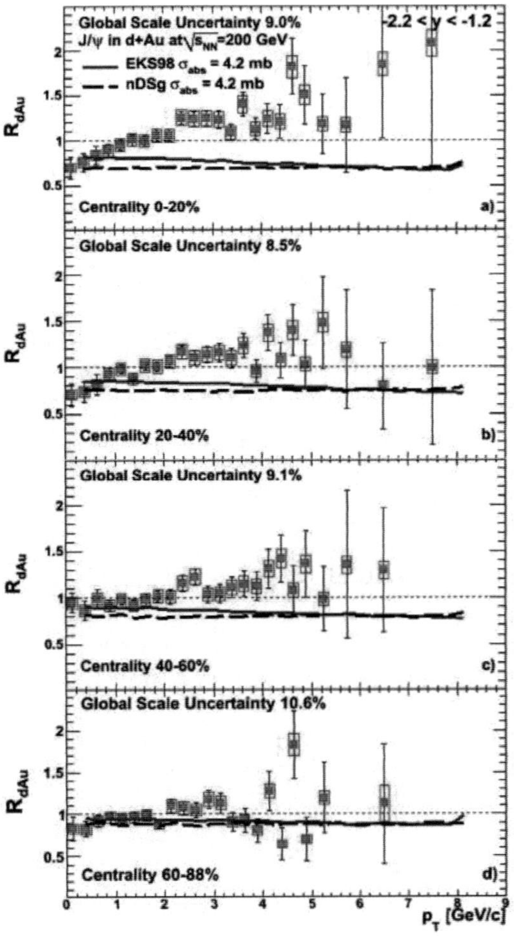

Figure 2. (Color online) J/ψ R_{dAu} as a function of p_T for (a) central, (b) midcentral, (c) midperipheral, and (d) peripheral events in the interval -2.2 < y < -1.2. Curves are calculations by Ferreiro *et al.* [5]

Figure 1 shows R_{dAu}, as a function of p_T over three rapidity ranges in d+Au collisions at $\sqrt{s_{NN}}$ = 200 GeV using the PHENIX detector. These measurements provide a large improvement in statistical precision and p_T reach over the previously published PHENIX d + Au results [6,7]. The R_{dAu} vs p_T displays similar behavior at mid- and forward rapidity, showing suppression at low p_T with a gradual increase to a value consistent with 1.0. The R_{dAu} at backward rapidity has a different distribution with p_T, showing a more rapid increase from suppression to a value of 1.0 and transitioning to $R_{dAu} > 1.0$ above 2 GeV/c. These trends are greater for central collisions, while the peripheral collisions show R_{dAu} consistent with 1.0 across all rapidities, for example, at backward rapidity as seen at Fig. 2.

The theory models in the figure 1,2 are from Kopeliovich *et al.* [8,9] and Ferreiro *et al.* [5,10]. They include shadowing as well as breakup and Cronin effect where the breakup cross section is dependent on the p_T and rapidity of the J/ψ. The shape of the p_T dependence is in good agreement with the data at mid- and forward rapidity, but the theory shows a greater overall level of suppression than is seen in the data. At backward rapidity, there is a disagreement with the shape of the p_T distribution. More discussion and details on the theory model comparison is available at [2], which also includes discussion of initial parton energy loss in the nucleus.

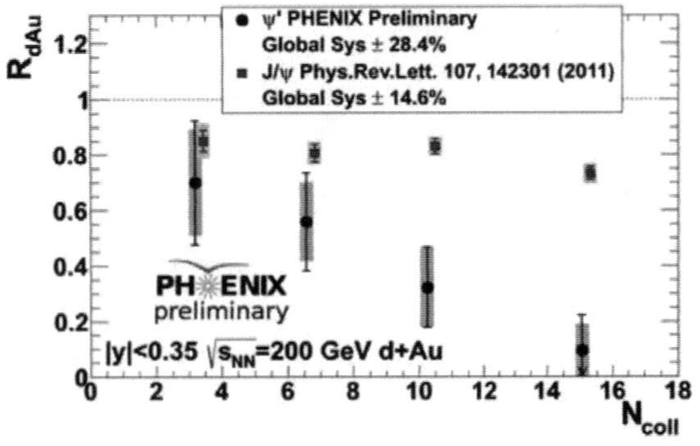

Figure 3. (Color online) The ψ' nuclear modification factor, R_{dAu}, as a function of N_{coll}. Also included are the previously measured J/ψ R_{dAu} [1].

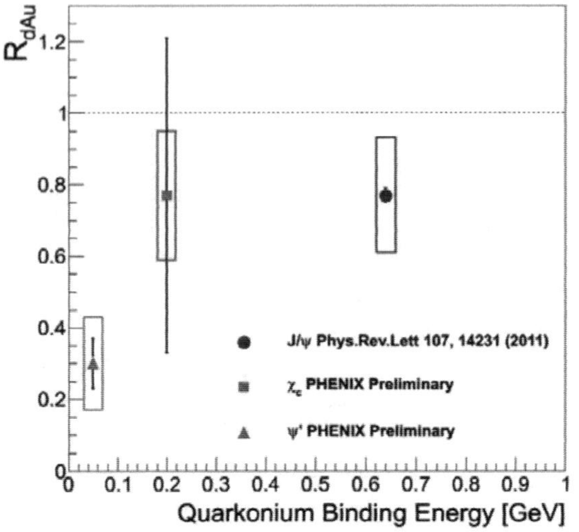

Figure 4. (Color online) The ψ', χ_c, and J/ψ R_{dAu} for 0-100% centrality integrated d+Au collisions as a function of the quarkonia binding energy.

Recently the PHENIX got preliminary result on ψ' and χ_c measurement at mid rapidity [4]. Figure 3 shows R_{dAu} vs N_{coll} of J/ψ and ψ'. Interestingly, we see mores suppression for ψ' than J/ψ as N_{coll} increases. This gives a clue to understand final status effects on ψ'. Since ψ' has weaker binding energy and larger nuclear size, ψ' can be dissociated or broken up easily. Particles in the nucleus and co-mover around ψ' can affect the dissociation though we need more study to understand it. Figure 4 shows charmonia binding energy dependence on R_{dAu}. And we see a more suppression trend in case of weak binding energy. We would estimate any rapidity dependence on the trend once we archive ψ' and χ_c measurement at forward rapidity. Recently two Forward Silicon Vertex Trackers (FVTX) are installed at the PHENIX and working successfully. They are expected to improve mass resolution so reveal ψ' resonance apart from J/ψ at forward rapidity.

2.2. $\Upsilon(1S+2S+3S)$ *measurement*

Because the Υ mass is heavier than J/ψ, the nuclear effects on the gluon distribution can be studied in different kinematic regions with the Υ measurement. At forward rapidity (the deuteron going direction) and the same collision energy of $\sqrt{s_{NN}} = 200$ GeV, the average momentum fraction of the gluon in the gold nucleus that is sampled for Υ production is $\langle x_2 \rangle \approx 1 \times 10^{-2}$, whereas that for J/Ψ production samples $\langle x_2 \rangle \approx 3 \times 10^{-3}$. Figure 5

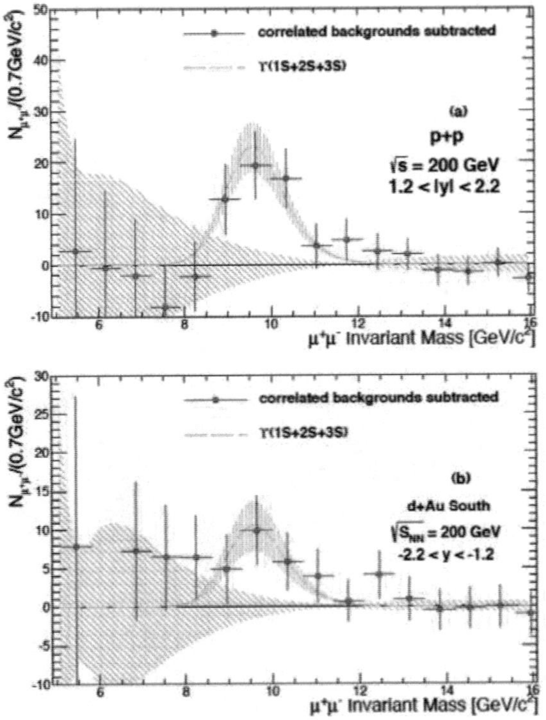

Figure 5. (Color online) The mass distributions are drawn after subtraction of all the correlated backgrounds and Drell-Yan process. The fitted peak curves represent only $\Upsilon(1S+2S+3S)$. The shaded bands (green vertical shading) around the fitted peak curves represent fitting uncertainties and those around zero yield (blue slanted shading) represent systematic uncertainties.

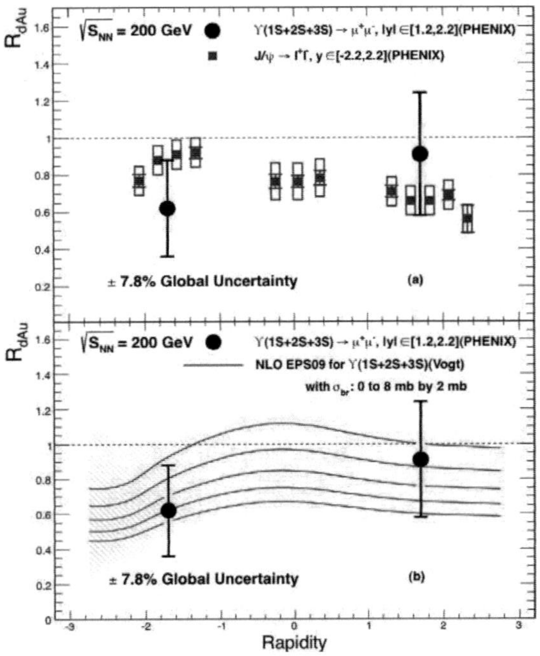

Figure 6. (Color online) Nuclear modification factors, R_{dAu}, are shown as a function of rapidity. For comparison, the upper panel (a) shows R_{dAu} for the J/Ψ [3] as well as for the Υ. The lower panel (b) shows theoretical predictions of nuclear modification based on a NLO EPS09 combined with a breakup cross section, with $\sigma_{br} = 0$ to 8 mb in 2-mb steps from top to bottom [12].

shows $\Upsilon(1S+2S+3S)$ distribution after subtraction of combinatorial background and physical backgrounds of Drell-Yan process and correlated backgrounds from open heavy flavors [3]. Drell-Yan process is estimated by NLO calculation. And for d+Au collision, additional isospin effects of deuteron, nuclear shadowing, energy loss effects are included [11]. Υ peak is composed of 3 Gaussians as 1 Gaussian is corresponding to 1 Υ state. Figure 6 shows Υ R_{dAu} and comparisons to (a) J/Ψ [1] and (b) EPS09 shadowing model + breakup cross section [12]. The results show moderate suppression trend at backward rapidity and the comparison to shadowing model gives some possibility to breakup effect on Υ but does not result in any definitive constraint on the breakup cross section given the large experimental uncertainties. More comparisons to gluon saturation models and to models including initial state energy loss would also be of interest. Currently, Υ R_{AuAu} analysis is actively ongoing at mid- and forward rapidities at the PHENIX. Comparison to the coming R_{AuAu} results will be also interesting to study hot nuclear medium effects on the Υ production.

3. Forward Silicon Vertex Trackers

The FVTX is installed at the PHENIX experiment at 2011 and has undergone successful commissioning and operations during the 2012 and 2013 RHIC Runs. The FVTX cover same η range to muon arm, $1.2 < |\eta| < 2.2$ so improve background rejection, detection resolution and extend the vertex capability of the PHENIX Silicon Vertex Tracker (VTX) to forward and backward rapidities. Figure 7 shows the half of the vertex detectors in the cage. They

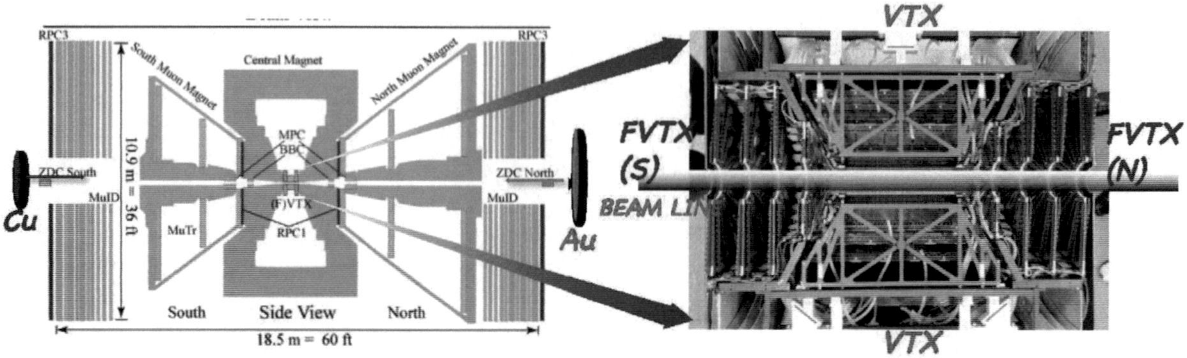

Figure 7. (Color online) Side view of PHENIX forward detectors & close view of the vertex trackers.

are sitting at the closest location to interaction point (IP). The FVTX spans z location of 18.5 $cm < |z| < 38\ cm$ and 2π azimuthal angle with 4 stations for each arm. The FVTX composed of 384 wedges, and strips of wedge keep 75 μm pitch in radial direction and 3.75̊ along ϕ direction, so is most sensitive to r-direction.

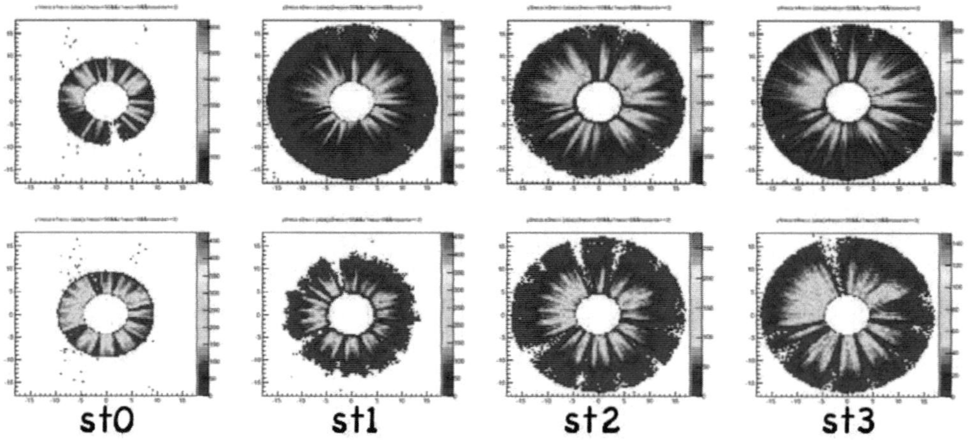

Figure 8. (Color online) Stereo plot of hits for each FVTX station, 200 GeV p-p run.

At first working year of 2012, the PHENIX took a considerable data of $\sqrt{s} = 200$ GeV and 510 GeV p+p collisions and $\sqrt{s_{NN}} = 200$ GeV Cu+Au collision As shown at Fig. 8, the FVTX was alive > 95% from alive area of run 12 data. For now, the FVTX self alignment is completed and global alignment between the FVTX-VTX-Muon Tracker is ongoing. With the vertex detectors, we expect to measure primary and secondary vertices, improve mass resolution by measuring open angle precisely and reject hadron backgrounds. At the following section, more details about detector performance and working principle, etc., which we use for analyzing signals, are shown.

3.1. DCA$_R$ measurement & Background rejection

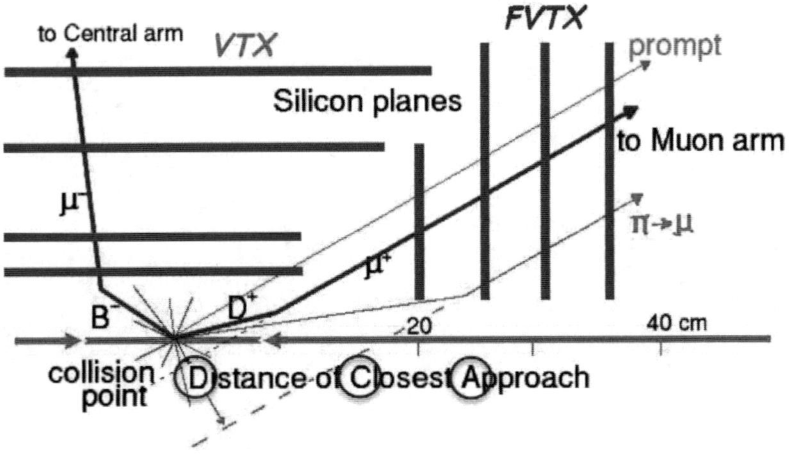

Figure 9. (Color online) Schematic for DCA measurement.

One of important parameters, which can be measured with the vertex detectors, is distance of closest approach (DCA). As seen by the schematic view of Figure 9, DCA is defined as the closest distance when we project the track into primary vertex, and measured by using the primary and secondary decay vertices. The DCA distribution from hadron decay muons would be broader than prompt muons since the hadrons, such as pions and kaons, fly away from the primary vertex before they decay whereas prompt particles decay shortly once they are produced. Moreover, the DCA distribution is different each other along its process and decay time so the shape can be used to identify each of them.

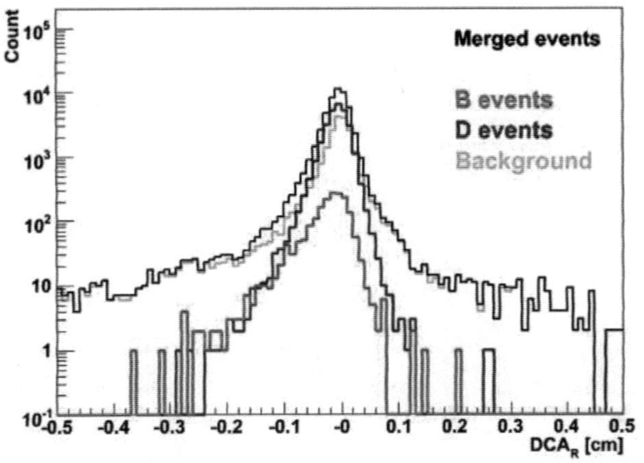

Figure 10. (Color online) DCA$_R$ distribution for B and D mesons and the backgrounds.

Figure 10 shows simulated DCA$_R$, DCA in the r direction for B, D mesons and the hadron backgrounds. The expected resolution for the prompt particle is 200 μm. Since decay muon from relatively short-lived B and D mesons has narrower shape than decay muon from long-lived hadrons and the shapes are different between B, D mesons, we can extract a signal by fitting the shape or improve signal to background ratio efficiently requiring certain DCA limit. We can apply same method to extract the prompt Drell-Yan signal.

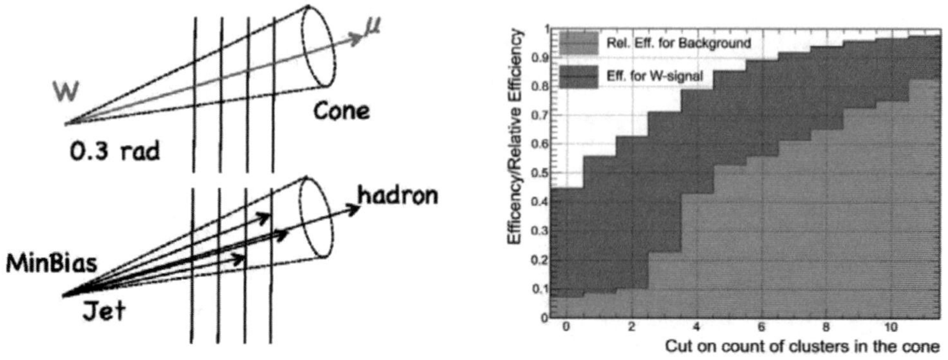

Figure 11. (Color online) Schematic for W signal and jet background tracks in a cone (left) and simulated efficiencies along the cut on the clusters (right).

Besides the DCA shape, isolation cut can be applied with FVTX to exclude the jet background. If particles are produced by jet, they make many hits/clusters around them whereas W or Drell-Yan process don't make such clusters as seen at Fig 11. So giving a limit on the clusters/hits around certain geometrical range, we can achieve efficient background rejection from the jet.

3.2. Opening angle measurement & mass resolution improvement

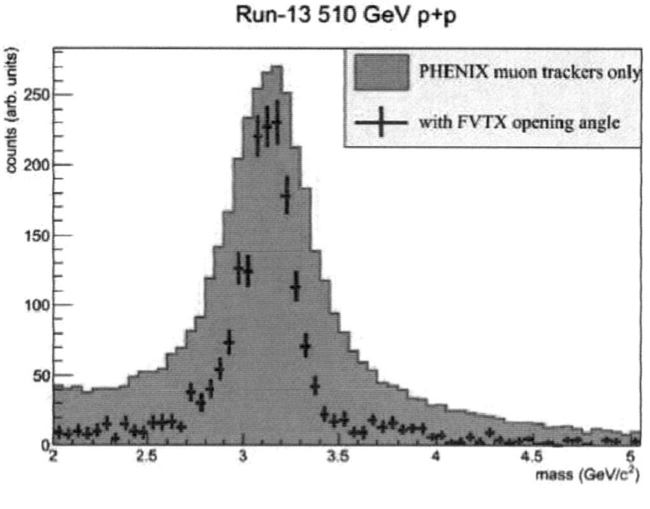

Figure 12. (Color online) Dimuon distribution of $\sqrt{s}$ = 510 GeV p+p collision at Run13. Red color is from muon tracker only and blue crosses are with the FVTX refitting. Counts are re-scaled and arbitrary.

The FVTX can measure precise opening angle of dimuon and improve invariant mass resolution by refitting the track with it. Figure 12. shows the reconstructed dimuon mass distribution from a set of Run 13 data and we see the J/ψ peak gets narrower with the FVTX opening angle measurement (blue crosses) than the muon arm alone (red). We also expect the reduction of hadron backgrounds, which happens when hadron decays in the muon tracker volume. If the hadron decays at the inside of the tracker volume, they can be reconstructed as high momentum particles since they make false straight-like trajectory but they can be excluded by requiring good χ^2 values through the FVTX and the muon tracker.

4. References

[1] A. Adare *et al.* (PHENIX Collaboration) Phys. Rev. Lett. **107**, 142301 (2011).
[2] A. Adare *et al.* (PHENIX Collaboration) Phys. Rev. C **87**, 034904 (2013).
[3] A. Adare *et al.* (PHENIX Collaboration) Phys. Rev. C **87**, 044909 (2013).
[4] D. McGlinchey "Recent Heavy Quarkonia Results from PHENIX", QM 2012 proceedings.
[5] E. G. Ferreiro, F. Fleuret, J. P. Lansberg, N. Matagne, and A. Rakotozafindrabe, Few Body Syst. **53**, 27 (2012).
[6] A. Adare *et al.* (PHENIX Collaboration) Phys. Rev. C **77**, 024912 (2008).
[7] S. S. Adler *et al.* (PHENIX Collaboration) Phys. Rev. Lett. **96**, 012304 (2006).
[8] B. Kopeliovich, I. Potashnikova, and I. Schmidt, Nucl. Phys. A **864**, 203 (2011).
[9] B. Z. Kopeliovich, I. K. Potashnikova, and I. Schmidt, Phys. Rev. C **82**, 024901 (2010).
[10] E. Ferreiro, F. Fleuret, J. Lansberg, and A. Rakotozafindrabe, Phys. Lett. B **680**, 50 (2009).
[11] R. B. Neufeld, I. Vitev, and B.-W. Zhang, Phys. Lett. B **704**, 590 (2011).
[12] R. Vogt, Phys. Rev. C **81**, 044903 (2010).

Journal of Physics: Conference Series **458** (2013) 012009 doi:10.1088/1742-6596/458/1/012009

Energy dependence of J/ψ production in Au+Au collisions at $\sqrt{s_{NN}}$ = 39, 62.4 and 200 GeV from the STAR experiment

Wangmei Zha (for the STAR collaboration)

University of Science and Technology of China, Heifei City, Anhui Province, China
Brookhaven National Laboratory, Upton, NY, USA

E-mail: wangmei@rcf.rhic.bnl.gov

Abstract. In this article, we report on the measurements of J/ψ invariant yields as a function of transverse momentum at midrapidity ($|y|<1.0$) in Au+Au collisions at $\sqrt{s_{NN}}$ = 39 , 62.4 and 200 GeV taken in 2010 by STAR with full Time-of-Flight detector and Barrel Electromagnetic Calorimeter detector in operation. Centrality dependence of J/ψ production and nuclear modification factors are presented. Comparisons among different collision energies and model calculations are discussed.

1. Introduction

The Relativistic Heavy Ion Collider (RHIC) is built to search for the Quark-gluon Plasma (QGP) and to study its properties in laboratory through high energy heavy-ion collisions [1]. J/ψ suppression in heavy-ion collisions, due to color screening of quark and anti-quark, has been proposed as a signature of QGP formation [2]. Various measurements of J/ψ have been performed in different collision systems and at different energies, and indeed a suppression of J/ψ production has been observed [3, 4, 5, 6]. A similar pattern of suppression was found at SPS and RHIC, even though the temperature and energy density reached in these collisions are significantly different. Furthermore, a stronger suppression at forward rapidity compared to midrapidity was observed at RHIC [7]. These observations indicate that effects other than color screening contribute to J/ψ production. The regeneration of J/ψ from the statistical coalescence of charm quarks was suggested to explain the similar suppressions at SPS and RHIC [8]. Despite the increase in collision energy, the increased regeneration contribution from the larger charm quark density could compensate for the additional suppression due to color-screening. This also explains a stronger suppression at forward rapidity at RHIC where the charm quark density is lower compared to midrapidity. In this paper, we further study the collision energy dependence of J/ψ production under these two competing mechanisms.

We present the measurement of the J/ψ production at midrapidity with the STAR experiment in Au+Au collisions at $\sqrt{s_{NN}}$ = 39, 62.4 and 200 GeV using data collected during RHIC year 2010 run and study the nuclear modification factors at these energy points.

2. Experiment and Analysis

The STAR experiment is a large-acceptance multi-purpose detector which covers full azimuth and pseudorapidity of $|\eta|<1$ [9]. The Au+Au data were obtained using a minimum-bias trigger by selecting on coincidences in the Vertex Position Detector (VPD) and the Zero Degree Calorimeter (ZDC). The total number of minimum-bias events used in analysis is 182 million, 94 million, 304

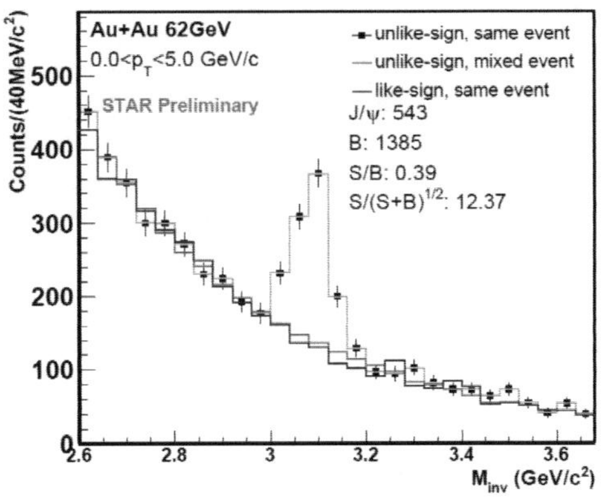

Figure 1. Invariant mass distribution of e^+e^- pairs in 0 - 60% central Au+Au collisions at 62.4 GeV.

million for 39, 62.4 and 200 GeV, respectively. In this analysis, the J/ψ is reconstructed through its decay into electron-positron pairs, J/ψ → e^+e^- (branching ratio B_{e+e-} = 5.9%). The primary detectors used in this analysis are the Time Projection Chamber (TPC) [10], the Time-of-Flight (TOF) [11], and the Barrel Electromagnetic Calorimeter (BEMC) [12]. The TPC provides tracking and particle identification via the ionization energy loss (dE/dx) of charge particles. In year 2010, STAR has installed 100% of TOF trays (which measure the time of flight and velocity of particles) at midrapidity. This detector, combined with TPC, can clearly identify electrons by rejecting hadrons in the low and intermediate p_T range. The BEMC is a lead-scintillator calorimeter and has been used to improve the electron identification at high p_T. We use two methods to estimate the combinatorial background in this analysis:

a. **Mixed-Event**: Events are categorized according to the z-position of event vertex and the centrality of event. Electrons from one event are paired with positrons from other random events from an event pool with similar global features.

b. **Like-Sign**: Electrons (or positrons) of the same charge sign are paired within same event.

Figure 1 shows the invariant mass distribution of e^+e^- paris in 0 - 60% central Au+Au collisions at 62.4 GeV. For the results reported in this paper, we use Mixed-Event method for the combinatorial background estimation and the Mixed-Event background is normalized to the Like-Sign in a mass range of 2.0 - 4.0 GeV/c^2. Good Signal-to-Background ratios (for 62.4 GeV, S/B = 0.39; for 39 GeV, S/B = 0.62) are observed.

3. Results

Figure 2 shows the J/ψ invariant yield as a function of p_T for 0 - 60% central Au+Au collisions at 39, 62.4 and 200 GeV. As expected, the J/ψ invariant yields are larger in Au+Au collisions at larger center-of-mass energy.

There are several different ways to quantify the suppression of J/ψ production. Since no p+p reference baselines at $\sqrt{s_{NN}}$ = 39 or 62.4 GeV were measured at RHIC, we first discuss the J/ψ R_{CP} which defined as follows:

$$R_{CP} = \frac{\dfrac{dN/dy}{<N_{coll}>}(\text{central})}{\dfrac{dN/dy}{<N_{coll}>}(\text{peripheral})} \quad (1)$$

where $<N_{coll}>$ is the average number of nucleon-nucleon collisions in relative centrality bins. R_{CP} is a ratio of J/ψ production in central collisions (where QGP formation is expected) to peripheral collisions (which naively should be a superposition of p+p interactions).

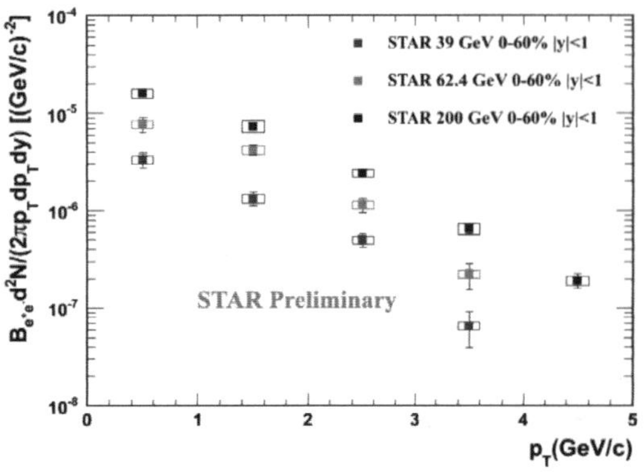

Figure 2. J/ψ invariant yields for Au+Au collisions (centrality 0 - 60%) at 39, 62.4 and 200 GeV as a function of p_T. The error bars depict the statistical uncertainties. The boxes represent the systematic uncertainties.

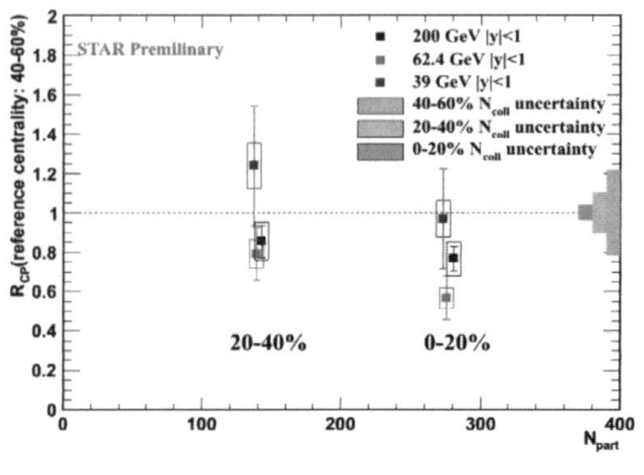

Figure 3. J/ψ R_{CP} results with respect to 40 - 60% (peripheral) for Au+Au collisions at 39, 62.4 and 200 GeV as a function of the number of participating nucleons. The error bars represent the statistical uncertainties. The boxes represent the systematic uncertainties. The shaded bands around unity on the right-hand axis represent the normalization uncertainty from the average number of binary nucleon-nucleon collisions (<N_{coll}>) in different centrality bins.

The R_{CP}, as a function of the average number of participants (<N_{part}>) for Au+Au collisions at 39, 62.4 and 200 GeV , are shown in Fig. 3. Note that the peripheral bin selection for Au+Au collisions at these three energy points is 40 - 60% centrality. Significant suppression is observed in central Au+Au collisions at 62.4 GeV which is similar as at 200GeV.

R_{AA} is another way to quantify the suppression of J/ψ production. It is defined as follows:

$$R_{AA} = \frac{1}{T_{AA}} \frac{d^2N_{AA}/dp_T dy}{d^2\sigma_{pp}/dp_T dy} \tag{2}$$

where $d^2N_{AB}/dp_T dy$ is the invariant J/ψ yield in A+A collisions and $d^2\sigma_{pp}/dp_T dy$ is the J/ψ cross section in p+p collisions. The nuclear overlap function is defined as T_{AB} = <N_{coll}>/σ_{inel}^{pp} , and takes into account the inelastic cross section in p+p collisions (σ_{inel}^{pp}) and the number of nucleon-nucleon collisions in A+A collisions (<N_{coll}>). The STAR experiment has no J/ψ cross section measurement for p+p collisions at 39 and 62.4 GeV. There are several p+p measurements from fixed target p+A experiment [13, 14, 15] and from Intersecting Storage Ring (ISR) collider experiments [16, 17] near these two energy points. However, the p_T shapes from ref. [16] and ref. [17] at 63 GeV are inconsistent with each other and the cross section measurements at 39 GeV are comparable to (or even larger than) that at 63 GeV. Therefore, we use Color Evaporation Model (CEM) prediction as our p+p reference baselines for $\sqrt{s_{NN}}$ = 39 and 62.4 GeV [18] because CEM calculation describes the p_T and rapidity distribution in p+p 200 GeV collisions [18]. Figure 4 shows the R_{AA} of J/ψ as a function of N_{part} for Au+Au collisions at 39, 62.4 and 200 GeV. The theoretical calculation curves, which contain two main components: initial suppression and regeneration, are from reference [7]. As the collision energy increases the QGP temperature increases, thus the J/ψ color screening (initial suppression) becomes more significant. However, in this theoretical calculation, the regeneration contribution increases with collision energy due to the increase in the charm pairs production, and nearly compensates the additional suppression arising from higher temperature. Significant suppression of J/ψ production is observed in Au+Au collisions from 39 to 200 GeV with respect to N_{coll} scaled p+p yields. No significant energy dependence is observed for R_{AA} with uncertainties. Theoretical calculation describes the data within error bars.

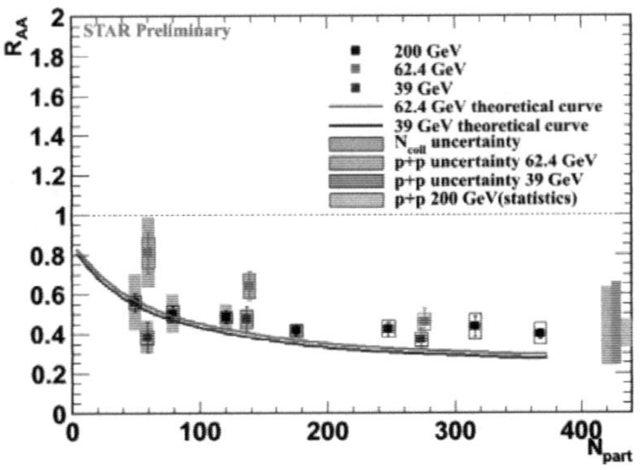

Figure 4. J/ψ R_{AA} results as a function of N_{part} Au+Au collisions at 39, 62.4 and 200 GeV . The theoretical curves are from reference [7].The error bars represent the statistical uncertainties. The boxes represent the systematic uncertainties. The shaded bands indicates the uncertainties from <N_{coll}>. The boxes on the vertical axis represent the uncertainties from CEM estimation for 39 and 62.4 GeV and the statistical uncertainty for 200 GeV p+p baseline.

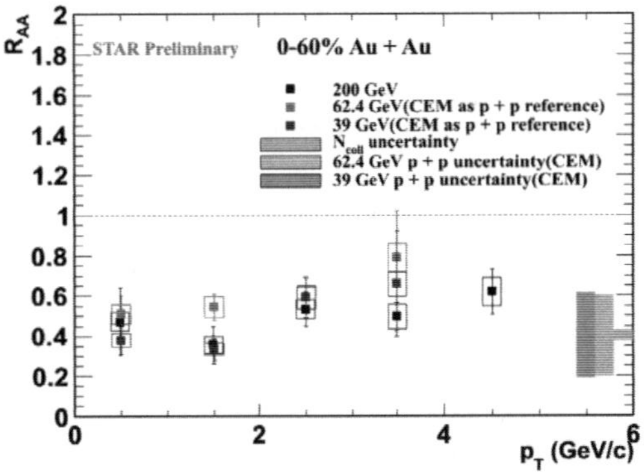

Figure 5. J/ψ R_{AA} results for Au+Au collisions (centrality 0 - 60%) at 39, 62.4 and 200 GeV as a function fo p_T. The error bars represent the statistical uncertainties. The boxes represent the systematic uncertainties. The boxes on the vertical axis represent the uncertainties from CEM estimation for 39 and 62.4 GeV and <N_{coll}> uncertainties.

Figure 5 shows the J/ψ R_{AA} as a function of p_T for Au+Au collisions at 39, 62.4 and 200 GeV. The CEM estimation is used as reference baseline for 39 and 62.4 GeV. Significant suppression is observed from 39 to 200 GeV. Similar suppression pattern within uncertainties is observed for these energies.

4. summary

In summary, we report on recent STAR measurements of J/ψ production in Au+Au collisions at 39, 62.4 and 200 GeV at midrapidity. Significant suppression of J/ψ production, with respect to number of binary collisions scaled p+p yields, is observed for these three energies. No significant energy dependence of nuclear modification factor (both for R_{AA} and R_{CP}) is found within uncertainties. Model calculations, which include direct suppression and regeneration, describe the centrality dependence of R_{AA} within uncertainties. Precise p+p reference measurements at 39 and 62.4 GeV are needed for further understanding of the J/ψ interactions with the nuclear matter at those energies.

References

[1] Harrison M, Ludlam T and Ozaki S 2003 *Nucl. Instr. Meth.* A **499** 235
[2] Matsui T and Satz H 1986 *Phys. Lett.* B **178** 416
[3] Lansberg J P 2006 *Int. J. Mod. Phys.* A **21** 3857
[4] Abreu M C et al 1999 *Phys. Lett.* B **449** 128
[5] Arnaldi R et al 2007 *Phys. Rev. Lett.* **99** 132302
[6] Atomssa E T 2009 *Eur. Phys. J.* C **61** 683
[7] Zhao X and Rapp R 2010 *Phys. Rev.* C **82** 064905 (private communication)

[8] Adare A et al 2007 *Phys. Rev. Lett.* **98** 232301
[9] Ackermannn K H et al 2003 *Nucl. Instr. Meth.* A **499** 624
[10] Anderson M et al 2003 *Nucl. Instr. Meth.* A **499** 659
[11] Llope W J et al 2012 *Nucl. Instr. Meth.* A **661** S110
[12] Beddo M et al 2003 *Nucl. Instr. Meth.* A **499** 725
[13] Alexopoulos T et al 1997 *Phys. Rev.* D **55** 3927
[14] Schub M H et al 1995 *Phys. Rev.* D **52** 1307
[15] Gribushin A et al 2000 *Phys. Rev.* D **62** 012001
[16] Clark A G et al 1978 *Nucl. Phys.* B **142** 29
[17] Kourkounelis C et al 1980 *Phys. Lett.* B **91** 481
[18] Nelson R and Vogt R et al 2012 *arXiv: 1210.4610v1* (private communication)

Journal of Physics: Conference Series **458** (2013) 012010 doi:10.1088/1742-6596/458/1/012010

J/ψ Measurements with the ALICE Experiment at the LHC

Christoph Blume, for the ALICE Collaboration
Institut für Kernphysik, University of Frankfurt,
Max-von-Laue-Str. 1, 60438 Frankfurt am Main, Germany

E-mail: blume@ikf.uni-frankfurt.de

Abstract. We review recent results on J/ψ production measured by the ALICE collaboration at the LHC. For pp collisions at $\sqrt{s} = 7$ TeV yields and spectra of inclusive and prompt J/ψ, as well as results on their polarization and the charged particle multiplicity dependence of yields are presented. The nuclear modification factor R_{AA} obtained for inclusive J/ψ at mid- ($|y| < 0.9$) and forward-rapidities ($2.5 < y < 4$), covering the p_T range down to 0, for centrality selected Pb-Pb collisions are discussed. Also, first results on the J/ψ v_2 at forward-rapidities are shown.

1. Introduction

The understanding of J/ψ production processes in elementary collisions still poses a challenge to theoretical models. The main question is to what extent $q\bar{q}$ octet states have to be included in the calculation [1, 2]. A consistent description of the measured cross sections and polarization has to be achieved within a given model approach, such as the Color Singlet Model (CSM), Color Evaporation Model (CEM), or the Non-Relativistic QCD model (NRQCD). It is important to distinguish between the different sources contributing to inclusive J/ψ measurements. Roughly speaking, at low p_T about 10% of these originate from the weak decay of b-hadrons, while the 90% prompt J/ψ are either directly produced (~50%) or are due to feed down from heavier states (~40%), such as ψ' or χ_c. The J/ψ from b-decays can be measured by exploiting the fact that the secondary decay vertex will have a displacement relative to the main interaction vertex and allows thus a direct determination of the beauty cross section. Another interesting aspect is the question whether Multi-Parton Interactions (MPI) affect the J/ψ production in pp collisions. This can be addressed by measuring the dependence of the yield as a function of the underlying charged particle multiplicity. Finally, pp collisions serve as an important reference measurement to heavy-ion collisions in order to identify any medium effects.

Quarkonia in general are a unique probe for the properties of the hot and dense medium produced in high energy heavy-ion collisions. It has been suggested very early on that J/ψ production should be suppressed in AA collisions relative to pp due to screening of the $q\bar{q}$ potential by free color charges [3]. Indeed, such a suppression beyond what is expected due to Cold Nuclear Matter (CNM) effects has been observed experimentally for J/ψ at the SPS and at RHIC [4, 5, 6]. The CNM effects include nuclear absorption and shadowing of parton distributions and can be determined from pA collisions. A surprising aspect of these measurements is the observation that the amount of suppression is very similar at SPS and

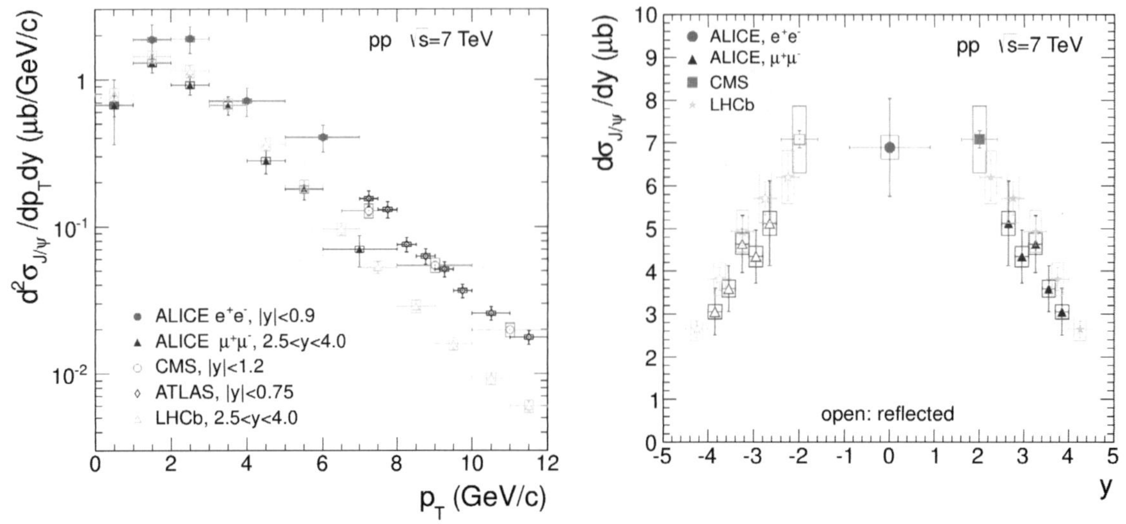

Figure 1. The differential J/ψ cross section for pp collisions at $\sqrt{s} = 7$ TeV as a function of transverse momentum $d^2\sigma_{J/\psi}/dp_T dy$ measured at midrapidity ($|y| < 0.9$) and at forward rapidity ($2.5 < y < 4$) (left panel) and as a function of rapidity $d\sigma_{J/\psi}/dy$ [11]. The ALICE results are compared to data from the other LHC experiments [12, 13, 14], obtained in similar rapidity ranges.

RHIC, despite the clearly different energy densities. In addition, a larger suppression at forward rapidities than at mid-rapidity has been seen at RHIC. These findings have led to the idea that also regeneration mechanisms might play a role at higher energies where the number of produced $c\bar{c}$ pairs is large [7, 8]. This regeneration might happen throughout the evolution of the Quark-Gluon Plasma (QGP) [9] or statistically at hadronization [10] and counteracts the suppression within a QGP.

ALICE is capable of measuring J/ψ at mid-rapidity ($|y| < 0.9$) with the detectors of the central barrel (ITS, TPC, TRD) via the di-electron decay channel. Electrons are identified using the particle identification capabilities of TPC, TRD, and TOF. In addition, secondary vertices of $B \rightarrow J/\psi$ decays can be localized with the help of the ITS. At forward rapidities ($2.5 < y < 4$) J/ψ are detected in the muon spectrometer via their di-muon decay channel. Hadrons are suppressed by an absorber of ten interaction lengths thickness and the muons passing through it are tracked by several layers of pad chambers. A trigger on single and di-muons allows the collection of a high statistics J/ψ sample. The acceptance covers in both the central and forward case, the p_T range down to 0.

2. pp Collisions

Figure 1 shows the J/ψ spectra measured for pp collisions at $\sqrt{s} = 7$ TeV [11]. The presented data correspond to an integrated luminosity of 5.6 nb^{-1} (15.6 nb^{-1}) for the di-electron (di-muon) measurement. A good agreement of the p_T spectrum at forward rapidity with the corresponding measurement by the LHCb collaboration [13] is found. Due to the coverage of the low p_T region, the data at mid-rapidity complement the results by the ATLAS and CMS experiments [12, 14]. Therefore they allow to fill the gap in the J/ψ rapidity distribution (right panel of Fig. 1) and make it possible to determine the integrated cross section.

The same measurement has been performed for pp collisions at $\sqrt{s} = 2.76$ TeV, although

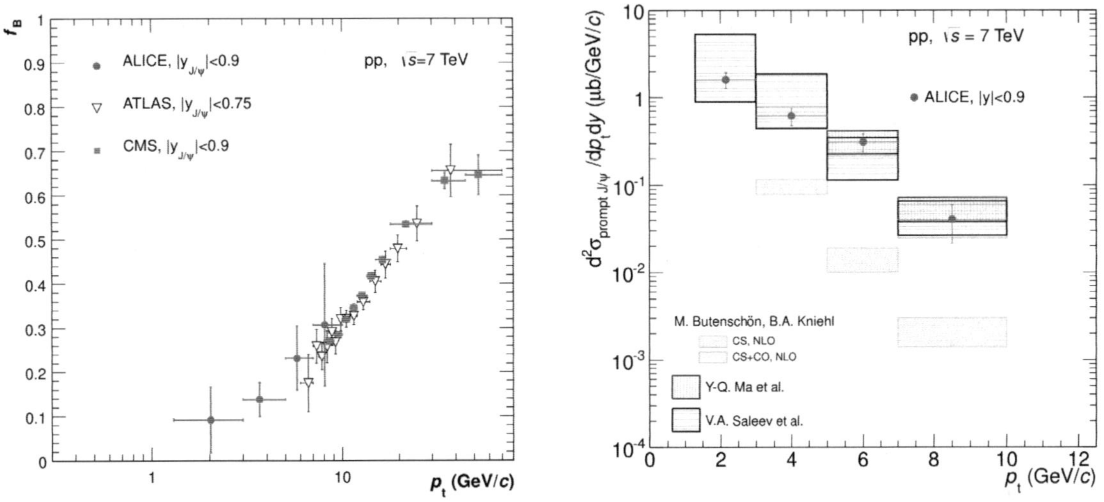

Figure 2. Left: The fraction of J/ψ from the decay of b-hadrons as a function of $p_{\rm T}$ [16] together with results from ATLAS [14] and CMS [17] in pp collisions at $\sqrt{s}$ = 7 TeV. Right: The differential cross section of prompt J/ψ as a function of $p_{\rm T}$ compared to results from ATLAS [14] and CMS [17] at mid-rapidity and to theoretical calculations [18, 19, 20].

only a limited integrated luminosity is available in the di-electron case due to the brevity of the run (1.1 $\rm nb^{-1}$ for $J/\psi \to e^+e^-$, 19.9 $\rm nb^{-1}$ for $J/\psi \to \mu^+\mu^-$) [15]. Since these data serve as a reference for the determination of the nuclear modification factor $R_{\rm AA}$ in heavy-ion collisions, they currently limit the achievable statistical and systematic errors on $R_{\rm AA}$.

The fraction of J/ψ originating from decays of beauty hadrons $f_{\rm B}$ is extracted from fits with simulated templates to distributions of the pseudo-proper decay length

$$x = \frac{c\, L_{\rm xy}\, m_{J/\psi}}{p_{\rm T}^{J/\psi}} \quad \text{with} \quad L_{\rm xy} = \vec{L}\, \vec{p}_{\rm t}^{\,J/\psi} / p_{\rm T}^{J/\psi}, \tag{1}$$

where $\vec{L}$ is the vector pointing from the primary to the decay vertex. The left panel of Fig. 2 shows the resulting $p_{\rm T}$ dependence of $f_{\rm B}$ at mid-rapidity in pp collisions at $\sqrt{s}$ = 7 TeV. Due to its unique acceptance at very low $p_{\rm T}$, ALICE determines $f_{\rm B}$ down to $p_{\rm T}$ = 1.3 GeV/c. At higher $p_{\rm T}$ a good agreement with data by the ATLAS [14] and CMS [17] collaborations is found. Combining the measurement of $f_{\rm B}(p_{\rm T})$ with the $p_{\rm T}$ spectrum of inclusive J/ψ (left panel Fig. 1) yields the corresponding spectrum of prompt J/ψ (right panel of Fig. 2). The data match the NLO NRQCD calculations which include color singlet as well as octet contributions [18, 19, 20]. However, the theory is below the measurements by roughly an order of magnitude, if only the color singlet part is considered.

Another observable that can help to discriminate between the different theory approaches is the J/ψ polarization. ALICE measures it via the distributions of the polar angle θ and the azimuthal angle ϕ of the decay muons at forward rapidities ($2.5 < y < 4$) in the $p_{\rm T}$ range 2 - 8 GeV/c [21]

$$W(\cos\theta) \propto \frac{1}{3+\lambda_\theta}(1+\lambda_\theta\cos^2\theta) \quad \text{and} \quad W(\phi) \propto 1 + \frac{2\lambda_\phi}{3+\lambda_\theta}\cos 2\phi. \tag{2}$$

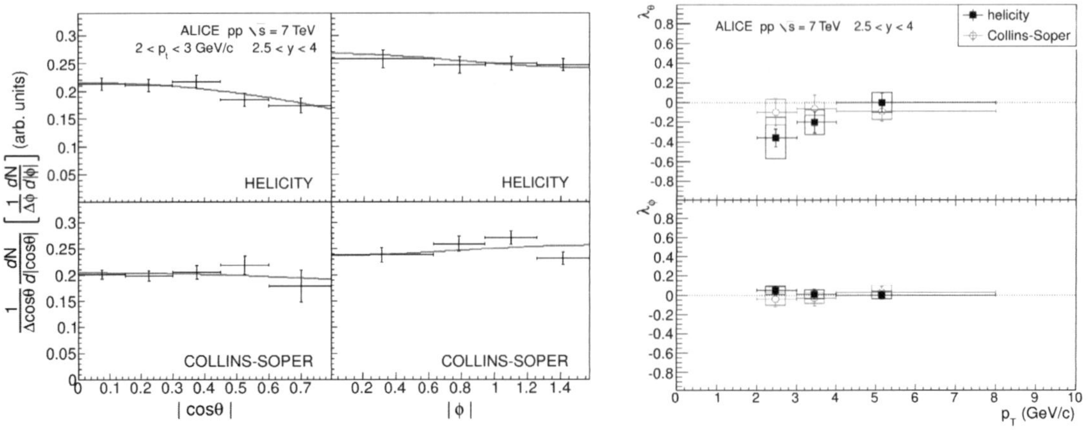

Figure 3. Left: The acceptance corrected angular distributions for the J/ψ decay muons in the range $2 < p_{\mathrm{T}} < 3$ GeV/c. The simultaneous fit to the results in the Collins-Soper and helicity frames is also shown. Right: λ_θ and λ_ϕ as a function of p_{T} for inclusive J/ψ, measured in the helicity (closed squares) and Collins-Soper (open circles) frames [21].

An example for the fits, which are performed in the Collins-Soper and the helicity frame, is shown in the left panel of Fig. 3. The p_{T} -dependence of the resulting values for λ_θ and λ_ϕ is shown in the right panel of Fig. 3. All values are found to be close to zero. It should be pointed out that these fits are for inclusive J/ψ, i.e. they contain the contributions from ψ'-, χ_{c}-, and b-decays.

The interplay between processes on soft and hard processes can be investigated by studying the dependence of the J/ψ yield on the charged particle multiplicity of the underlying event. The latter is dominated by soft processes, which usually require a modeling of MPI in order to achieve a proper theoretical description, and thus could probe the impact parameter of the pp collision. ALICE has measured the relative J/ψ yield as a function of $\mathrm{d}N_{\mathrm{ch}}/\mathrm{d}\eta$ up to four times the mean multiplicity of $\langle \mathrm{d}N_{\mathrm{ch}}/\mathrm{d}\eta \rangle = 6.01 \pm 0.01(\mathrm{stat.})^{+0.20}_{-0.12}(\mathrm{syst.})$ [22]. As shown in Fig. 4 an almost linear increase of the J/ψ yield is observed, both at mid-rapidity and at forward rapidities, while $\mathrm{d}N_{\mathrm{ch}}/\mathrm{d}\eta$ is determined in both cases around mid-rapidity ($|\eta| < 1$). Such a behavior is not seen in Monte Carlo models as Pythia6.4, if only J/ψ from hard productions processes are considered. In this case the J/ψ yield is essentially independent from the charged particle multiplicity of the underlying event. The experimental finding suggests that MPI affect also processes on a harder scale, such as J/ψ production [23], and thus introduce an impact parameter dependence even in elementary pp collisions [24].

3. Pb-Pb Collisions

The nuclear modification factor, defined as

$$R_{\mathrm{AA}} = \frac{\mathrm{d}^2 N^{\mathrm{AA}}_{\mathrm{J}/\psi}/(\mathrm{d}p_{\mathrm{T}}\mathrm{d}y)}{\langle T_{\mathrm{AA}} \rangle \, \mathrm{d}^2 \sigma^{\mathrm{pp}}_{\mathrm{J}/\psi}/(\mathrm{d}p_{\mathrm{T}}\mathrm{d}y)}, \tag{3}$$

where $\langle T_{\mathrm{AA}} \rangle$ is the nuclear overlap function as determined by Glauber calculations, has been measured by ALICE for inclusive J/ψ at mid-rapidity ($|y| < 0.9$) and forward rapidity ($2.5 < y < 4$) for Pb-Pb collisions at $\sqrt{s_{\mathrm{NN}}} = 2.76$ TeV [25, 26]. A clear suppression relative to pp collisions is observed in both cases. While at forward rapidities R_{AA} quickly drops to a value

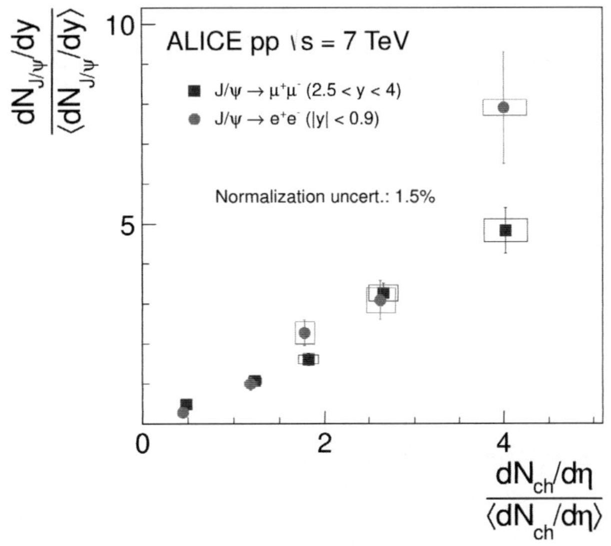

Figure 4. J/ψ yield $\mathrm{d}N_{\mathrm{J}/\psi}/\mathrm{d}y$ as a function of the charged particle multiplicity densities at mid-rapidity $\mathrm{d}N_{\mathrm{ch}}/\mathrm{d}\eta$. Both values are normalized by the corresponding value for minimum bias pp collisions ($\langle \mathrm{d}N_{\mathrm{J}/\psi}/\mathrm{d}y\rangle$, $\langle \mathrm{d}N_{\mathrm{ch}}/\mathrm{d}\eta\rangle$). Shown are measurements at forward rapidities ($2.5 < y < 4$) and at mid-rapidity ($|y| < 0.9$) [22].

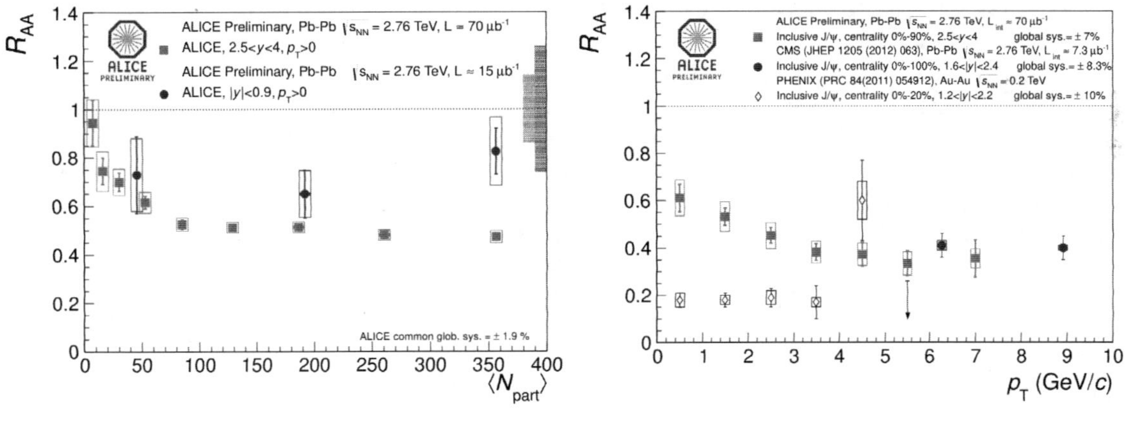

Figure 5. Left: The nuclear modification factor R_{AA} as a function of the number of participating nucleons $\langle N_{\mathrm{part}}\rangle$, measured in centrality selected Pb-Pb collisions at $\sqrt{s_{\mathrm{NN}}} = 2.76$ TeV at mid-rapidity ($|y| < 0.9$) and forward rapidity ($2.5 < y < 4$) [25, 26]. Right: R_{AA} at forward rapidities ($2.5 < y < 4$) as a function of p_{T} in comparison to date by PHENIX [6] and CMS [27].

of 0.5 and then is independent of centrality for $N_{\mathrm{part}} > 100$, there seems to be an indication that R_{AA} is higher at mid-rapidity (see left panel of Fig. 5). However, the current systematic errors, especially on the pp reference, are still too large to draw a firm conclusion. Since the ALICE acceptance allows to measure J/ψ down to $p_{\mathrm{T}} = 0$, the p_{T} dependence of R_{AA} can be studied in this low transverse momentum region. This is shown in the right panel of Fig. 5 for forward rapidities. A significant decrease of R_{AA} with increasing p_{T} is observed. While R_{AA} is

Journal of Physics: Conference Series **458** (2013) 012010 doi:10.1088/1742-6596/458/1/012010

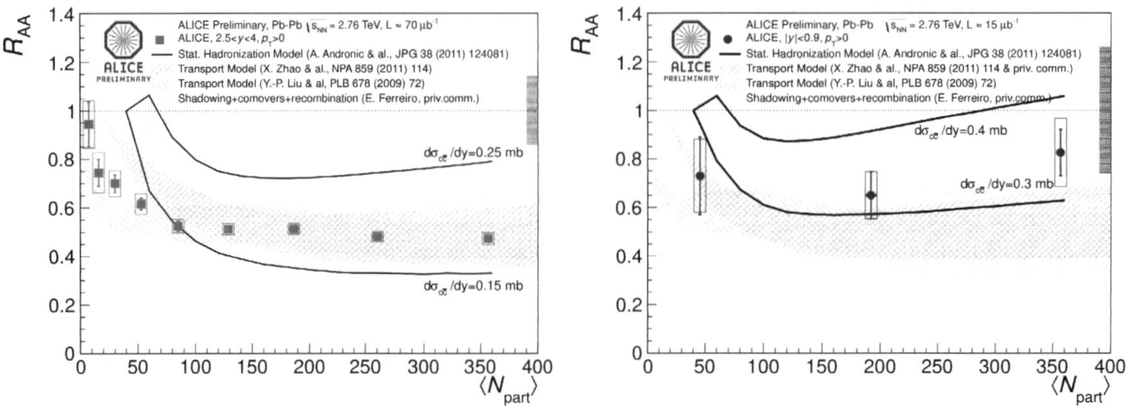

Figure 6. The nuclear modification factor R_{AA} as a function of the number of participating nucleons $\langle N_{\mathrm{part}} \rangle$, measured in centrality selected Pb-Pb collisions at $\sqrt{s_{\mathrm{NN}}} = 2.76$ TeV at forward rapidity ($2.5 < y < 4$, left panel) and at mid-rapidity ($|y| < 0.9$, right panel) [25, 26]. Shown are comparisons to several model predictions [28, 29, 30, 31].

about 0.6 for $p_{\mathrm{T}} < 1$ GeV/c, it drops to values below 0.4 at higher p_{T} and thus agrees with the measurement by the CMS collaboration in this p_{T} region [27]. Also shown in this figure is a comparison with PHENIX results for Au-Au collisions at $\sqrt{s_{\mathrm{NN}}} = 200$ GeV [6], which indicates a sizable rise of R_{AA} at lower p_{T} with increasing center-of-mass energy.

A possible explanation for the increase of R_{AA} with $\sqrt{s_{\mathrm{NN}}}$ might be that at higher energies an additional contribution from J/ψ re-creation is playing a stronger role, due to the higher number of $c\bar{c}$ pairs present in the fireball. In fact, models that include this effect in addition to suppression mechanisms provide a reasonable description of the data (see Fig. 6). The transport models include CNM effects (shadowing and absorption by hadrons), suppression in the hot medium due to color screening, feed down from b-hadrons, and continuous $c\bar{c}$ recombination throughout the evolution of the plasma phase [29, 30]. The statistical hadronization model assumes that all J/ψ are dissolved inside the plasma and that they are only formed at the hadronization stage by distributing the $c\bar{c}$ quarks formed in initial hard scatterings according to statistical weights [28]. Another model ascribes the suppression to nuclear shadowing and comover absorption, but also has to include recombination in order to match the data [31]. However, for a final interpretation a precise understanding of CNM effects is mandatory and is addressed with the recent p-Pb data.

Figure 7 shows the J/ψ v_2 measurement by ALICE for the 20% - 60% central Pb-Pb collisions at $\sqrt{s_{\mathrm{NN}}} = 2.76$ TeV [32]. It is determined in the region $2.5 < y < 4$ by fitting the data with the following equation

$$\frac{\mathrm{d}N_{\mathrm{J}/\psi}}{\mathrm{d}\Delta\phi} \propto 1 + 2v_2^{\mathrm{obs}} \cos(2\,\Delta\phi) \quad \text{with} \quad \Delta\phi = \phi_{\mu\mu} - \Psi_{\mathrm{EP},2}, \tag{4}$$

where $\phi_{\mu\mu}$ is the azimuthal angle of the di-muon pair and $\Psi_{\mathrm{EP},2}$ the event plane angle. $\Psi_{\mathrm{EP},2}$ is measured with a 2% resolution using the VZERO detectors. In the range $2 < p_{\mathrm{T}} < 6$ GeV/c an indication for a non-zero v_2 is observed, as shown in the right panel of Fig. 7, although with a significance of only 2.7 σ. However, the magnitude of the effect is in agreement with transport model calculations that include regeneration effects [33, 34].

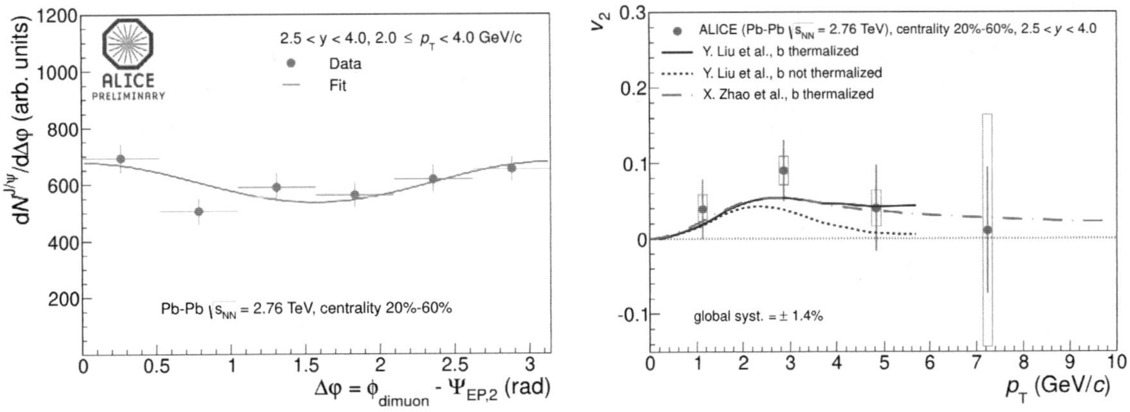

Figure 7. Left: Example of the J/ψ v_2 signal extraction in $\Delta\phi$ bins. Right: The inclusive J/ψ v_2 measured in the 20% - 60% most central Pb-Pb collisions at $\sqrt{s_{\rm NN}} = 2.76$ TeV [32], compared to predictions from parton transport models [33, 34].

4. Conclusions

Transverse momentum spectra and rapidity distributions of inclusive J/ψ have been measured down to $p_{\rm T} = 0$ for pp collisions at $\sqrt{s} = 2.76$ TeV and 7 TeV. A good agreement with NRQCD calculations is observed. The same applies to the spectra of prompt J/ψ. No significant J/ψ polarization in the forward region is seen in pp collisions at $\sqrt{s} = 7$ TeV. An almost linear increase of the J/ψ yield with the charged particle multiplicity of the underlying event is observed, indicating an impact parameter dependence of J/ψ production in pp collisions.

The nuclear modification factor $R_{\rm AA}$ has been measured for Pb-Pb collisions at $\sqrt{s_{\rm NN}} = 2.76$ TeV at forward and mid-rapidity for $p_{\rm T} > 0$. A clear indication for less suppression at low $p_{\rm T}$ than at high $p_{\rm T}$ is found and the $R_{\rm AA}$ at low $p_{\rm T}$ is clearly higher at LHC than at RHIC. Also, a first hint of a non-zero v_2 has been observed. These observations might indicate that at LHC energies regeneration effects are effective in addition to the suppression mechanisms seen at lower energies.

References

[1] N. Brambilla *et al.*, Eur. Phys. J. **C71**, 1534 (2011).
[2] J.P. Lansberg, Eur. Phys. J. **C61**, 693 (2009).
[3] T. Matsui and H. Satz, Phys. Lett **B 178**, 416 (1986).
[4] B. Alessandro et al. (NA50 Collaboration), Eur. Phys. J. **C 39**, 335 (2005).
[5] A. Adare et al. (PHENIX Collaboration), Phys. Rev. Lett. **98**, 232301 (2007).
[6] A. Adare et al. (PHENIX Collaboration), Phys.Rev. **C 84**, 054912 (2011).
[7] P. Braun-Munzinger and J. Stachel, Phys. Lett. **B 490**, 196 (2000).
[8] R.L. Thews, M. Schroedter, and J. Rafelski, Phys. Rev. **C 63**, 054905 (2001).
[9] X. Zhao and R. Rapp, Phys. Lett. **B 664**, 253 (2008).
[10] A. Andronic, P. Braun-Munzinger, K. Redlich, and J. Stachel, Phys. Lett. **B 652**, 259 (2007).
[11] K. Aamodt et al. (ALICE Collaboration), Phys. Lett. **B 704**, 442 (2011), errat.: Phys. Lett. **B 718**, 692 (2012).
[12] V. Khachatryan et al. (CMS Collaboration), Eur. Phys. J. **C 71**, 1575 (2011).
[13] R. Aaij et al. (LHCb Collaboration), Eur. Phys. J. **C 71**, 1645 (2011).
[14] G. Aad et al. (ATLAS Collaboration), Nucl. Phys. **B 850**, 387 (2011).
[15] B. Abelev et al. (ALICE Collaboration), Phys. Lett. **B 718**, 295 (2012).
[16] B. Abelev et al. (ALICE Collaboration), JHEP **11**, 065 (2012).
[17] S. Chatrchyan et al. (CMS Collaboration), J. High Energy Phys. **2**, 011 (2012).
[18] M. Butenschoen and B.A. Kniehl, Phys. Rev. Lett. **106**, 022003 (2011).

[19] Y.-Q. Ma, K. Wang, and K. Chao, Phys. Rev. Lett. **106**, 042002 (2011).
[20] V.A. Saleev, M.A. Nefedov and A.V. Shipilova, Phys. Rev. **D 85**, 074013 (2012).
[21] B. Abelev et al. (ALICE Collaboration), Phys. Rev. Lett. **108**, 082001 (2012).
[22] B. Abelev et al. (ALICE Collaboration), Phys. Lett. **B 712**, 165 (2012).
[23] S. Porteboeuf and R. Granier de Cassagnac, Nucl. Phys. **B 214**, 181 (2011).
[24] L. Frankfurt, M. Strikman, D. Treleani, and C. Weiss, Phys. Rev. Lett. **101**, 202003 (2008).
[25] B. Abelev et al. (ALICE Collaboration), Phys. Rev. Lett. **109**, 072301 (2012).
[26] R. Arnaldi et al. (for the ALICE Collaboration), arXiv:1211.2578.
[27] S. Chatrchyan et al. (CMS Collaboration), J. High. Energy Phys. **1205**, 063 (2012).
[28] A. Andronic, P. Braun-Munzinger, K. Redlich, and J. Stachel, J. Phys. **G 38**, 124081 (2011).
[29] X. Zhao and R. Rapp, Nucl. Phys. **A 859**, 114 (2011).
[30] Y.-P. Liu, Z. Qu, N. Xu, P.-F. Zhuang, Phys. Lett. **B 678**, 72 (2009), and Priv. Comm.
[31] E. Ferreiro, Priv. Comm.
[32] E. Abbas et al. (ALICE Collaboration), arXiv:1303.5880.
[33] Y. Liu, N. Xu, and P. Zhuang, Nucl. Phys. **A 834**, 317C (2010), and Priv. Comm., arXiv:0910.0959.
[34] X. Zhao, A. Emerick, and R. Rapp, arXiv:1210.6583.

Journal of Physics: Conference Series **458** (2013) 012011 doi:10.1088/1742-6596/458/1/012011

Quarkonium Results in PbPb Collisions at CMS

M Calderón de la Barca Sánchez[1], for the CMS Collaboration
[1] Physics Department, UC Davis. One Shields Avenue, Davis CA 95616, USA
E-mail: mcalderon@ucdavis.edu

Abstract. We summarize the results from the study of charmonium and bottomonium via the dimuon decay channel in PbPb collisions with the CMS experiment. We discuss the observation of sequential suppression of the Υ states. We present preliminary results of prompt J/ψ and ψ' production, as well as of non-prompt J/ψ's coming from the weak decay of b-quarks. This latter measurement is sensitive to b-quark energy loss. We discuss the results and compare to model predictions.

1. Introduction

Quarkonium suppression in a hot colored medium is a longstanding signature of quark-gluon plasma (QGP) formation. It is expected that suppression will arise due to two main effects. The first mechanism proposed was color Debye screening [1]. In the vacuum, a heavy quark-antiquark bound state, such as charmonium or bottomonium, are kept together by the QCD color field. Simulations of QCD on the lattice show that the color fields take the form of a color-flux tube, which gives rise to an approximately linear potential. Since the quarks are massive and move at low velocities, one can use the framework of non-relativistic QCD. The spectroscopy of charmonia and bottomonia can be obtained this way. In a hot QCD medium, the expectation is that the additional color charges around the heavy quark-antiquark pair will produce color fields which screen the potential between them. The Debye screening radius depends on the temperature, and this then leads to a suppression of excited quarkonium states according to their average radius. This picture has been augmented with recent studies which find that there is also a broadening of the quarkonium spectral functions due to the interaction of soft gluons in the QGP medium with the quarkonium state. The effects of gluo-dissociation and Landau damping can be encapsulated in the quark-antiquark potential as an imaginary part, with the real part governing the screening behavior. For a review of this subject, see Ref. [2]. More recently, attempts to calculate the potential directly in the lattice have been done, including the real and imaginary parts, i.e. screening and Landau damping/gluo-dissociation. These studies motivate the continued study of quarkonia in relativistic heavy ion collisions, since an observation of suppression could then be connected to the properties studied in the lattice. In this way, we could estimate the temperature reached in the medium. Furthermore, a suppression due to hot nuclear matter effects in the bottomonium case would be a clean signature of deconfinement. (The picture is complicated in the case of charmonium due to the competing effect of recombination, which is negligible for the case of bottomonium due to the small bottom cross section even at LHC energies.) In this paper, we present a summary of some key results from the CMS in the field of quarkonium studies in relativistic heavy ion collisions. The manuscript is organized as follows: We briefly describe the detector and analysis preliminaries, then we discuss the results from prompt charmonium. Next, we present the results from non-prompt J/ψ which originate mainly from B meson decays. We then discuss bottomonium measurements, followed by a summary of our findings.

2. CMS Detector and Data

A detailed description of the CMS detector can be found in Ref. [3]. The main detectors used in this analysis are the Muon detectors and the silicon tracker. The solenoidal superconducting magnet of 6m internal diameter provides a $B = 3.8$ T field. It houses the silicon pixel and silicon strip tracker, as well as electromagnetic and hadronic calorimeters. The tracker can measure charged tracks in the kinematic range $|\eta| < 2.5$. Muons are detected via drift tubes, cathode strip chambers and resistive plate chambers in the range $|\eta| < 2.4$. The granularity of the tracker working in tandem with the strong bending B field allow for measuring muon transverse momenta (p_T) with resolution in the range $1 - 2\%$ for the analyses presented here. The muons need a minimum total momentum in the range $3 - 5$ GeV/c to reach the muon stations. This places limits on the acceptance of J/ψ mesons. For mid-rapidity, we can measure J/ψ in the range $p_T > 6.5$ GeV/c, while at forward rapidity, we can measure it down to $p_T > 3$ GeV/c. For the Υ mesons, the higher mass is transformed to a higher muon momentum in the decay process, allowing us to measure Υ mesons down to $p_T > 0$ for all rapidity.

To estimate the collision centrality, we measure the transverse energy (E_T) in a hadron calorimeter placed in the forward region (HF, $2.9 < |\eta| < 5.2$). We use a Glauber model to simulate the N_{part} distributions based on fits to the HF E_T measurements, as described in Ref. [4].

3. J/ψ Production

In the invariant mass spectrum of dimuons, CMS can measure J/ψ mesons and estimate the prompt and non-prompt contributions [5]. We distinguish these two categories as follows. The prompt component originates from directly produced J/ψ plus those which arise from the secondary decays of charmonium excited states, such as the ψ' and χ_c. These excited state transitions are sufficiently fast that the muons we observe are indistinguishable from primary muons. The non-prompt component arises mainly from the decay of beauty hadrons, chiefly B mesons. These weak decays are longer lived, with lifetimes of order $c\tau \approx 490\mu$m (for B^+). By reconstructing the invariant mass, the position of the secondary vertex in the transverse plane L_{xy}, and the pair p_T, we can do a simultaneous fit of the invariant mass spectrum together with the pseudo-decay length $\ell_{J/\psi} = L_{xy} m_{J/\psi}/p_T$. This allows us to estimate the contributions of prompt and non-prompt J/ψ separately.

3.1. Prompt J/ψ

We have measured the production of prompt J/ψ in both PbPb and pp collisions at $\sqrt{s}$=2.76 TeV. To quantify the modifications observed in going from pp to PbPb, we scale the yields measured in pp using the Glauber model from Ref. [4] and use this to normalize the PbPb measurements, constructing the nuclear modification factor R_{AA}. Figure 1 shows R_{AA} of prompt J/ψas a function of N_{part}, p_T, and y.

For the top 5% most central collisions in the kinematic range $p_T > 6.5$ GeV/c we observe a suppression of a factor of ~ 5. For the centrality-integrated data, we do not observe a significant p_T or rapidity dependence of the suppression in our measured kinematics. Comparing the data to the model calculations from Zhao and Rapp [6], given by the various lines in the left panel of Fig. 1, shows a qualitative agreement with the observed trend of increasing suppression, i.e. decreasing R_{AA}, with increasing N_{part}. In this model, for our kinematics, the mechanism responsible for the suppression are the hot nuclear matter effects, screening and gluo-dissociation, discussed previously.

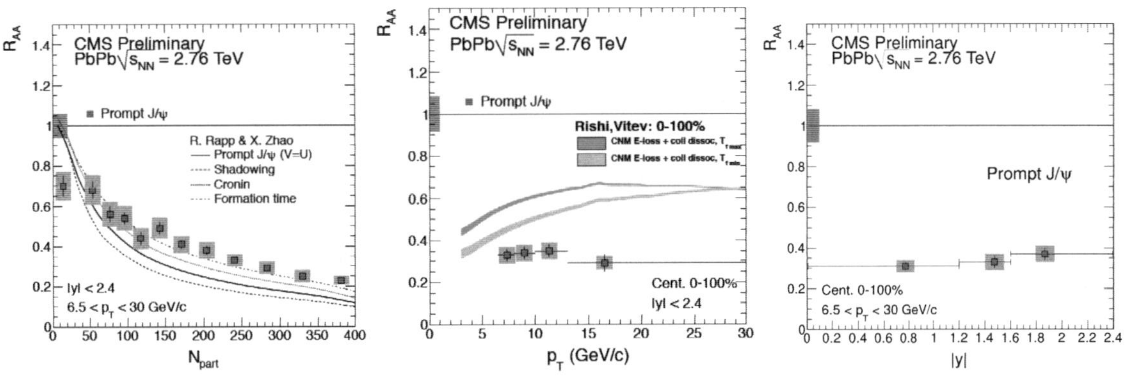

Figure 1. The nuclear modification factor, R_{AA}, for prompt J/ψ as a function of $N_{\rm part}$ (left), p_T(center) and $|y|$ (right). The kinematic range for the measurement is $6.5 < p_T < 30$ GeV/c, and $|y| < 2.4$. The data show a clear suppression with increasing $N_{\rm part}$, but no p_T or y dependence. For discussion of the model comparisons, see text.

We see no evidence for a contribution from statistical recombination, which is expected to contribute to the J/ψ yield at lower p_T. In contrast, a model from Sharma and Vitev [7] based on cold-nuclear-matter energy loss plus collisional dissociation of J/ψ mesons expects less suppression with increasing p_T, in disagreement with the observed level of suppression that remains constant in our measurement region.

We have also measured the nuclear modification factor for the ψ' meson. We find that in the kinematic region $p_T > 6.5$ GeV/c and $|y| < 1.6$ the ψ' is suppressed much more than the J/ψ. Our preliminary result in this kinematic region integrated over all centralities is $R_{AA}(\psi(2S)) = 0.11 \pm 0.03$ (stat.) $\pm\ 0.02$(syst.) $\pm\ 0.02$(pp). In the forward rapidity region, $1.6 < |y| < 2.4$ and lower p_T (down to 3 GeV/c) we find an enhancement of the ψ'. However, the current statistics from the pp measurement hinder the statistical significance of the forward y measurement. With pp data from 2013 this situation will be ameliorated.

3.2. Non-prompt J/ψ production

The non-prompt contribution to the inclusive J/ψ yield was also extracted from the PbPb and pp data. The nuclear modification factor R_{AA} is shown in Figure 2. The dependence of R_{AA} on centrality, p_T and y is shown in the left, middle and right panel, respectively.

We find an increasing suppression for non-prompt J/ψ with increasing $N_{\rm part}$. The suppression reaches a factor 2.5 ($R_{AA} \approx 0.4$) for the top 5% most central events. When integrated over centrality, there is a hint of increasing suppression with increasing p_T or y. As mentioned earlier, the non-prompt J/ψ production is dominated by the decays of B mesons. Therefore, suppression of the non-prompt J/ψ yield is expected to be a sensitive probe to energy loss of b-quarks. Due to the large mass of the b ($m_b \sim 4.2$ GeV/c^2), the radiative energy loss of b quarks is expected to be suppressed, when compared to light parton gluon Bremsstrahlung, at forward angles of order $\theta \lesssim m/E$, where θ is the angle between the radiated gluon and the direction of the propagating quark, m is the quark mass of the heavy quark, and E its Energy. Measurements of non-photonic electron R_{AA} at RHIC indicated that, in addition to radiative energy loss, b quark production also had a significant modification due to collisional energy loss. In the comparison of the R_{AA} vs. p_T data to model calculations, we find that the model from Vitev, based on Ref. [8], with only radiative energy loss does not give enough suppression to be consistent with our measurement (light blue band). The addition of collisional energy loss (dark blue) is in much better quantitative agreement with our data. A calculation

Journal of Physics: Conference Series **458** (2013) 012011 doi:10.1088/1742-6596/458/1/012011

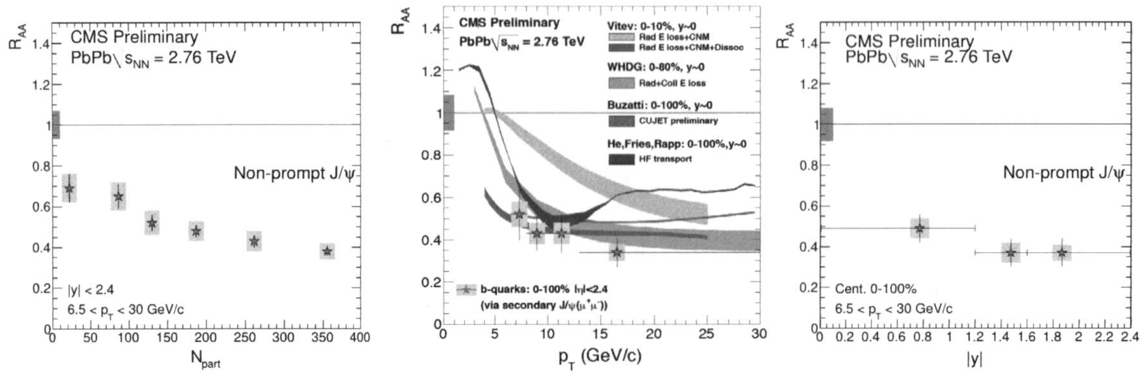

Figure 2. The nuclear modification factor, R_{AA}, for non-prompt J/ψ, which mainly come from B meson decays, as a function of N_{part} (left), p_T(center) and $|y|$ (right). The kinematic range for the measurement is $6.5 < p_T < 30$ GeV/c, and $|y| < 2.4$. For discussion of the experiment and model comparisons, see text.

in the energy-loss framework developed by Wicks, Horowitz, Djordjevic and Gyulassy, including both the radiative and collisional energy-loss mechanisms, is also shown. The framework was developed for comparing to RHIC data, and is based on pQCD weak-coupling and AdS/CFT strong-coupling drag energy loss models [9]. The model parameters were set at RHIC and then kept constant to predict the R_{AA} at LHC (green band). We see that this model is in good quantitative agreement with our data. Note that our data are given in terms of the p_T of the measured J/ψ while the calculations are given in terms of the p_T of the B meson, which is always higher than the daughter J/ψ p_T. Therefore, one would expect a slight shift of the model curves to lower p_T to account for this. We also compared our data to calculations from the CUJET model [10] and He et al. [11], which also are in good agreement with our data at low p_T. The energy-loss model from He et al., which is based on a heavy-quark diffusion and hadronization picture, expects less suppression with increasing p_T which is in disagreement with our highest measured p_T R_{AA}.

4. Bottomonium

We have measured the Υ family in both the *pp* and PbPb collision systems. Figure 3 shows the invariant mass of dimuon pairs measured in *pp* (left) and PbPb (center and right). The excellent resolution of the CMS detector for muons allows a clear separation of all 3 states [12]. The data are fit in order to extract various parameters of interest, in particular the raw (not acceptance corrected) single ratios $\Upsilon(nS)/\Upsilon(1S)$ in both collision systems and the double ratio of these raw single ratios between the PbPb and *pp* system $(\Upsilon(nS)/\Upsilon(1S))_{\text{PbPb}}/(\Upsilon(nS)/\Upsilon(1S))_{pp}$. The raw single ratios cancel uncertainties arising from mechanisms which affect the excited states and the ground state in the same way. An example of such a mechanism is the uncertainty arising from the parton distribution functions. For Υ production, the dominant production channel is gluon-gluon fusion, hence the gluon PDF is the one that plays the most important role. Since this PDF enters in the same way in the production of the bottomonium final state, namely in the initial state entrance channel, any modifications will affect the ground state and the excited states in the same way, thus they will cancel in the single ratio. An important experimental effect which does not cancel in the raw single ratio is a difference in the acceptance for excited states compared to the ground state. This difference in acceptance arises from the fact that the muons chosen for the analysis must satisfy ${p_T}^{\mu} > 4$ GeV/c. Since we look for two muons in order to reconstruct the Υ mesons, and the p_T threshold is close to half of the Υ mass, it is appreciably

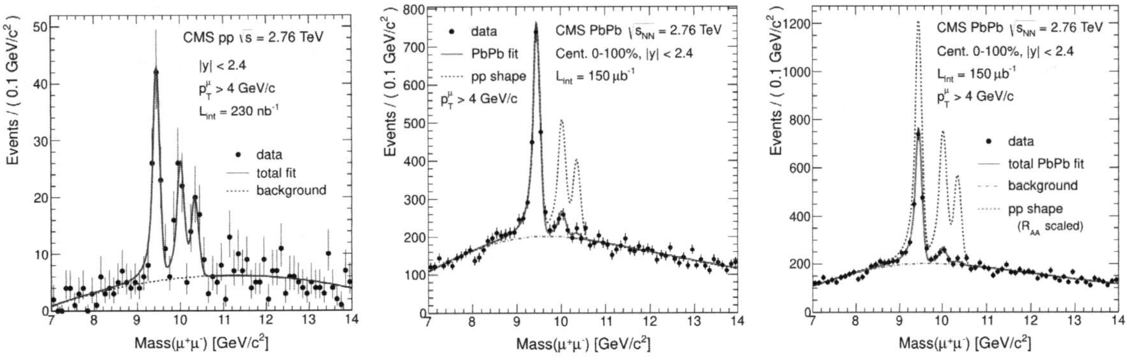

Figure 3. The invariant mass distribution of dimuons. The pp data are shown in the left panel. The center and right panel both show the PbPb data. The pp and PbPb data are fit, with the result shown on the plot. The pp fit is shown in the center and right panels to graphically illustrate the double ratio $(\Upsilon(nS)/\Upsilon(1S))_{\mathrm{PbPb}}/(\Upsilon(nS)/\Upsilon(1S))_{pp}$ (center) and the nuclear modification factor scaling R_{AA} (right). See text for details.

easier for a state at higher mass to have its muon daughters satisfy the p_T threshold compared to a lower-mass Υ meson. This acceptance effect will cancel in the double ratio between PbPb and pp. This makes the double ratio a very robust observable for quantifying modifications of the excited states with a very small systematic uncertainty. This is illustrated graphically in the center panel of Fig. 3. The shape of the invariant mass fit to the Υ states in the pp collision system is overlaid on the PbPb data. In the single ratio, the Υ(1S) yield is the reference, and in the double ratio the PbPb and pp ratios are then compared. By construction, the double ratio for the 1S is 1, so the 1S peaks in PbPb and pp have the same height in the figure. The interesting observation then is to see what happens to the excited states. If the excited-to-ground state ratio in PbPb was the same as in pp, the blue dotted line (pp shape) would also match the PbPb data for the excited states. Clearly it does not. Therefore, we conclude that the excited states are modified in PbPb. Furthermore, we surmise that this modification is not arising from nuclear PDFs, because they would affect the excited states in the same way as the ground state. It remains to be seen whether this is strictly a hot nuclear matter effect by doing this measurement in the pPb collision system.

In the right panel, we illustrate graphically a different scaling. To compare absolute yields for each state between pp and PbPb, we again use a Glauber model to estimate the nuclear overlap and the number of collisions, N_{coll}. Using this information, we scale the measured yields in pp and compare them directly to the data. The blue dotted line in the right panel shows this expected pp yield, which is the denominator in the construction of the nuclear modification factor R_{AA}. One can see in this way that the inclusive yield for all Υ states in PbPb collisions shows suppression compared to the binary-scaled pp yield.

The results for the nuclear modification factor, R_{AA} of Υ mesons are shown in Fig. 4. With the 150 μb^{-1} of integrated luminosity sampled in 2011, we are able to split the Υ(1S) and Υ(2S) data into 7 centrality bins. The Υ(1S) data are shown as the red filled squares in the left panel of Fig. 4, and as the green diamonds in the center and right panels. The Υ(2S) data are shown as the green filled circles in the left panel, and as the blue filled circles in the center and right panels. The small panel to the right of the left panel shows the centrality integrated R_{AA} values:

- $R_{AA}(\Upsilon(1\mathrm{S})) = 0.56 \pm 0.08(\mathrm{stat.}) \pm 0.07(\mathrm{syst.})$
- $R_{AA}(\Upsilon(2\mathrm{S})) = 0.12 \pm 0.04(\mathrm{stat.}) \pm 0.02(\mathrm{syst.})$
- $R_{AA}(\Upsilon(3\mathrm{S})) < 0.1$ at 95% C.L.

Journal of Physics: Conference Series **458** (2013) 012011 doi:10.1088/1742-6596/458/1/012011

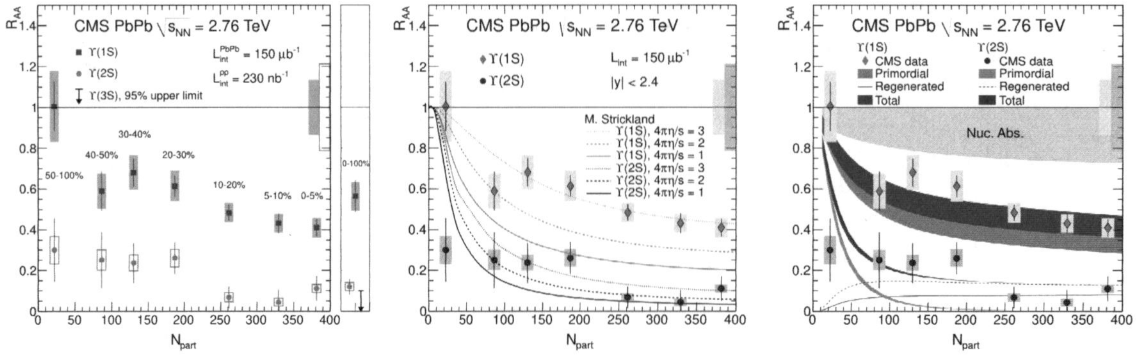

Figure 4. The nuclear modification factor, R_{AA}, for Υ mesons as a function of N_{part}. (Left) The red filled squares show our results for $\Upsilon(1S)$ and the green filled circles are for $\Upsilon(2S)$. The small panel to the right of this figure shows the centrality integrated values. The upper limit for the $\Upsilon(3S)$ is shown as the arrow in this panel. (Center) The same data for the $\Upsilon(1S)$ and $\Upsilon(2S)$ are shown in as green diamonds and blue circles, respectively. The data are compared to a model calculation from Ref. [15]. (Right) The Υ data are the same as in the center panel. The data are compared to a model calculation from Ref. [16]. See text for details.

This is the first measurement of $\Upsilon(2S)$ R_{AA}. In addition, the R_{AA} values we observe are ordered in the expected way: smaller R_{AA} for the states with the smaller binding energy and larger radius.

In studying the data as a function of centrality, we find that the $\Upsilon(1S)$ and $\Upsilon(2S)$ suppression increases with N_{part}. For all the centrality bins, we observe the suppression of the $\Upsilon(2S)$ state to be larger than the suppression of the $\Upsilon(1S)$. In the most peripheral bin, the $\Upsilon(1S)$ nuclear modification factor is consistent with unity, while that for the $\Upsilon(2S)$ remains low. However, it should be noted that this bin includes a wide impact parameter range (50-100% of the total cross section), and it is expected that most of the events where an Υ will be produced will be biased towards larger N_{coll} and hence smaller impact parameter.

Our measurements are of inclusive production of Υ mesons, which include both the directly produced Υ mesons and those which are the daughters of excited states. For example, the χ_b meson can decay via $\chi_b \to \Upsilon(1S)+\gamma$. The χ_b is less tightly bound and has a larger average radius than the $\Upsilon(1S)$. If the χ_b states are suppressed in the QGP, this will reduce the contribution from one of the sources of $\Upsilon(1S)$ and will lead to a suppression even if the directly produced $\Upsilon(1S)$ mesons do not melt in the QGP or are otherwise modified. This contribution from feed-down of higher excited states has been measured by CDF at lower $\sqrt{s}$ and by LHCb at higher $\sqrt{s}$ [13, 14]. Both experiments find that the feed-down contribution to the $\Upsilon(1S)$ yield is $\approx 50\%$. However, this contribution has only been measured at high p_T. If the contribution is the same at lower p_T, a suppression of all the excited states down to $R_{AA} = 0$ but with no suppression of $\Upsilon(1S)$ would result in $R_{AA} \approx 0.5$ for the $\Upsilon(1S)$. Under this hypothesis, our data for the most central events for the $\Upsilon(1S)$ is consistent with the suppression affecting only the excited states. This scenario illustrates the importance of including the effects of feed-down when making comparisons to our inclusive measurements.

Two such comparisons are shown in the center and right panels of Fig. 4. In the center panel, we compare our results to a model calculation by Strickland and Bazow [15]. In this model, the heavy-quark potential is taken to be the heavy-quark internal energy as calculated in the lattice. (The calculation was also done with the free energy and found to be inconsistent with our data. The free-energy calculation is not shown.) In addition, the model also includes the imaginary part of the potential (the Landau damping and gluo-dissociation effects discussed previously). These ingredients naturally lead to sequential melting. In addition to including the effects of feed-down, under the assumption that the contribution from excited states is as measured by CDF in the entire p_T range, the model includes a dynamical expansion using the framework of anisotropic hydrodynamics. The model was run under three different choices for the value of the shear viscosity to entropy ratio, $4\pi\eta/S = 1$, 2, and 3. For each of these choices, the initial temperature was also adjusted in order to match the existing measurements of multiplicity. The temperatures used were in the range $552 < T_0 < 580$ MeV. The highest temperature corresponds to the lowest shear viscosity to entropy ratio, and these are shown as the solid lines in Fig. 4, center. The red line is for the Υ(1S) and the blue line is for the Υ(2S). The Υ(1S) data seems to be best described by the calculation with the lowest temperature and highest shear viscosity to entropy ratio in this model. There is some tension to describe the Υ(2S) data with this same set of parameters, although with the statistical uncertainty for the Υ(2S) data none of the choices can be ruled out to more than 2σ.

The right panel of Fig. 4 shows a comparison to a model calculation from Emerick et al. [16]. The authors of this calculation considered two scenarios. In one scenario, the binding energy of the states was varied with temperature, such that at large temperature the binding was small. This was dubbed the "Weak-Binding Scenario". The second scenario assumed that there is no variation of the binding energy with temperature. The bound states remain with the same masses and binding energies as in the vacuum throughout their dynamical evolution in the QGP. This was called the "Strong-Binding Scenario". As an example of the differences, the bottomonium spectral function is much wider in the Weak-Binding compared to the Strong-Binding scenario. In comparing this to lattice data using the so-called euclidean correlator, which is the integral of the spectral function over some kernel, it was found that the Strong-Binding scenario results in a better agreement with the finding that the correlators do not change very much with temperature. This has been taken as a strong motivation for preferring the Strong-Binding scenario. In this model, both Landau damping and gluo-dissociation become the leading break-up mechanisms, with the former being more important for weakly bound states and the latter for states with large binding energy. The dynamical evolution is done in the framework of kinetic theory. The primordial suppression, including feed-down contributions, is shown as the red band in Fig. 4, right, for the Υ(1S). For the Υ(2S) it is shown as the magenta band. In this model, the primordial Υ(2S) yield is completely suppressed in the most central collisions. This model also includes a parameterization of cold-nuclear matter effects, based on preliminary results from d+Au collisions from STAR [17]. In addition, the model incorporates a contribution from regeneration of Υ mesons via the coalescence of uncorrelated b-$\bar{b}$ pairs produced in the same event. The solid black line shows this contribution for the Υ(1S), and the dotted black line shows the corresponding contribution for the Υ(2S). The total nuclear modification factor, including all contributions, is shown as the blue band for the Υ(1S) and as a brown band for the Υ(2S). Given the small b-$\bar{b}$ cross section, the total suppression is not too different from the primordial result in this model. There is still a large uncertainty in the nuclear absorption, underlining the need for results from d+Au and p+Pb collisions. In this model the initial temperature reached in the QGP is $T_0 \approx 610$ MeV.

5. Summary

We presented the results from Quarkonia production measured by CMS in the heavy ion collision environment. We have measured charmonia at high-p_T in the mid-rapidity region. We found that the J/ψ mesons are suppressed, and the ψ' mesons at mid-rapidity are suppressed more than the J/ψ. This pattern is as expected from the sequential melting picture. The non-prompt J/ψ measurement allows us to study b-quark energy loss. Calculations incorporating radiative and collisional energy loss are consistent with our preliminary measurements. In the bottomonium sector, we also observe a clear ordering of the suppression of the three Υ states with binding energy. A plot summarizing the observed nuclear modification factors for all the quarkonia states discussed in this manuscript is shown in Fig. 5. The horizontal axis indicates

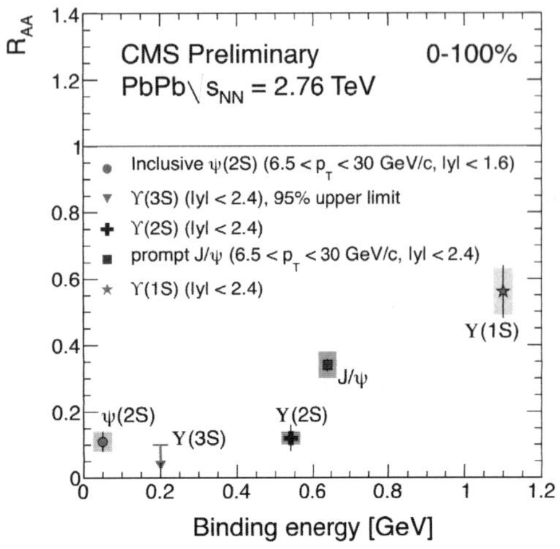

Figure 5. Nuclear modification factor, R_{AA}, for the quarkonia measured by CMS in heavy ion collisions shown as a function of the binding energy of the state. A sequential suppression pattern is observed. See text for details.

the binding energy of the given quarkonium state. The ψ' is the state with the smallest binding energy, and the $\Upsilon(1S)$ is the most tightly-bound state. The binding energy is simply taken to be the vacuum binding energy, i.e. the mass difference between the quarkonium state and the open heavy-flavor continuum, given by twice the mass of the D meson for the charmonium states and twice the mass of the B meson for the bottomonium family. We observe a pattern of sequential suppression of quarkonium states. This pattern is consistent with the expectation that quarkonia will melt in a hot quark-gluon plasma, with the least tightly bound states melting first. The temperatures reached in the plasma in the two model calculations used for comparison to our data are in the range 550 – 610 MeV. We look forward to measurements of quarkonia in p+Pb collisions, which will be important to distinguish cold and hot nuclear matter effects.

References

[1] T. Matsui and H. Satz, Phys. Lett. B **178**, 416 (1986)
[2] N. Brambilla, S. Eidelman, B. K. Heltsley, R. Vogt, G. T. Bodwin, E. Eichten, A. D. Frawley and A. B. Meyer *et al.*, Eur. Phys. J. C **71**, 1534 (2011) [arXiv:1010.5827 [hep-ph]].
[3] S. Chatrchyan *et al.* [CMS Collaboration], JINST **3**, S08004 (2008).
[4] S. Chatrchyan *et al.* [CMS Collaboration], JHEP **1108**, 141 (2011) [arXiv:1107.4800 [nucl-ex]].
[5] CMS Collaboration, J/ψ results from CMS in PbPb collisions with $150\mu b^{-1}$ data, PAS HIN-12-014.
[6] X. Zhao and R. Rapp, Nucl. Phys. A **859**, 114 (2011) [arXiv:1102.2194 [hep-ph]].
[7] R. Sharma and I. Vitev, Phys. Rev. C 87, **044905** (2013) [arXiv:1203.0329 [hep-ph]].
[8] I. Vitev, J. Phys. G **35**, 104011 (2008) [arXiv:0806.0003 [hep-ph]].
[9] W. A. Horowitz, AIP Conf. Proc. **1441**, 889 (2012) [arXiv:1108.5876 [hep-ph]].
[10] A. Buzzatti and M. Gyulassy, arXiv:1207.6020 [hep-ph].
[11] M. He, R. J. Fries and R. Rapp, Phys. Rev. C **86**, 014903 (2012) [arXiv:1106.6006 [nucl-th]].
[12] S. Chatrchyan *et al.* [CMS Collaboration], Phys. Rev. Lett. **109**, 222301 (2012) [arXiv:1208.2826 [nucl-ex]].
[13] T. Affolder *et al.* [CDF Collaboration], Phys. Rev. Lett. **84**, 2094 (2000) [hep-ex/9910025].
[14] R. Aaij *et al.* [LHCb Collaboration], JHEP **1211**, 031 (2012) [arXiv:1209.0282 [hep-ex]].
[15] M. Strickland and D. Bazow, Nucl. Phys. A **879**, 25 (2012). [arXiv:1112.2761 [nucl-th]].
[16] A. Emerick, X. Zhao and R. Rapp, Eur. Phys. J. A **48**, 72 (2012) [arXiv:1111.6537 [hep-ph]].
[17] R. Reed [STAR Collaboration], J. Phys. Conf. Ser. **270**, 012026 (2011) [Nucl. Phys. A **855**, 440 (2011)].

Journal of Physics: Conference Series **458** (2013) 012012 doi:10.1088/1742-6596/458/1/012012

Static potential and singlet free energy at non-zero temperature

A. Bazavov and P. Petreczky

Physics Department, Brookhaven National Laboratory, Upton, NY 11973, USA

Abstract. We study Wilson loops at non-zero temperature and extract the static quark potential from them. The extracted potentials are larger than the singlet free energies and do not show screening for $T < 190$ MeV. We also calculate the singlet free energy and find that only at high temperatures the potential approaches the singlet free energy

1. Introduction

Quarkonium suppression was proposed by Matsui and Satz as a signature of formation of deconfined medium in heavy ion collisions [1]. The basic idea behind this proposal is that color screening in the deconfined medium will modify the heavy quark potential, eventually leading to the dissolution of the heavy quarkonium states. The problem of dissolution of quarkonium states at high temperatures could be formulated in terms of spectral functions. Early attempts to calculate spectral functions on the lattice have been presented in Refs. [2]. However, extraction of the spectral functions from lattice results on Euclidean correlation functions is quite difficult [3] and one should also be careful with cutoff effects in the spectral functions extracted from the lattice [4]. Furthermore, the Euclidean time correlators may not be sensitive to the melting of the bound states at high temperatures due to the fact that the Euclidean time extent is limited to $< 1/(2T)$ (see e.g. discussions in Ref. [5]). The effective field theory framework for heavy quark bound states, namely pQNRCD could be a useful tool for calculating quarkonium spectral functions [6]. The effective field theory approach allows to rigorously define the concept of the static quark anti-quark potential both at zero and non-zero temperatures. One of the main outcomes of the effective field theory analysis is the finding that at non-zero temperature the potential has also an imaginary part, which has important consequences for the dissolution of the quarkonium states. While pNRQCD is formulated in the weak coupling framework it is possible to extend it to the non-perturbative regime. For example, if the binding energy is the smallest scale in the problem all the other scales, like the thermal scales, the inverse size of the bound state and Λ_{QCD} can be integrated out. In this case the potential should be determined non-perturbatively and is identical to the energy of a static $Q\bar{Q}$ pair. If one further neglects the dipole interactions one gets the generalization of the simple potential model to the case of high temperatures [7]. However, one still needs to specify the potential. In the past model considerations based on lattice calculations of the so-called singlet free energy have been used (see e.g. discussion in Ref. [7]). In Ref. [8] it has been suggested to extract the energy of a static $Q\bar{Q}$ pair using the spectral decomposition of the temporal Wilson loops at non-zero temperature. In this contribution we discuss the calculation of the static quark anti-quark energy in 2+1 flavor QCD based on this idea. We also compare the static energy with the singlet free energy.

Journal of Physics: Conference Series **458** (2013) 012012 doi:10.1088/1742-6596/458/1/012012

2. Numerical results

In lattice QCD calculations the static $Q\bar{Q}$ energy is extracted from Wilson loops $W(r,\tau)$. At large Euclidean time separations the exponential decay of the Wilson loops is governed by the static energy or potential, $W(r,\tau) \sim \exp(-V(r)\tau)$. More generally one can write a spectral decomposition for the Wilson loops [8]

$$W(r,\tau) = \int_{-\infty}^{\infty} d\omega \sigma(r,\omega) e^{-\omega\tau}. \tag{1}$$

At zero temperature the spectral function is proportional to $\delta(\omega - V(r))$ plus a sum of delta functions corresponding to the excited states (hybrid potentials). At non-zero temperature the delta function becomes a Lorentzian with the width related to the imaginary part of the potential. In Ref. [8] the potential at non-zero temperature was extracted by inverting Eq. (1) via the maximum entropy method (MEM). This procedure overestimated both the real and the imaginary parts of the potential. The problem is similar to the problem of extracting meson spectral functions [2, 3], where the width of the bound state peaks is mostly an artifact of MEM. Refined methods to extract the potential have been presented in Ref. [9] but the extraction of the imaginary part of the potential remains very challenging.

The usual problem that appears in the calculation of the potential is that the Wilson loops become noisy at large spatial separations r. To deal with this problem smeared gauge fields are used in spatial links when constructing Wilson loops on the lattice. Alternatively, one can fix the Coulomb gauge and calculate the correlation functions of two temporal Wilson lines separated by distance r without connecting them by spatial links [10]. At zero temperature where one only interested in the energy levels both choices are equally good, and merely correspond to different choices of static meson interpolating operators. The same should be true at non-zero temperature provided the imaginary part of the potential is not too large.

We have calculated the correlation functions of temporal Wilson lines in 2+1 flavor QCD using the Highly Improved Staggered Quark (HISQ) action [11] and improved gauge action with the physical value of the strange quark mass and light quark masses corresponding to pion mass of 160 MeV in the continuum limit. The calculations have been performed in Coulomb gauge on $48^3 \times 16$ and $48^3 \times 12$ lattices for bare gauge coupling $\beta = 10/g^2 = 7.5$ as well as on $24^3 \times 6$ lattices at various gauge couplings that have been used in the study of the chiral and deconfinement transition by HotQCD [12]. As in Ref. [12] the lattice spacing was set by the r_1 scale extracted from the static potential and using the value $r_1 = 0.3106$ fm. We also used the renormalization constants for the Wilson line correlators determined in Ref. [12]. On $24^3 \times 6$ lattices we also calculated on-axis smeared and unsmeared Wilson loops. We used iterative APE and HYP smearing with 5, 10 and 20 smearing steps for APE smearing and 1,5 and 10 smearing steps for HYP smearing. It should be noted that for $\tau T = 1$ the correlator gives the so-called singlet free energy [13], $F_1(r,T) = -T \ln W(r, \tau = 1/T)$.

The correlators of the Wilson lines as well as smeared Wilson loops show exponential τ-dependence, except close to $\tau T \simeq 1$, where the correlators show a slight increase. This increase is due to the contribution of a backward propagating state $\sim \exp(-E_{back}(1/T - \tau))$, that arises from the fact that static quarks propagate in gauge field background that is periodic in τ. Similar behavior has been observed in the Wilson loop calculations at non-zero temperature in pure gauge theory [8] as well as in full QCD calculations of bottomonium spectral functions within the non-relativistic formulation [16]. For the largest two values of N_τ considered here the backward propagating state does not cause any problem and we get stable results for the potential by performing single exponential fits in the τ-interval around the mid-point $\tau T = 1/2$. However, the results are quite noisy for $rT > 1$ in this case. Thus to explore the potential at larger distances we use $24^3 \times 6$ lattices for which statistical errors are small. Here the results are sensitive to how the fits are done. We performed three type of fits. First we used only $\tau T = 1/3$ and $\tau T = 1/2$ to

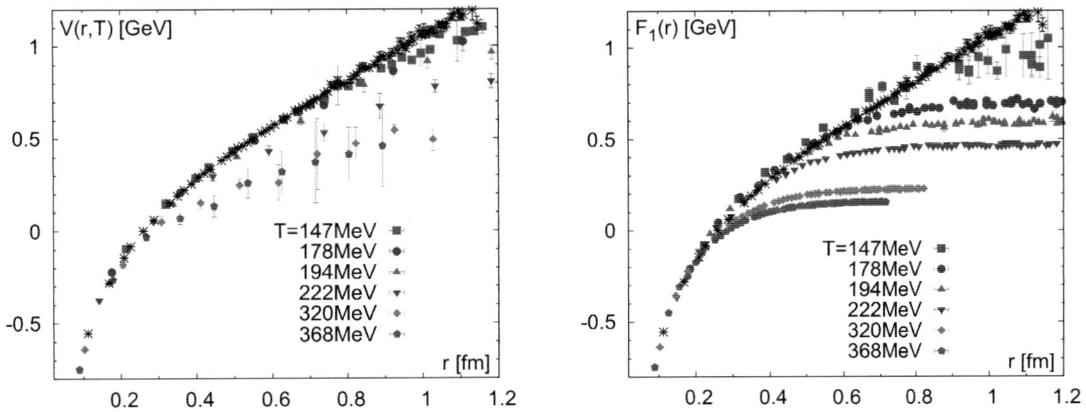

Figure 1. The static quark anti-quark potential (left) and the singlet free energy in Coulomb gauge (right) calculated on $24^3 \times 6$ lattice. The zero temperature potential is shown as black crosses.

extract the potential. Then we determined the backward propagating contribution by performing fits with $\tau T = 1$ and $\tau T = 2/3$. Subtracting the backward propagating contribution from the correlator we extracted the potential using $\tau T = 1/2$ and $\tau T = 2/3$ points. This gives our central value of the extracted potential. Finally, we performed single exponential fits for $\tau T = 1/2$ and $\tau T = 2/3$ which gave us the lower value of the potential. While the Coulomb gauge results and smeared Wilson loops give similar results excited state contributions are visible in τ-dependence of the unsmeared Wilson loops.

Our numerical results for the static quark anti-quark potential at different temperatures extracted from Coulomb gauge Wilson line correlators are shown in Fig. 1. The errors for the temperature dependent potential shown in the figure are mostly systematics and are estimated as described above. We also calculated the singlet free energy in Coulomb gauge and the results are shown in Fig. 1. Qualitatively the temperature dependence of the singlet free energy obtained in our calculations is similar to the temperature dependence of the singlet free energy obtained earlier with the p4 action [14, 15] but there quantitative differences. For the lowest temperature both the potential and the singlet free energy agree with the zero temperature result. For $T = 178$MeV the singlet free energy is very different from the zero temperature potential, while the difference is small for the finite temperature potential. Furthermore, the in-medium potential does not show screening at distances smaller or equal to 1 fm. We fitted the potential with screened Coulomb plus linear form which gave us estimates for the screening length of about 2 fm. Screening effects become apparent in the potential at $T = 194$MeV and happen at distances of about 1fm. At the same temperature the screening effects in the singlet free energy set in at distance of about 0.6fm. At higher temperatures screening effects in the potential set in at smaller and smaller distances and the difference between the potential and the singlet free energy becomes smaller. This is expected as at very high temperatures the singlet free energy should be equal to the potential [6]. At all temperatures the potential is larger than the singlet free energy, i.e. it seems to approach the singlet free energy from above. At the highest two temperatures our results are in qualitative agreement with the findings of the recent quenched study [9].

To study to what extent our conclusions depend on the choice of the correlators we compare the singlet free energy obtained from unsmeared and smeared Wilson loops as well as from Coulomb gauge correlators. The results are shown in Fig. 2. At low temperatures the results

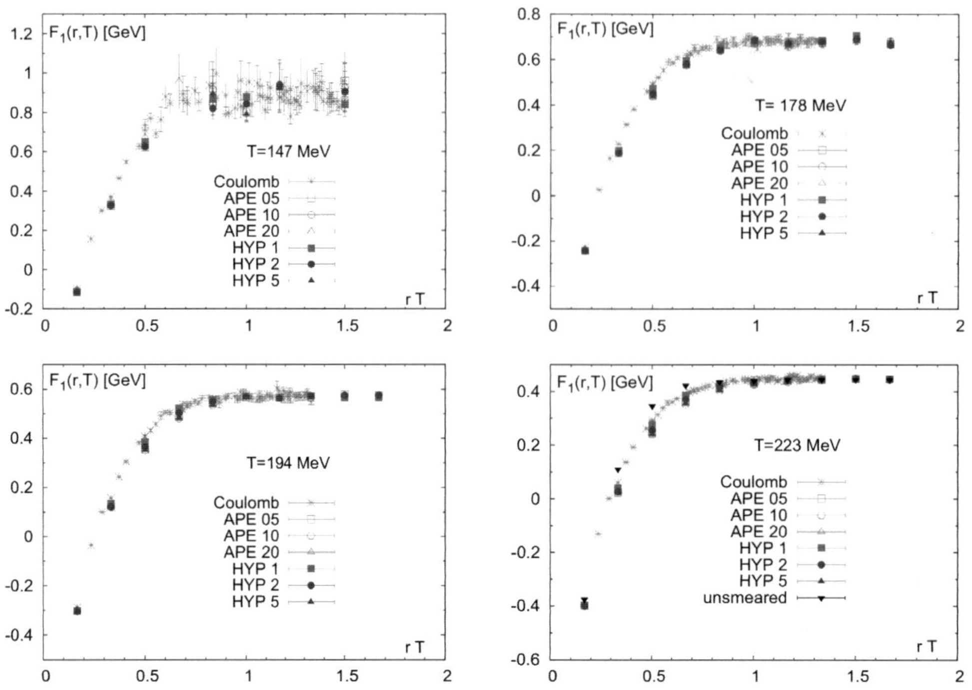

Figure 2. The singlet free energy at $T = 147$ MeV, 178 MeV, 194 MeV and 220 MeV calculated for different choices of the correlators.

obtaibed using unsmeared Wilson loops were too noisy and therefore we do not show them here. Smeared Wilson loops give results which are slightly smaller than Coulomb gauge results at intermediate distances, except for one iteration of HYP smearing. There is no dependence on the smearing if more than one smearing steps are used. The difference between the Coulomb gauge result and the results obtained using smeared Wilson loops is about the same or smaller than the errors of the extracted potential (c.f. Fig. 1). For $T = 222$MeV we also show F_1 obtained from unsmeared Wilson loops, which turns out to be systematically larger than F_1 obtained in Coulomb gauge at intermediate distances. This could be due to the intersection divergences of cyclic Wilson loops [17]. At large distances the singlet free energy obtained from various correlators converges to the same value as expected.

3. Conclusion

We have calculated the correlation functions of Wilson lines at non-zero temperature for different τ and extracted the temperature dependent potential. The temperature dependent potential is always larger than the singlet free energy, approaching it from above. We do no see screening effects present in the potential for $T < 190$MeV. The value of the extracted temperature potential at large distance is very close to the value of the phenomenological potential used in Ref. [7]. Our results are not very sensitive to the specific choice of the correlation function as long as Coulomb gauge Wilson line correlators and smeared Wilson loops are considered.

Acknowledgments

This work was supported by U.S. Department of Energy under Contract No. DE-AC02-98CH10886. The numerical simulations have been performed at NERSC and on BlueGene/L computers at the New York Center for Computational Sciences (NYCCS) at Brookhaven National Laboratory.

[1] T. Matsui and H. Satz, Phys. Lett. B **178** (1986) 416.
[2] T. Umeda et al., Eur. Phys. J. C **39S1** (2005) 9; M. Asakawa and T. Hatsuda, Phys. Rev. Lett. **92**, 012001 (2004); S. Datta et al., Phys. Rev. D **69**, 094507 (2004)
[3] I. Wetzorke et al., Nucl. Phys. Proc. Suppl. **106**, 510 (2002).
[4] F. Karsch, et al., Phys. Rev. D **68**, 014504 (2003)
[5] P. Petreczky, J. Phys. G **39**, 093002 (2012)
[6] N. Brambilla, et al, Phys. Rev. D **78**, 014017 (2008); Phys. Rev. D **82**, 074019 (2010)
[7] P. Petreczky, C. Miao and A. Mocsy, Nucl. Phys. A **855**, 125 (2011)
[8] A. Rothkopf et al, PoS LAT **2009**, 162 (2009); Phys. Rev. Lett. **108**, 162001 (2012);
[9] Y. Burnier and A. Rothkopf, Phys. Rev. D **86** (2012) 051503; arXiv:1304.4154 [hep-ph]
[10] C. Aubin et al., Phys. Rev. D70, 094505 (2004)
[11] E. Follana et al., Phys. Rev. D75, 054502 (2007)
[12] A. Bazavov et al., Phys. Rev. D **85**, 054503 (2012); J. Phys. Conf. Ser. **230**, 012014 (2010)
[13] O. Kaczmarek et al., Phys. Lett. B **543**, 41 (2002)
[14] P. Petreczky, J. Phys. G **37**, 094009 (2010)
[15] O. Kaczmarek, PoS CPOD **07**, 043 (2007) [arXiv:0710.0498 [hep-lat]].
[16] G. Aarts et al., Phys. Rev. Lett. **106**, 061602 (2011)
[17] M. Berwein, N. Brambilla, J. Ghiglieri and A. Vairo, JHEP **1303** (2013) 069 [arXiv:1212.4413 [hep-th]].

Journal of Physics: Conference Series **458** (2013) 012013 doi:10.1088/1742-6596/458/1/012013

Measurements of the nuclear modification factor and the elliptic azimuthal anisotropy of heavy flavours with ALICE

Denise Moreira de Godoy for the ALICE Collaboration

Department of Nuclear Physics, University of Sao Paulo, 187 R Matao Street, Sao Paulo, SP 66318, Brazil

E-mail: denise.m.godoy@gmail.com

Abstract. Heavy quark production is sensitive to the transport properties of the Quark-Gluon Plasma created in heavy-ion collisions at high energies. The production yields of charmed mesons and heavy-flavour decay leptons in most central collisions allow to study the parton energy loss mechanism of heavy quarks propagating through the dense and colored medium. Concurrently, measurements of the elliptic azimuthal anisotropy of D mesons and heavy-flavour leptons in semi-central collisions allow to study the degree of thermalization of heavy quarks within the strongly coupled, flowing medium. We present ALICE measurements of the nuclear modification factor (R_{AA}) and elliptic flow coefficient (v_2) of heavy-flavour decay electrons and D mesons reconstructed via their hadronic decays at mid-rapidity ($|y| < 1$) and muons from heavy-flavour hadron decays at forward rapidity ($2.5 < y < 4$) in Pb-Pb collisions at $\sqrt{s_{\mathrm{NN}}} =$ 2.76 TeV. Several theoretical predictions are confronted with these measurements.

1. Introduction

In heavy-ion collisions at ultrarelativistic energies sufficiently high temperature and/or energy density can be achieved to form the Quark-Gluon Plasma (QGP), the state of matter predicted by Quantum Chromodynamics (QCD) in which quarks and gluons are deconfined. Heavy flavours, charm and beauty, produced in the initial stage of collisions via hard parton scattering processes, propagate through the strongly interacting medium and lose energy via elastic scattering and gluon radiation mechanisms. The resulting modification of the transverse momentum spectra of heavy flavours in heavy-ion collisions with respect to proton-proton collisions is a sensitive probe of the parton energy loss within the QGP. The nuclear modification factor R_{AA} is defined as:

$$R_{\mathrm{AA}}(p_{\mathrm{T}}) = \frac{1}{\langle T_{\mathrm{AA}}\rangle} \frac{dN_{AA}/dp_{\mathrm{T}}}{d\sigma_{pp}/dp_{\mathrm{T}}}, \tag{1}$$

where dN_{AA}/dp_{T} is the differential yield in nucleus-nucleus collisions, $d\sigma_{pp}/dp_{\mathrm{T}}$ is the differential cross section in proton-proton collisions, and $\langle T_{\mathrm{AA}}\rangle$ is the nuclear thickness function in nucleus-nucleus collisions, which is evaluated with a Glauber Monte Carlo calculation. One expects R_{AA} $= 1$ in absence of initial and final state effects in nuclear collisions.

Another experimental observable used in studies of the QGP properties is the elliptic azimuthal anisotropy of produced particles. The azimuthal particle production can be represented as a Fourier series with respect to the reaction plane, which is defined by the impact parameter direction and the beam direction [1]:

$$E\frac{d^3N}{d^3p} = \frac{1}{2\pi}\frac{d^2N}{p_T dp_T dy}\left[1 + \sum_{n=1}^{\infty} 2v_n \cos\left[n(\varphi - \Psi_{RP})\right]\right], \tag{2}$$

where φ is the azimuthal angle of the emitted particle and Ψ_{RP} is the reaction plane angle. The second harmonic v_2, the so-called elliptic flow or elliptic azimuthal anisotropy, is sensitive to the initial geometry of the heavy-ion collisions and can be connected to the properties of the created medium via hydrodynamics [1]. The measurement of v_2 of heavy-flavour hadrons at low transverse momentum provides a way to test whether heavy quarks take part in the collective expansion of the medium. v_2 of heavy-flavour hadrons at high transverse momentum is interpreted as a path length dependence of heavy-quark energy loss within the created dense medium.

2. ALICE experiment and data analysis

The ALICE detector [2] consists of a central barrel (-0.9 $< \eta <$ 0.9) located inside a 0.5 T magnetic field oriented along the LHC beam direction, a muon spectrometer (-4.0 $< \eta <$ -2.5) and several forward detectors. The Inner Tracking System (ITS), the Time Projection Chamber (TPC), and the Time-Of-Flight (TOF) detectors are located in the central barrel and provide charged particle identification. The ITS and TPC detectors are also used for track reconstruction and, consequently, the momentum measurement. The starting time of the collision, which is required for the particle time-of-flight determination, is measured with the T0 detector. The central barrel also includes the Electromagnetic Calorimeter (EMCal), which provides measurement of electrons, photons, neutral hadrons, and jets, and also serves as a trigger detector. The V0 detector (-3.7 $< \eta <$ -1.7 and 2.8 $< \eta <$ 5.1) and the Zero Degree Calorimeters (ZDC) are used for event selection and triggering. The event plane, which is an experimental estimate of the reaction plane, is determined from the azimuthal distribution of tracks in the TPC or counts in the V0 scintillator tiles.

Electrons were identified in Pb-Pb collisions at $\sqrt{s_{NN}}$ = 2.76 TeV based on information from the TPC, TOF and EMCal detectors. The contribution of electrons from non-heavy flavour sources was estimated with two different techniques [3]. The first one used a cocktail of the non-heavy-flavour sources of electrons, and the second one was based on a selection of the invariant mass of e^-e^+ pairs. This contribution was then subtracted from the inclusive electron spectrum. In the case of the v_2 analysis, the subtraction of the non-heavy flavour electron v_2 from the inclusive electron v_2 used as input the heavy-flavour decay electron to background ratio. The cross section in pp collisions for $p_T < 8$ GeV/c was obtained by scaling the measured cross section [4] at $\sqrt{s}$ = 7 TeV [3] to $\sqrt{s}$ = 2.76 TeV based on FONLL [5] predictions. For $p_T > 8$ GeV/c, the cross section in pp collisions was obtained by FONLL calculations directly since the pp measurement was not available.

Charm mesons in Pb-Pb collisions at $\sqrt{s_{NN}}$ = 2.76 TeV were reconstructed through their hadronic decay channels $D^0 \to K^-\pi^+$, $D^+ \to K^-\pi^+\pi^+$, $D^{*+} \to D^0\pi^+ \to K^-\pi^+\pi^+$, $D_s^+ \to \phi\pi^+ \to K^-K^+\pi^+$, and their charge conjugates. The raw yields of D mesons were extracted from fits to invariant mass distributions. The D-meson contribution from weak decays of B mesons was estimated by using FONLL predictions and an hypothesis on the nuclear modification factor of B meson spectra in the medium. This contribution was then subtracted from the D-meson raw yield in order to obtain the yield of prompt D mesons [6]. The pp reference was computed, as for heavy-flavour decay electrons, by means of a FONLL-driven scaling of the cross sections measured at $\sqrt{s}$ = 7 TeV. FONLL predictions were also used for a p_T extrapolation of the measured cross sections to the highest p_T bin where the pp measurement was not available.

Journal of Physics: Conference Series **458** (2013) 012013 doi:10.1088/1742-6596/458/1/012013

The inclusive muon identification in Pb-Pb collisions at $\sqrt{s_{\rm NN}} = 2.76$ TeV was performed with the muon spectrometer. The background contribution was evaluated with Monte Carlo simulations in pp collisions and, in Pb-Pb collisions, from the extrapolation to forward rapidity of the π and K yields measured at mid-rapidity [7]. The measured cross section in pp collisions at $\sqrt{s} = 2.76$ TeV was used for the reference.

3. Results

Figure 1 shows $R_{\rm AA}$ and v_2 of heavy-flavour decay electrons as a function of p_T in the 0-10% and 20-40% centrality classes, respectively, in Pb-Pb collisions at $\sqrt{s_{NN}} = 2.76$ TeV [8]. The observed suppression of electrons from heavy-flavour decays for $p_{\rm T} > 3$ GeV/c indicates a significant energy loss of heavy quarks: charm and beauty. The contribution from beauty hadron decays dominates the measured yield of heavy-flavour decay electrons at high $p_{\rm T}$ in pp collisions [9]. The observed non-zero v_2 at low p_T might indicate that charm quarks participate in the collective motion of the medium. The measured $R_{\rm AA}$ and v_2 of electrons from heavy-flavour decays are compared to theoretical models [10–13]. The comparisons indicate that the simultaneous description of $R_{\rm AA}$ and v_2 of heavy-flavour decay electrons presents a challenge for theoretical models.

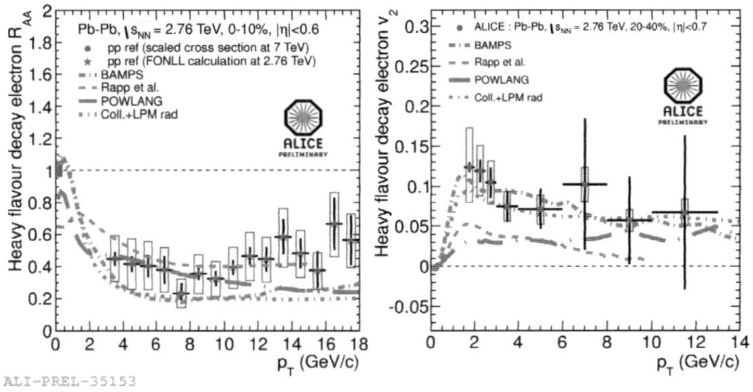

Figure 1. $R_{\rm AA}$ and v_2 of electrons from heavy-flavour decays as a function of $p_{\rm T}$ in the 0-10% and 20-40% centrality classes, respectively, in Pb-Pb collisions at $\sqrt{s_{NN}} = 2.76$ TeV. Results are compared to theoretical predictions [10–13]. Vertical error bars are the statistical uncertainties, horizontal error bars indicate the bin widths and empty boxes are the total systematic uncertainties. The filled box centered at $R_{\rm AA} = 1$ is the normalization uncertainty.

The nuclear modification factors of prompt D^0, D^+, and D^{*+} mesons in Pb-Pb collisions at $\sqrt{s_{\rm NN}} = 2.76$ TeV as a function of p_T for the centrality classes 0-20% and 40-80% were published in [6]. Figure 2 shows $R_{\rm AA}$ of prompt D^0, D^+, D^{*+}, and D_s^+ as a function of p_T in the 0-7.5% centrality class measured from the high statistics data sample collected in 2011 [8]. The nuclear modification factors of the measured prompt D mesons are compatible within the uncertainties.

The relative yield of D_s^+ mesons with respect to non-strange D meson could be enhanced in Pb-Pb collisions at intermediate p_T if charm quarks hadronize via recombination [14]. However, no conclusion can be drawn on the expected enhancement of D_s^+ within the current uncertainties.

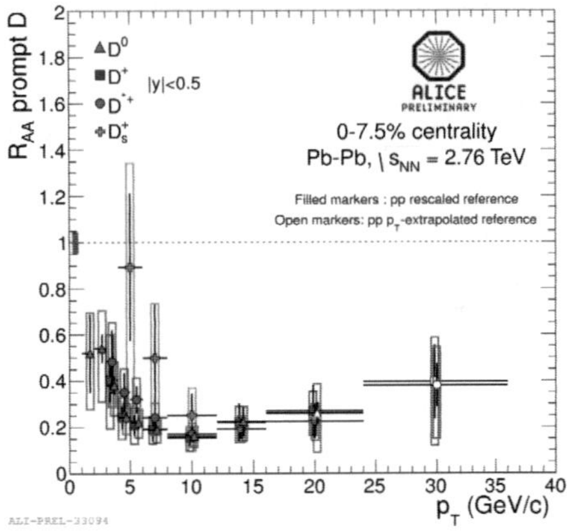

Figure 2. R_{AA} of prompt D^0, D^+, D^{*+}, and D_s^+ as a function of p_{T} in the 0-7.5% Pb-Pb collisions at $\sqrt{s_{NN}} = 2.76$ TeV. Vertical error bars are the statistical uncertainties, horizontal error bars indicate the bin widths and empty boxes are the total systematic uncertainties. The filled box centered at $R_{\mathrm{AA}} = 1$ is the normalization uncertainty.

In the left-hand panel of Fig. 3, we show the average R_{AA} of prompt D^0, D^+, and D^{*+} mesons weighted by their statistical uncertainties as a function of p_T in the 0-7.5% centrality class in Pb-Pb collisions at $\sqrt{s_{NN}} = 2.76$ TeV. The right-hand panel of Fig. 3 shows v_2 of prompt D^0, D^+, and D^{*+} mesons in the 30-50% centrality class in Pb-Pb collisions at $\sqrt{s_{NN}} = 2.76$ TeV. A strong suppression of prompt D mesons is observed over the measured p_{T} range, which indicates that charm quarks lose energy within the created medium. The suppression is similar to that observed for charged particles [6], with a possible indication, not conclusive within the current uncertainties, of $R^D_{\mathrm{AA}} > R^{\mathrm{charged}}_{\mathrm{AA}}$. The measured values of v_2 at low p_{T}, similar to those of light hadrons, although with large uncertainties, may indicate that low momentum charm quarks take part in the collective motion of the system.

The measurements are compared to theoretical models based on parton energy loss in the deconfined medium [13–17].

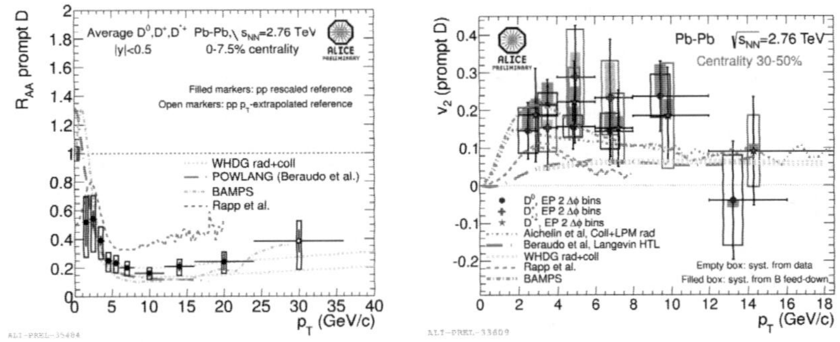

Figure 3. R_{AA} and v_2 of prompt D mesons as a function of p_{T} in the 0-7.5% and 30-50% centrality classes, respectively, in Pb-Pb collisions at $\sqrt{s_{NN}} = 2.76$ TeV. Measurements are compared to theoretical predictions [13–17]. Vertical error bars are the statistical uncertainties, horizontal error bars indicate the bin widths and empty boxes are the total systematic uncertainties. The filled box centered at $R_{\mathrm{AA}} = 1$ is the normalization uncertainty.

The nuclear modification factors as a function of the number of participants of prompt D mesons in $|y| < 0.5$ and $6 < p_T < 12$ GeV/c [6] and J/Ψ from B meson decays (non-prompt J/Ψ) measured by the CMS Collaboration in $|y| < 1.2$ and $p_T > 6.5$ GeV/c [18,19] in Pb-Pb collisions at $\sqrt{s_{NN}} = 2.76$ TeV are shown in Fig. 4. The larger suppression of prompt D mesons with respect to non-prompt J/Ψ is an indication that charm quarks lose more energy than beauty quarks in the deconfined medium.

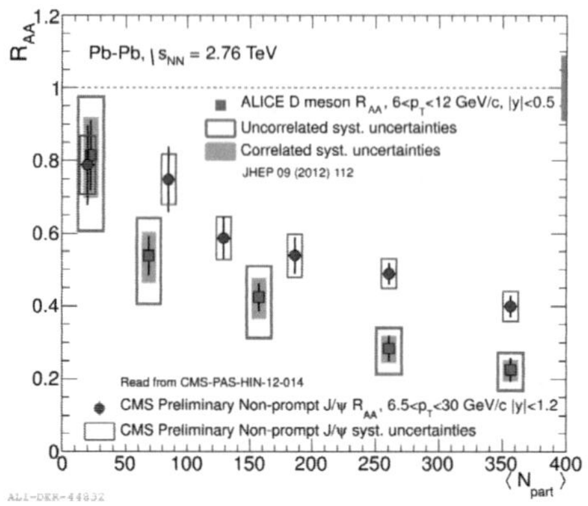

Figure 4. Centrality dependence of R_{AA} of prompt D mesons in $|y| < 0.5$ and $6 < p_T < 12$ GeV/c [6] and non-prompt J/Ψ measured by the CMS Collaboration in $|y| < 1.2$ and $p_T > 6.5$ GeV/c [18,19] in Pb-Pb collisions at $\sqrt{s_{NN}} = 2.76$ TeV. Vertical error bars are the statistical uncertainties and boxes are the systematic uncertainties.

The production of muons from heavy-flavour decays at forward rapidity in pp and Pb-Pb collisions at $\sqrt{s_{NN}} = 2.76$ TeV was published in [7]. The nuclear modification factors of heavy-flavour decay muons in $2.5 < y < 4$ as a function of p_T in central (0-10%) and peripheral (40-80%) Pb-Pb collisions at $\sqrt{s_{NN}} = 2.76$ TeV are shown in Fig. 5. A larger suppression is observed in central collisions than in peripheral collisions.

The measurement in the 0-10% centrality class is compared to theoretical models [10,17,20–23]. The measured R_{AA} of muons from heavy-flavour decays can not be described by nuclear shadowing alone [20] in the measured p_T range. However, models based on parton energy loss in the colour charged medium can describe the data within the uncertainties.

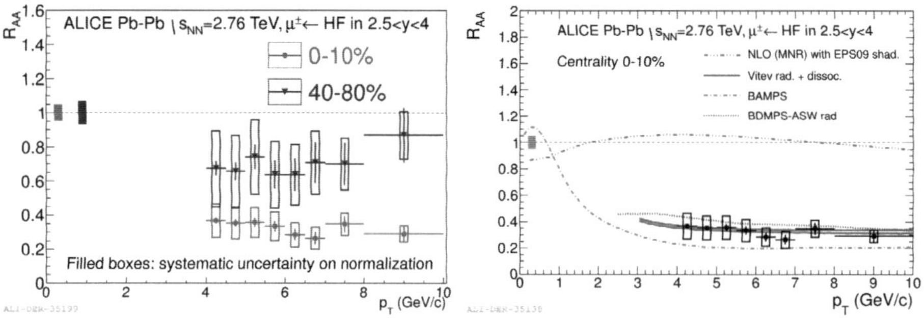

Figure 5. R_{AA} of muons from heavy-flavour decays in $2.5 < y < 4$ as a function of p_T in central (0-10%) and peripheral (40-80%) Pb-Pb collisions at $\sqrt{s_{NN}} = 2.76$ TeV [7]. Result in the 0-10% centrality class is compared to theoretical predictions [10, 17, 20–23]. Vertical error bars are the statistical uncertainties, horizontal error bars indicate the bin widths and empty boxes are the total systematic uncertainties. The filled boxes centered at $R_{AA} = 1$ are the normalization uncertainties.

4. Conclusions

The nuclear modification factor and the elliptic azimuthal anisotropy of heavy-flavour decay leptons and prompt D mesons have been measured in Pb-Pb collisions at $\sqrt{s_{\rm NN}} = 2.76$ TeV by the ALICE collaboration. The observed suppression of heavy-flavour decay products is similar for the three studied decay channels and indicates energy loss of heavy quarks in the created medium. The observed non-zero v_2 of heavy-flavour decay electrons and prompt D mesons at low p_T provides a first hint for collective motion of the heavy quarks in the QGP. Several models were compared to the measured $R_{\rm AA}$ and v_2 of heavy-flavour decay leptons and prompt D mesons. The models based on parton energy loss in the deconfined medium can describe the data within the uncertainties, but the simultaneous description of heavy-flavour $R_{\rm AA}$ and v_2 is challenging.

References

[1] A. M. Poskanzer and S. A. Voloshin, Phys. Rev. C **58** (1998) 1671.
[2] K. Aamodt *et al.* [ALICE Coll.], JINST **3** (2008) S08002.
[3] B. Abelev *et al.* [ALICE Coll.], Phys. Rev. D **86** (2012) 112007.
[4] R. Averbeck, N. Bastid, Z. Conesa del Valle, P. Crochet, A. Dainese and X. Zhang, arXiv:1107.3243 [hep-ph].
[5] M. Cacciari, S. Frixione, N. Houdeau, M. L. Mangano, P. Nason and G. Ridolfi, JHEP **1210** (2012) 137.
[6] B. Abelev *et al.* [ALICE Coll.], JHEP **09** (2012) 112.
[7] B. Abelev *et al.* [ALICE Coll.], Phys. Rev. Lett. **109** (2012) 112301.
[8] Z. C. del Valle [ALICE Coll.], arXiv:1212.0385 [nucl-ex].
[9] B. Abelev *et al.* [ALICE Coll.], Physics Letters B **721** (2013) 13.
[10] J. Uphoff, O. Fochler, Z. Xu and C. Greiner, Physics Letters B **717** (2012) 430.
[11] M. He, R. J. Fries and R. Rapp, arXiv:1208.0256 [nucl-th].
[12] M. Monteno, W. M. Alberico, A. Beraudo, A. De Pace, A. Molinari, M. Nardi and F Prino, J. Phys. G. **38** (2011) 124144.
[13] P. B. Gossiaux, R. Bierkandt and J. Aichelin, Phys. Rev. C **79** (2009) 044906.
[14] M. He, R. Fries, Rainer J. and R. Rapp, Phys. Rev. Lett. **110** (2013) 112301.
[15] W. A. Horowitz and M. Gyulassy, J. Phys. G **38** (2011) 124114.
[16] M. Monteno, W. M. Alberico, A. Beraudo, A. De Pace, A. Molinari, M. Nardi and F Prino, Eur. Phys. J. **C 71** (2011) 1666.
[17] O. Fochler, J. Uphoff, Z. Xu and C. Greiner, J. Phys. G **38** (2011) 124152.
[18] S. Chatrchyan *et al.* [CMS Coll.], JHEP **05** (2012) 063.
[19] CMS Coll., CMS-PAS-HIN-12-14 (2012).
[20] K.J. Eskola, H. Paukkunen and C.A. Salgado, JHEP **04** (2009) 065.
[21] R. Sharma, I. Vitev and B. Zhang, Phys. Rev. C **80** (2009) 054902.
[22] Y. He, I. Vitev and B. Zhang, Physics Letters B **713** (2012) 224.
[23] N. Armesto, A. Dainese, C. A. Salgado and U. A. Wiedemann, Phys. Rev. D **71** (2005) 054027.

29th Winter Workshop on Nuclear Dynamics (WWND2013) IOP Publishing
Journal of Physics: Conference Series **458** (2013) 012014 doi:10.1088/1742-6596/458/1/012014

Charm production asymmetry at the LHC

E.R. Cazaroto[1], V.P. Goncalves[2], F.S. Navarra[1] and M. Nielsen[1]

[1] Instituto de Física, Universidade de São Paulo, C.P. 66318, 05315-970, São Paulo, SP, Brazil
[2] Instituto de Física e Matemática, Universidade Federal de Pelotas, C.P. 354, CEP 96010-900, Pelotas, RS, Brazil.

E-mail: navarra@if.usp.br

Abstract. In this work we use the meson cloud model to study the energy dependence of the asymmetry in D meson production in $p-p$ collisions. We find a good agreement with the recent data from LHCb on D^-/D^+ asymmetry. Although small, this asymmetry may shed light on the role played by the charm meson cloud of the proton.

1. Introduction

It is well known from experiments [1, 2, 3, 4, 5, 6] that the production of D^+/D^- in hadronic collisions is not exactly symmetric. For large x_F (Feynman momentum) it is clearly observed that there are more D^-'s than D^+'s produced in pp, $\Sigma^- p$ and $\pi^- p$ collisions. The recent data from COMPASS collaboration [7] have confirmed that charm production asymmetry is also observed in γp collisions, and the very recent data from LHCb collaboration [8] on pp collisions at 7 TeV show a very small (but different of zero) asymmetry. The origin of the observed asymmetries is still an open question and cannot be explained with usual perturbative QCD (pQCD) or with the string fragmentation model contained in PYTHIA. This fact motivated the construction of alternative models [9, 10, 11] that were able to obtain a reasonable description of the low energy data and make predictions for higher energies. Now we can, for the first time, compare the predictions of these models with the high energy data from LHCb and study the energy dependency of the production asymmetry.

2. Charm production asymmetry

The production asymmetry is quantified as:

$$A = \frac{N_{D^-} - N_{D^+}}{N_{D^-} + N_{D^+}} \tag{1}$$

where N_i represents the number of mesons i produced as a function of, e.g., the Feynman x_F variable. In pQCD the most relevant elementary processes of charm production are $q+\bar{q} \to c+\bar{c}$ and $g+g \to c+\bar{c}$. At high energies the distribution of gluons is much larger than the one of any other parton and the second process dominates. After being produced the c and $\bar{c}$ quarks fragment independently and the resulting D^+ and D^- (also D^0 and $\bar{D}^0$) mesons will have the same rapidity y, p_T and x_F distributions. Consequently, in calculations using pQCD there is no production asymmetry ($A = 0$). This is indeed true for the bulk of charm production. Experimental results on hadronic collisions show that differences between the D^+ and D^- x_F

Journal of Physics: Conference Series **458** (2013) 012014 doi:10.1088/1742-6596/458/1/012014

distributions appear at large x_F, with D^- being harder. Given the valence quark content of the proton $p(uud)$ and of the $D^-(d\bar{c})$, a natural explanation of the observed effect is that the $\bar{c}$ is "dragged" by the projectile valence d quark, forming the, somewhat harder, $d\bar{c}$ bound state. This process, that has been called "recombination" (or coalescence), is a non-perturbative process. Recombination models have been first proposed long time ago [12, 13, 14] and then used more recently [15, 16] to study the accumulated experimental data and to make predictions for the RHIC collisions.

A possible way to implement the idea of recombination is to use models in which the proton is a fluctuating object. In this approach, before colliding with the target the proton projectile fluctuates into a " charm baryon - charm meson " virtual pair. The charmed meson can be liberated during the interaction, being present in the final state of the collision. This type of fluctuation happens all the time. Sometimes the proton fluctuates into a virtual pair "pion - neutron" and then the virtual pair recombines back to its original form, the proton. It was shown [17] that this type of fluctuation is quite relevant to the understanding of hadron structure. This mechanism, in which the "meson cloud" plays a major role (see [18] for light mesons) is ususally called the Meson Cloud Model (MCM). A simple and accurate description of charm asymmetry production at lower energies ($\sqrt{s} \simeq 10 - 40$ GeV) within the framework of the MCM can be found in [19].

In Fig. 1 we show an illustration of D^+/D^- production in the meson cloud approach. Before the collision, the proton projectile fluctuates into a virtual meson-baryon (MB) pair. Fig. 1 and 1 b show the situation in which the meson (the baryon) scatters with the target producing a system X of particles. The system X contains all sorts of particles, including the mesons D^+ and D^-, however there is no asymmetry since the D^+ and the D^- are identically distributed. In what follows we shall call it "indirect production" (I) of $D^\pm$. The production asymmetry happens when the proton fluctuates into a pair $\Sigma_c^{++}D^-$ and the D^- is liberated by the interaction with the target, as shown in Fig. 1c. We shall call it "direct production" (D) of D^-. This last mechanism is, of course, very much suppressed but it is the responsible for the asymmetry because only D^- is produced in this process. This happens because the valence quark content of the proton (uud) only allows its fluctuation into a baryon-meson pair where the meson is always D^- ($d\bar{c}$) and never D^+ ($\bar{d}c$). The meson and the baryon in the cloud have fractional momentum y_M and y_B, with distributions called $f_{M/MB}(y_M)$ and $f_{B/MB}(y_B)$ respectively (we shall use for them the short notation f_M and f_B). Of course, by momentum conservation, $y_M + y_B = 1$ and these distributions are related by [17, 20]:

$$f_M(y) = f_B(1-y) \tag{2}$$

The "splitting function" $f_M(y)$ represents the probability density to find a meson with momentum fraction y of the total cloud state MB. With f_M and f_B we can compute the differential cross section of D production. In the reaction $pp \to D^-X$ the differential cross section of D^- production is given by:

$$\frac{d\sigma^{pp\to DX}}{dx_F} = \Phi_0 + \Phi_I + \Phi_D \tag{3}$$

where Φ_0 and Φ_I refer respectively to "bare" and indirect contributions to D meson production and x_F is the fractional longitudinal momentum of the outgoing meson. By "bare" contribution we mean the process in which the hadron projectile does not fluctuate into any cloud state. The bare contribution is responsible for most of the production and does not generate any asymmetry. The term Φ_D represents the direct process depicted in Fig. 1c and is given by [21, 20]:

$$\Phi_D = \frac{\pi}{x_F} f_D(x_F)\, \sigma^{\Sigma p} \tag{4}$$

Journal of Physics: Conference Series **458** (2013) 012014 doi:10.1088/1742-6596/458/1/012014

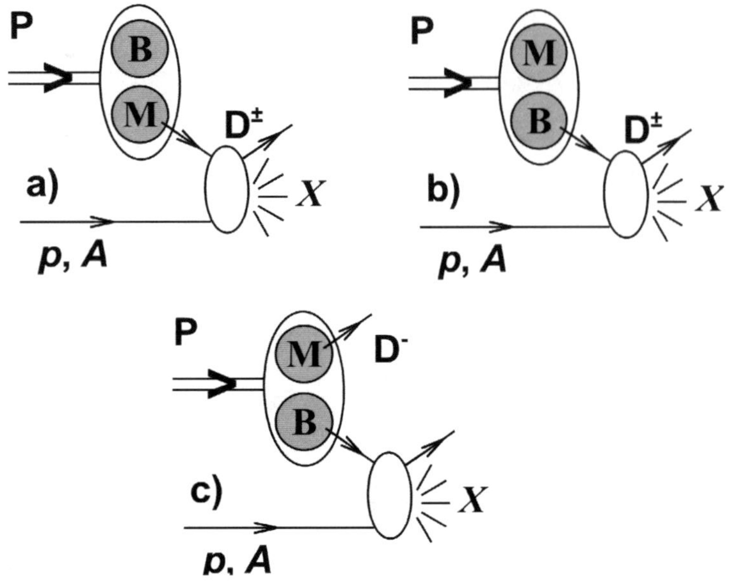

Figure 1. pp collision in which the projectile is in a MB state. Figs. a) and b) show the "indirect" $D^{\pm}$ production and c) the "direct" D^{-} production.

where $f_D \equiv f_{D^-/\Sigma_c^{++}D^-}$ and $\sigma^{\Sigma p}$ is the total $p\,\Sigma_c^{++}$ cross section. The term Φ_I in Eq. (3) can generate some asymmetry, however we can neglect its contribution when compared with the asymmetry generated by the term Φ_D. Thus the D^+/D^- production asymmetry is given by [22]:

$$A^D(x_F) = \frac{\frac{d\sigma^{D^-}(x_F)}{dx_F} - \frac{d\sigma^{D^+}(x_F)}{dx_F}}{\frac{d\sigma^{D^-}(x_F)}{dx_F} + \frac{d\sigma^{D^+}(x_F)}{dx_F}} \simeq \frac{\Phi_D}{\Phi_T^D} \qquad (5)$$

where Φ_T^D is the total $D^+ + D^-$ x_F distribution. The denominator of this expression can be replaced by a parametrization of the experimental data [1, 2, 3, 4, 5, 6], being given by [22]:

$$\Phi_T^D = \frac{d\sigma^{D^-}(x_F)}{dx_F} + \frac{d\sigma^{D^+}(x_F)}{dx_F} \simeq 2\,\sigma_0^D\,(1-x_F)^{n_D} \qquad (6)$$

where $n_D = 5$ for D mesons [22]. Integrating the above expression we obtain the total cross section for charged charm meson production $\sigma^{D^{\pm}} = 1/3\,\sigma_0^D$. Assuming isospin symmetry the charged and neutral (σ^{D^0}) production cross sections are equal. Neglecting the contribution of other (heavier) charm states we can relate the D meson production cross section to the total $c-\bar{c}$ production cross section, $\sigma_{c\bar{c}}$, in the following way:

$$\sigma^{D^{\pm}} = \sigma^{D^0} = \frac{1}{3}\sigma_0^D = \frac{1}{2}\sigma_{c\bar{c}}\ . \qquad (7)$$

From the above relation we can extract the parameter σ_0^D from the experimentally measured $\sigma_{c\bar{c}}$:

$$\sigma_0^D = 1.5\,\sigma_{c\bar{c}}\ . \qquad (8)$$

Inserting (4) and (6) into (5) the asymmetry becomes:

$$A^D(x_F) = \frac{\pi\,\sigma^{\Sigma p}}{2\,\sigma_0^D}\,\frac{f_D(x_F)}{x_F\,(1-x_F)^{n_D}}\ . \qquad (9)$$

The behavior of the asymmetry, Eq. (9), is controlled by the splitting function $f_D(x_F)$. For the splitting function we will use the Sullivan approach [17, 20]. The fractional momentum distribution of a pseudoscalar meson M in the Fock state $|MB'\rangle$ (of a baryon $|B\rangle$) is given by [17, 20]:

$$f_M(y) = \frac{g^2_{MBB'}}{16\pi^2}\, y \int_{-\infty}^{t_{max}} dt\, \frac{[-t + (m_{B'} - m_B)^2]}{[t - m_M^2]^2}\, F^2_{MBB'}(t) \tag{10}$$

where t and m_M are the four momentum square and the mass of the meson in the cloud state and

$$t_{max} = m_B^2\, y - m_{B'}^2\, \frac{y}{1-y}$$

is the maximum t, with m_B and $m_{B'}$ being the B and B' masses respectively. Following a phenomenological approach, we use for the baryon-meson-baryon form factor $F_{MBB'}$ the exponential form:

$$F_{MBB'}(t) = \exp\left(\frac{t - m_M^2}{\Lambda^2_{MBB'}}\right) \tag{11}$$

where $\Lambda_{MBB'}$ is the form factor cut-off parameter. Considering the particular case where $B = p$, $B' = \Sigma_c^{++}$ and $M = D^-$, we insert (10) into (9) to obtain the final expression for the asymmetry in our approach:

$$A^D(x_F) = \frac{N^D}{(1 - x_F)^{n_D}} \int_{-\infty}^{t_{max}} dt\, \frac{[-t + (m_\Sigma - m_p)^2]}{[t - m_D^2]^2}\, F^2_{pD\Sigma} \tag{12}$$

where

$$N^D = \frac{g^2_{pD\Sigma_c}\, \sigma^{\Sigma_c p}}{32\,\pi\, \sigma_0^D} \tag{13}$$

Noticing that $y = x_F$ in the above equations, we can see that in the limit $x_F \to 1$, $t_{max} \to -\infty$, and the integral in (12) goes to zero. In fact, it vanishes faster than the denominator and therefore $A \to 0$. This behavior does not depend on the cut-off parameter but it depends on the choice of the form factor. For a monopole form factor we may obtain asymmetries which grow even at very large x_F. Before presenting the results we emphasize that they depend only on two parameters: Λ and N. Whereas Λ affects the width and position of the maximum of the momentum distribution of the leading meson in the cloud (and consequently of the asymmetry), N is a multiplicative factor which determines the strength of the asymmetry.

3. Results and discussion

Although the recent data [8] are given in terms of the pseudo-rapidity η, in order to study the energy dependence it is more convenient to use the Feynman x_F variable, which for $\eta > 1$ is given by $x_F \simeq 2\, m_T\, e^\eta/\sqrt{s}$. If the transverse momentum of the final D meson is zero or very small we have $m_T \simeq m_D$.

In order to study the energy dependence of the asymmetry we shall consider the two energies where we have experimental data: $\sqrt{s_1} = 33$ GeV and $\sqrt{s_2} = 7000$ GeV (see Table 1). Let us consider the asymmetry ratio $R_A = A(\sqrt{s_2})/A(\sqrt{s_1})$. Using the definitions of the splitting function (10) in (4) and then in (5), many energy independent factors cancel out and we find:

$$R_A = \frac{A(s_2)}{A(s_1)} = \left(\frac{\sigma^\Sigma(s_2)}{\sigma^\Sigma(s_1)}\right) / \left(\frac{\sigma_0^D(s_2)}{\sigma_0^D(s_1)}\right) = \left(\frac{\sigma^{pp}(s_2)}{\sigma^{pp}(s_1)}\right) / \left(\frac{\sigma_{c\bar{c}}(s_2)}{\sigma_{c\bar{c}}(s_1)}\right) \tag{14}$$

where in the last step we have used (8) and assumed that $\sigma^\Sigma = \sigma^{\Sigma_c^{++}p} = \text{const}\,.\,\sigma^{pp}$. Moreover, we have neglected the energy dependence of n_D. The above ratios can be estimated with the

Table 1. Total pp cross section from [23] and total $c\bar{c}$ production cross section from [24] as a function of the energy. The first two lines refer to measurements and the last one show model calculations described in the corresponding references.

Energy (GeV)	σ_{pp} (mb)	$\sigma_{c\bar{c}}$ (mb)
33	40	0.04
7000	97	8
14000	110	11

recently obtained experimental data listed in Table 1. For $\sqrt{s_1} = 33$ GeV and $\sqrt{s_2} = 7000$ GeV we obtain $R_A = 1/75$. Therefore there is a strong decrease in the asymmetry when we increase the energy. This happens because the responsible for the asymmetry is meson emission, a non-perturbative process, which has a slowly growing cross section. In contrast, the symmetric processes are driven by the perturbative partonic interactions, which have strongly growing cross sections. With the data presented in Table 1 we can make the prediction for the order of magnitude of the D^+/D^- asymmetry in the forthcoming 14 TeV pp collisions. Using the last line of Table 1, setting $\sqrt{s_2} = 14$ TeV and susbtituting the numbers in (14) we find $R_A = 1/100$ showing the decreasing trend of the asymmetry.

Due to the lack of experimental data a direct comparison of D^+/D^- asymmetries in the same reaction, e.g. proton-proton, at different energies is difficult. However, we can relate and compare similar reactions. In [22] we related the two sets of data on asymmetries in proton-proton collisions: the low energy one taken by the SELEX collaboration [6] and the high energy one obtained by the LHCb collaboration [8]. In Ref. [22], using the definition (13) we made an estimate for N^D ($N^D \simeq 32$).

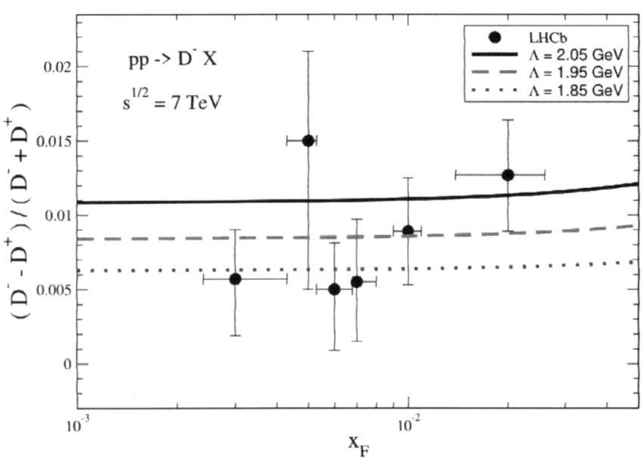

Figure 2. Comparison of the MCM asymmetry with experimental data [8] for D^-/D^+.

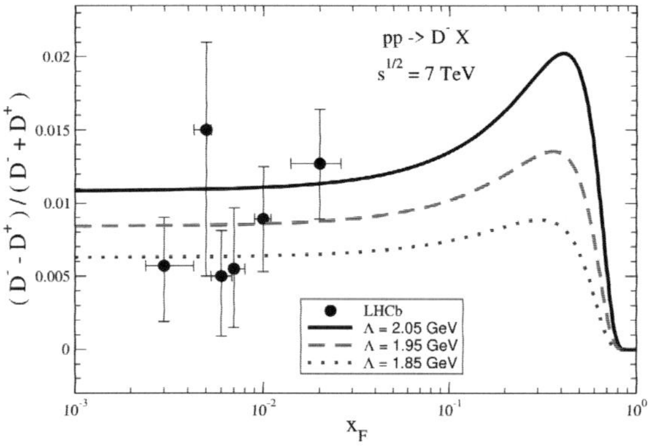

Figure 3. Comparison of the MCM asymmetry with experimental data [8] for D^-/D^+. Extension of Fig. 2 to the large x_F region.

Since we do not have data on D meson asymmetry at $\sqrt{s} = 33$ GeV we fixed the values of the cut off parameter by fitting the data of LHCb at $\sqrt{s} = 7$ TeV. In Fig. 2 we show the curves corresponding to the different cut off parameters. The values of Λ used to draw the curves are quite reasonable. It is interesting to amplify the region of higher x_F in Fig. 2, where we do not have experimental data. The amplifed curves are shown in Fig. 3. Having fixed all the parameters at the LHC energy, we can go back to the energy $\sqrt{s} = 33$ GeV and compute the asymmetry. This is shown in Fig. 4. Comparing Figs. 3 and 4 we draw the most important conclusion of this work: the asymmetry definitely decreases at increasing energies, reaching at most 2 % at $x_F \simeq 0.4$.

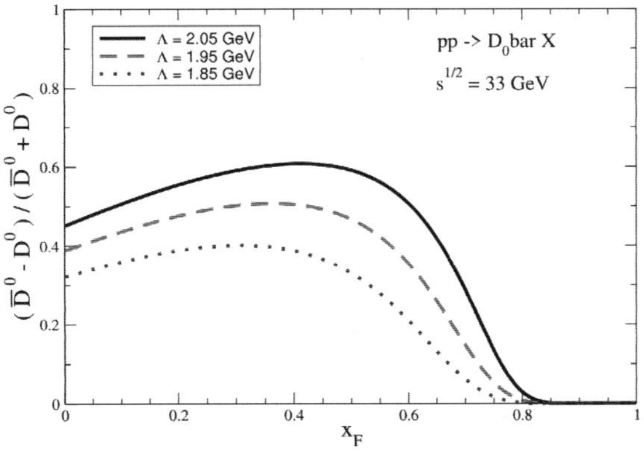

Figure 4. MCM prediction of the $\bar{D}^0/D^0$ asymmetry for $\sqrt{s} = 33$ GeV.

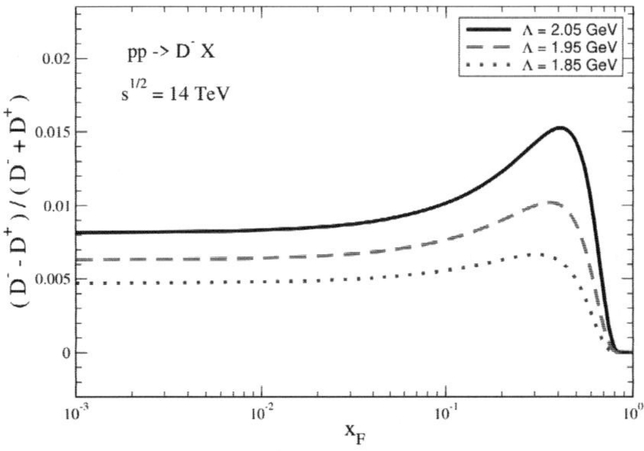

Figure 5. Prediction of the D^-/D^+ asymmetry for $\sqrt{s} = 14$ TeV.

Finally, in Fig. 5 we show our prediction for the D^-/D^+ asymmetry to be measured at $\sqrt{s} = 14$ TeV.

Acknowledgments

This work has been partially supported by the brazilian funding agencies CNPq and FAPESP.

References

[1] Adamovich M I *et al* (WA82 Collab.) 1993 *Phys. Lett.* B **306** 402
[2] Alves G A *et al* (E769 Collab.) 1996 *Phys. Rev. Lett.* **77** 2392
[3] Aitala E M *et al* (E791 Collab.) 1997 *Phys. Lett.* B **411** 230
[4] Adamovich M I *et al* (WA92 Collab.) 1997 *Nucl. Phys.* B **495** 3
[5] Adamovich M I *et al* (WA89 Collab.) 1999 *Eur. Phys. J.* C **8** 593
Adamovich M I *et al* (WA89 Collab.) 2000 *Eur. Phys. J.* C **13** 247
[6] Garcia F G *et al* (SELEX Collab.) 2002 *Phys. Lett.* B **528** 49
Iori M *et al* (SELEX Collab.) 1999 *Nucl. Phys.* B *Proc. Suppl.* **75** 16 (*Preprint* hep-ex/0009049)
Anjos J C and Cuautle E 2000 *Preprint* hep-ph/0005057
[7] Adolph C *et al* (COMPASS Collaboration) 2012 *Preprint* hep-ex/1211.1575
[8] Aaij R *et al* (LHCb Collaboration) 2013 *Phys. Lett.* B **718** 902 (*Preprint* hep-ex/1210.4112)
[9] Navarra F S, Nielsen M, Nunes C A A and Teixeira M 1996 *Phys. Rev.* D **54** 842
Paiva S, Nielsen M, Navarra F S, Durães F O and Barz L L 1998 *Mod. Phys. Lett.* A **13** 2715
Melnitchouk W and Thomas A W 1997 *Phys. Lett.* B **414** 134
[10] Brodsky S J, Hoyer P, Peterson C and Sakai N 1980 *Phys. Lett.* B **93** 451
[11] Vogt R, Brodsky S J and Hoyer P 1991 *Nucl. Phys.* B **360** 67
Vogt R and Brodsky S J 1995 *Nucl. Phys.* B **438** 261
Vogt R and Brodsky S J 1996 *Nucl. Phys.* B **478** 311
Gutierrez T and Vogt R 1999 *Nucl. Phys.* B **539** 189
[12] Das K P *et al* 1977 *Phys. Lett.* B **68** 459 (*Erratum* 1978 *Phys. Lett.* B **73** 504)
Chiu C B *et al* 1979 *Phys. Rev.* D **20** 211
[13] Hwa R 1980 *Phys. Rev.* D **22** 1593
Hwa R 1995 *Phys. Rev.* D **51** 85
[14] Kartvelishvili V G *et al* 1981 *Sov. J. Nucl. Phys.* **33** 434 (*Yad. Fiz.* 1981 **33** 832)
Likhoded A K *et al* 1983 *Sov. J. Nucl. Phys.* **38** 433 (*Yad. Fiz.* 1983 **38** 727)
[15] Cuautle E *et al* 1998 *Eur. Phys. J.* C **2** 473

Journal of Physics: Conference Series **458** (2013) 012014 doi:10.1088/1742-6596/458/1/012014

Braaten E *et al* 2002 *Phys. Rev. Lett.* **89** 122002
Mehen T 2004 *Acta Phys. Polon.* B **35** 121
Mehen T 2004 *AIP Conf. Proc.* **698** 508
Mehen T 2004 *J. Phys. G* **30** S295
Avila C, Magnin J and Mendoza-Navas L M 2003 *Preprint* hep-ph/0307358
[16] Rapp R and Shuryak E V 2003 *Phys. Rev.* D **67** 074036
[17] For a review see:
Speth J and Thomas A W 1998 *Adv. Nucl. Phys.* **24** 83
Kumano S 1998 *Phys. Rep.* **303** 183
Thomas A W 1983 *Phys. Lett.* B **126** 97
[18] Burkardt M, Hendricks K S, Ji C-R, Melnitchouk W and Thomas A W 2012 *Preprint* hep-ph/1211.5853 and references therein
[19] Carvalho F, Duraes F O, Navarra F S and Nielsen M 2001 *Phys. Rev. Lett.* **86** 5434
[20] Carvalho F, Durães F O, Navarra F S and Nielsen M 1999 *Phys. Rev.* D **60** 094015
Navarra F S, Nielsen M and Paiva S 1997 *Phys. Rev.* D **56** 3041
[21] Holtmann H, Szczurek A and Speth J 1996 *Nucl. Phys.* A **569** 631
Nikolaev N N, Schaefer W, Szczurek A and Speth J 1999 *Phys. Rev.* D **60** 014004
[22] Cazaroto E R, Goncalves V P, Navarra F S and Nielsen M 2013 *Preprint* hep-ph/1302.0035
[23] Fagundes D A, Menon M J and Silva P V R G 2012 *Preprint* hep-ph/1208.3456
[24] Nelson R E, Vogt R and Frawley A D 2013 *Phys. Rev.* C **87** 014908

29th Winter Workshop on Nuclear Dynamics (WWND2013) IOP Publishing
Journal of Physics: Conference Series **458** (2013) 012015 doi:10.1088/1742-6596/458/1/012015

Photons and Dilepton Measurements in PHENIX

Sky Rolnick
UC Riverside
E-mail: skymeson@gmail.com

Abstract.
Dileptons and direct photons are penetrating probes of the QGP and are emitted during the entire space-time evolution of heavy ion collisions. Recent measurements of direct photons indicate an excess of photons with $1 < p_T < 4$ GeV/c in Au+Au collisions beyond the N_{coll} scaled p+p yield. The large excess, interpreted as thermal radiation, can be fit to a variety of models consistent with high temperatures and low thermalization times. Measurements also indicate a large v_2 for $p_T < 4$ GeV/c where thermal photons dominate. Such a large v_2 is difficult to reconcile with current theoretical models. In these proceedings we will summarize the latest measurements of direct photons and dileptons produced at RHIC energies including recently released dielectron spectrum from p+p, dAu, and Au+Au collisions in PHENIX.

1. Introduction

Photons and dileptons are produced in copious amounts during heavy ion collisions and can be considered excellent probes to understand the hot dense medium. Information from almost every stage of the collision is carried through electromagnetic radiation and this allows us to better understand the space-time evolution of the system. Direct photons have now been measured in PHENIX for a variety of systems including p+p, d+Au, Cu+Cu, and Au+Au at a wide range of energies and momentum [1, 2]. Some of the exciting results from recent measurements in PHENIX include the larger than expected flow of direct and thermal photons, as well as the larger than expected enhancement of dielectrons in the low mass region. Both of these results have profound implications for understanding the medium itself and both are difficult to describe with current theoretical models.

Direct photons, as opposed to hadron decay photons, are of particular interest in Heavy Ion Physics because they do not interact strongly with the QGP, preserving their initial momentum. A large enhancement of direct photons has been measured in Au+Au and appears consistent with an initial temperature between $T_0 = 300 - 600$ MeV and a short thermalization time $\tau = 0.2 - 0.6$ fm/c [3]. Furthermore, large azimuthal anisotropies characterized by v_2, are measured in the particle emission and are thought to be a result of collective phenomena. For a strongly coupled system, even with very large pressure gradients, it would take time for flow to develop, so large v_2 would indicate later production time and lower momentum emission.

Dielectrons have also played a key role in understanding the QGP in heavy ion physics. A long history of measurements has painted a rather intriguing picture of what happens in such high energy densities found at RHIC. The rather significant enhancement observed in Au+Au at $\sqrt(s) = 200$ GeV of low mass dielectrons is also an indication of additional sources beyond known hadronic sources. The significant enhancement observed of low mass dielectrons is thought to be a possible signal for chiral symmetry restoration or perhaps medium modifications to the vector mesons [4]. Dielectrons also provide us with a detailed thermometer allowing us to resolve such things as temperature and multi-component flow as well as provide checks on a variety of production mechanisms. Understanding the electromagnetic component of an expanding medium will allow us to better understand the finer details of QCD itself.

2. Dielectrons in PHENIX

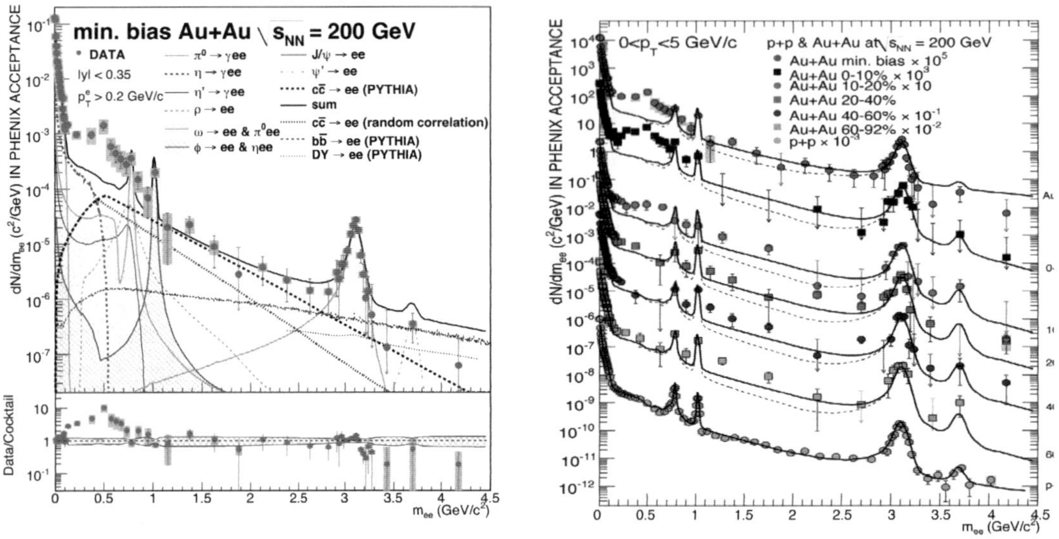

Figure 1. The invariant mass distributions of electron pairs (left), showing the enhancement in the low mass region. The enhancement as a function of centrality in which we see the majority of the enhancement being produced in central collisions (right).

The PHENIX detector is ideally suited for measuring photons and electrons. Previous measurements made by PHENIX in Au+Au collisions at $\sqrt{s_{NN}} = 200$ GeV have shown strong enhancement of electron pairs compared to the expected hadronic cocktail in the low mass region $m_{ee} = 0.15 - 0.75$ GeV$/c^2$ by a factor of $4.7 \pm 0.4(stat.) \pm 1.5(syst.)$ for minimum bias collisions [3]. For Au+Au, the mass spectra has been divided into p_T and centrality bins. It appears that the bulk of the enhancement can be described as originating from low p_T pairs in central events. The very low signal to background ratio of $\sim 1/200$ in the low mass region, results in large systematic uncertainty, particularly in the region of interest. Such a large enhancement was not expected and has posed a mystery to many theorists. Dielectrons are also useful for extracting direct photon yields as well.

3. Direct photons in PHENIX

There are several techniques used in PHENIX to measure direct photons using both real and virtual photons. For high momentum photons, $p_T > 5.0$ GeV$/c$, we are able to measure directly the energy deposited in the electromagnetic calorimeter (EMCal) and statistically subtract hadron decay photons from the inclusive photon yield to get at the direct photons. For lower momentum photons this subtraction becomes difficult due to the limited energy resolution of the EMCal at $p_T < 5.0$ GeV$/c$ and we must use either virtual photons that internally convert into e^+e^- pairs or measure real photons that externally convert in material into e^+e^- pairs. As a cross-check, we can make sure that both methods are in agreement with each other.

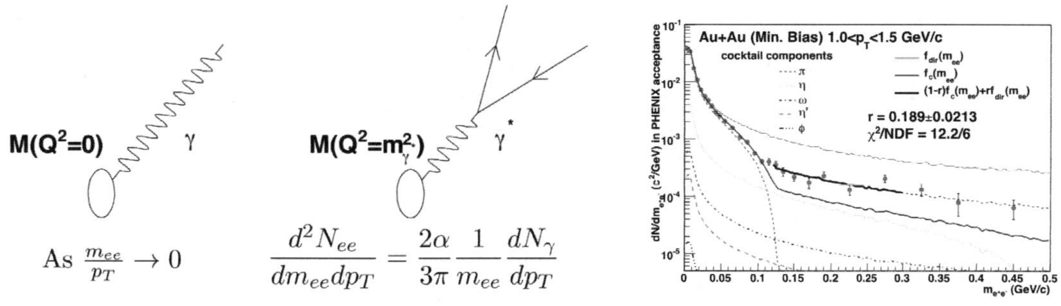

Figure 2. Processes which produce real photons (left) can also produce virtual photons which materialize as electron pairs (center). The invariant mass distribution of Dalitz decay of π^0, η and direct photons. The data is in red circles and clearly lies above known hadronic sources in blue. The excess is interpreted as arising from direct photons (right) [3].

Direct photons can be measured using the virtual photon method, which utilizes the relationship between real and virtual photons shown in Fig. 2. For our purposes we can assume any process which makes real photons will also have a fraction which manifest as electron pairs due to internal conversions. By selecting $p_T \gg m$ or in our case, a kinematic region of $m < 300$ MeV and $p_T > 1$ GeV/c, we can select a region where the dilepton signal is dominated by internal conversions. The clear enhancement in Au+Au above the hadronic cocktail can be fit with a two component function $F = (1-r)f_c + rf_d$ where f_c is the cocktail fraction and f_d is the contribution from direct photons. We should note that r is the only parameter allowed to vary in this fit, since the shapes of f_c and f_d are already determined. Typically, analysis is done above the π^0 mass to remove the majority of hadronic background. The excess is used to infer the yield of real direct photons by extrapolating to $m_{ee} = 0$.

$$r_\gamma = \frac{\gamma^*_{dir}(m > 0.15)}{\gamma^*_{inc}(m > 0.15)} = \frac{\gamma^*_{dir}(m = 0)}{\gamma^*_{inc}(m = 0)} = \frac{\gamma_{dir}}{\gamma_{inc}} \tag{1}$$

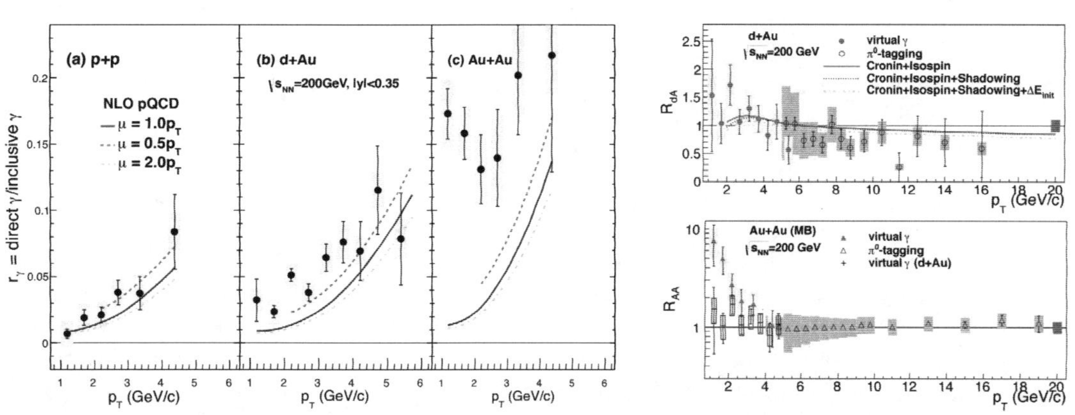

Figure 3. PHENIX has measured low p_T direct photon ratio in various collision systems using the virtual photon method. The theory curves represent NLO pQCD calculations for $\mu/p_T = 0.5$, 1, and 2 (left). Nuclear suppression factors for direct photons in d+Au and Au+Au using both real and virtual photon methods and comparing to CNM effects (right) [2].

Journal of Physics: Conference Series **458** (2013) 012015 doi:10.1088/1742-6596/458/1/012015

The direct photon ratio $r_\gamma = \frac{\gamma_{dir}}{\gamma_{inc}}$ has been measured using the the virtual photon method by PHENIX for p+p, Au+Au, Cu+Cu, and now d+Au at $\sqrt{s_{NN}}$=200 GeV. For p+p and d+Au collisions, the yield is consistent with NLO pQCD calculations, while a clear enhancement above predictions exists for Cu+Cu and Au+Au. Furthermore, nuclear suppression factors have been measured for d+Au and Au+Au in terms of R_{dA} and R_{AA}. CNM effect measured in d+Au do not explain the excess observed in Cu+Cu and Au+Au. This data supports the conclusion that additional sources arising from medium effects in Au+Au and Cu+Cu are not present in the baseline measurements of p+p and d+Au where a dense medium is not expected to be produced.

The direct photon yield can be determined from the product of r_γ and the inclusive yield $\gamma^{dir} = r_\gamma \gamma^{inc}$. The fits for p+p is based on a modified power law function $A_{pp}(1+p_T^2/b)^{-n}$ and for Au+Au we use an exponential plus the T_{AA} scaled p+p fit, i.e. $Ae^{-p_T/T} + T_{AA}A_{pp}(1+p_T^2/b)^{-n}$. Note that the only free parameters in the fit are the inverse slope parameter and the constant in front of the exponential. We observe a clear enhancement in Au+Au compared to the scaled p+p fit. This enhancement fits well to an exponential curve consistent with the view that it is thermal in origin.

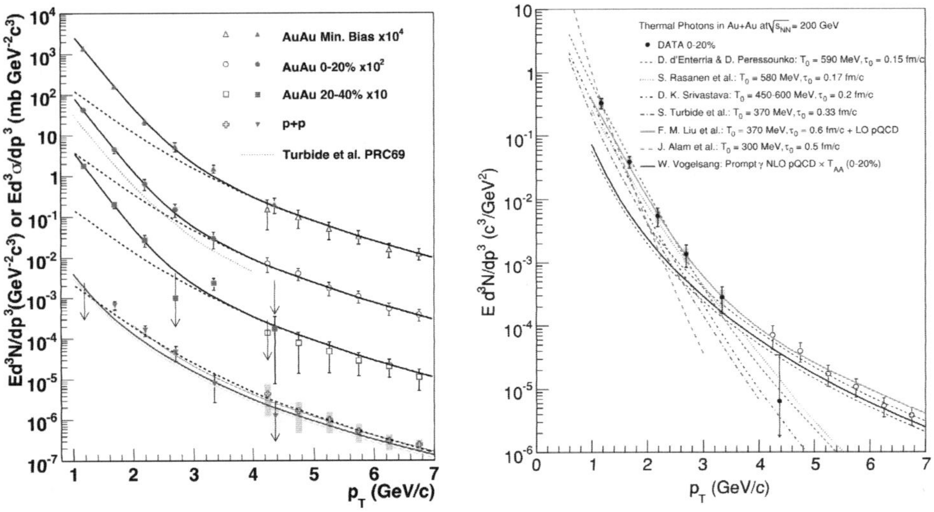

Figure 4. Invariant yields for direct photons in p+p and Au+Au (left) [6] On the right shows the same invariant yield in Au+Au but fit to a variety of theoretical models giving a range of temperatures and thermalization times [3]

4. Direct photon Elliptic Flow

A typical nucleus-nucleus collision starts with an initial spacial anisotropy, which turns into a momentum anisotropy due to large pressure gradients. Large azimuthal anisotropies are an indication of collective phenomena and cary information about where particles are produced in the collision. Since there are a variety of production mechanisms which overlap in transverse momenta, we can use this fact to distinguish between various production mechanisms. Sources originating early in the collision are typically hard processes from initial parton scattering and we should expect $v_2 = 0$. Later in the collision, softer processes such as thermal photons from the QGP and hadronic gas photons will contribute and we expect $v_2 > 0$. While thermal photons are obviously produced from a thermalized medium, initial state photons need not be in thermal equilibrium. There will also be a source from jet fragmentation and Bremsstrahlung photons which can have $v_2 < 0$ which is a reflection of the geometry and not dynamics [1].

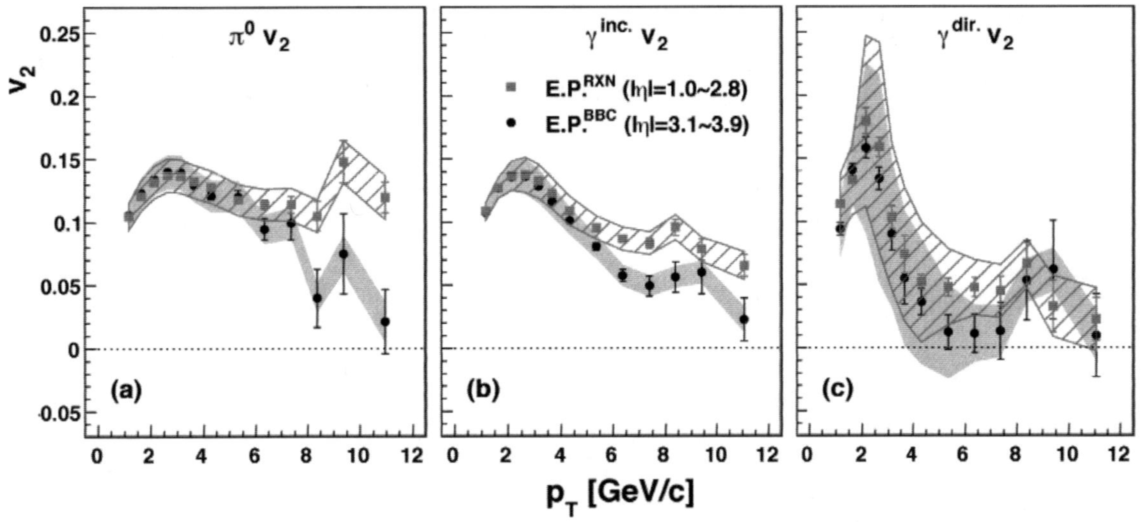

Figure 5. v_2 for pions, inclusive photons and direct photons [1].

We can begin be describing the steps in the analysis used to calculate direct photon v_2. The actual measurement of v_2 experimentally requires very good reaction plane resolution since we are essentially looking for an azimuthal asymmetry on an event-by-event basis.

$$R_\gamma = \frac{\gamma^{incl}(p_T)}{\gamma^{bg}(p_T)} = \frac{\epsilon(p_T) f(p_T) \cdot (\frac{N_\gamma^{incl}(p_T)}{N_\gamma^{\pi^0 tag}(p_T)})_{Data}}{(\frac{N_\gamma^{hadr}(p_T)}{N_\gamma^{\pi^0}(p_T)})_{Sim}} \tag{2}$$

$$v_2^{\gamma,obs} = \frac{v_2^{meas} - (N^{had}/N^{meas}) v_2^{had}}{1 - (N^{had}/N^{meas})} \tag{3}$$

$$v_2^{\gamma,dir} = \frac{R_\gamma(p_T) v_2^{\gamma,inc} - v_2^{\gamma,bg}}{R_\gamma(p_T) - 1} \tag{4}$$

Here $v_2^{\gamma,obs}$ is the observed inclusive v_2 before correcting for RP resolution, $v_2^{inc} = v_2^{obs}/\sigma_{RP}$ is for inclusive photons after corrections, $v_2^{\gamma,bg}$ is the hadronic background, and $v_2^{\gamma,dir}$ is for direct photons. Essentially we need to establish R_γ as the fraction of inclusive photons over decay photons, measure v_2 for inclusive photon yield correcting for hadron contamination, and subtract the hadron decay contributions from the inclusive photon to arrive at a direct photon v_2. It is worth noting that we only measure the π^0 v_2 which contributes about 80% of background. The other hadron v_2 is determined from KE_T/n_q scaling, where $KE_T = m_T - m_0$ [1].

We observe a significant direct photon signal with a rather large v_2 comparable to the inclusive photon and π^0 v_2 at low momentum. We also observe v_2 dropping to zero for high $p_T > 5$ GeV where hard processes dominate. This poses an intriguing question of how thermal photons could develop such a large v_2 when they are thought to originate from such early thermalization times.

5. Theory Comparisons

It can be difficult to reconcile our current understanding of the evolution with both large v_2 and short thermalization times. We can turn to two different theoretical models [7, 8] which

Journal of Physics: Conference Series **458** (2013) 012015 doi:10.1088/1742-6596/458/1/012015

both tend to under predict the data shown in Fig 6. Several other ideas are currently being developed which may help shed light on this puzzle, but are not discussed in this proceedings [9, 10, 11, 12].

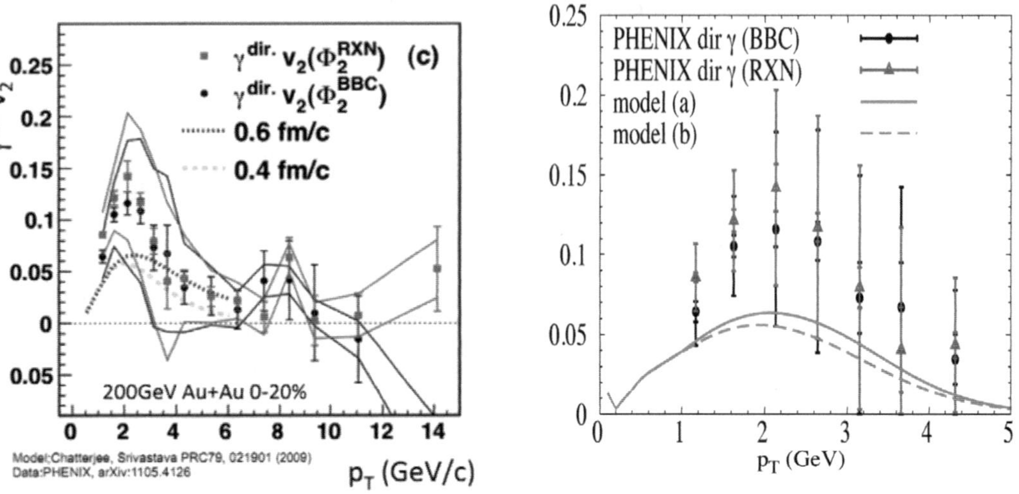

Figure 6. Theory and data comparison using models provided by Chatterjee, Srivastava [7] (left) and H. van Hees, C. Gale, and R. Rapp (right) [8]

6. Summary

Dileptons and photons have great potential to probe strongly interacting matter and explore general properties of the Quark Gluon Plasma. Many new results have been studied and reported on here. The dielectron mass spectrum and direct photon spectrum has been presented for $p+p$, $d+Au$ and $Au+Au$. Strong enhancement is observed in both dielectron spectra and direct photon spectra in central Au+Au collisions. In the case of direct photons, we also observe a large azimuthal asymmetry in the particle emission rates indicating a large flow component to the data. RHIC results are pushing forward our knowledge and understanding of the electromagnetic component of heavy ion collisions.

References

[1] A. Adare *et al.*, [PHENIXCollaboration], *Phys.Rev.Lett.* **109**, 122302, (2012) [arXiv:1105.4126[nucl-ex]]
[2] A. Adare et al., [PHENIXCollaboration], [arXiv:1208.1234]
[3] A. Adare et al., [PHENIXCollaboration], *Phys. Rev.* C **81**, 034911 (2010)
[4] R. Rapp, J. Wambach, H. van Hees. (2009) [arXiv:0901.3289 [hep-ph]]
[5] Turbide, Rapp, Gale, Phys. *Phys. Rev.* C **69**, 014903 (2004)[arXiv:0308085 [hep-ph]].
[6] A. Adare et al., [PHENIXCollaboration], *Phys. Rev. Lett.* **104**, 132301 (2010) [arXiv:0804.4168[nucl-ex]].
[7] R. Chatterjee, D. K. Srivastava, *Phys. Rev.* C **79** 021901 (2009)
[8] H. van Hees, C. Gale, R. Rapp, *Phys. Rev.* C **84**, 054906 (2011) [arXiv:1108.2131 [hep-ph]]
[9] V.S. Pantuev (2011) [arXiv:1105.4033 [nucl-ex]]
[10] Larry McLerran *et al.* [arXiv:1202.3679 [nucl-th]]
[11] G. Basar, D. Kharzeev, V.Skokov. [arXiv:1206.1334 [hep-ph]]
[12] Adam Bzdak, Vladimir Skokov [arXiv:1208.5502 [hep-ph]]

Journal of Physics: Conference Series **458** (2013) 012016 doi:10.1088/1742-6596/458/1/012016

Dilepton Measurements at STAR

Frank Geurts (for the STAR Collaboration)
Rice University, Houston, TX 77005, USA
E-mail: geurts@rice.edu

Abstract. In the study of hot and dense nuclear matter, created in relativistic heavy-ion collisions, dilepton measurements play an essential role. Leptons, when compared to hadrons, have only little interaction with the strongly interacting system. Thus, dileptons provide ideal penetrating probes that allow the study of such a system throughout its space-time evolution. In the low mass range ($M_{ll} < 1.1$ GeV/c^2), the dominant source of dileptons originates from the decay of vector mesons which may see effects from chiral symmetry restoration. In the intermediate mass range ($1.1 < M_{ll} < 3.0$ GeV/c^2), the main contributions to the mass spectrum are expected to originate from the thermal radiation of a quark-gluon plasma as well as the decays of charm mesons. In the high mass range ($M_{ll} > 3.0$ GeV/c^2), dilepton measurements are expected to see contributions from primordial processes involving heavy quarks, and Drell-Yan production.

With the introduction of the Time-of-Flight detector, the STAR detector has been able to perform large acceptance, high purity electron identification. In this contribution, we will present STAR's recent dielectron measurements in the low and intermediate mass range for RHIC beam energies ranging between 19.6 and 200 GeV. Compared to electrons, muon measurements have the advantage of reduced bremsstrahlung radiation in the surrounding detector materials. With the upcoming detector upgrades, specifically the muon detector (MTD), STAR will be able to include such measurements in its (di-)lepton studies. We will discuss the future dilepton program at STAR and the physics cases for these upgrades.

1. Introduction

Dileptons have long been proposed as one of the more crucial probes in the study of the hot and dense matter [1]. Leptons do not experience the strong force, and thus will have negligible final state interactions in a strongly interacting medium, with a mean free path that is much larger than the typical size of this medium. Electromagnetic probes such as the dielectrons are emitted throughout the evolution of a heavy-ion collision. A typical dielectron invariant mass spectrum, therefore, involves a plethora of sources, ranging from Dalitz decays, which dominate the low invariant mass range, to Drell-Yan pair production which dominates at masses above 3 GeV/c^2. In addition, dilepton contributions originate from vector mesons, open heavy-flavor decays, and thermal radiation emitted from a quark-gluon plasma (QGP).

The subsequent stages in the evolution of a hot and dense nuclear system can be identified with certain ranges in the dilepton invariant mass spectrum. In the high invariant mass range ($M_{ll} > 3$ GeV/c^2), dileptons from the decay of the heavy quarkonia such as the J/ψ and Υ mesons provide a means to study deconfinement effects in the hot and dense medium. These contributions are on top of a continuum from primordial Drell-Yan pair production. In the intermediate mass range ($1.1 < M_{ll} < 3.0$ GeV/c^2), the production of dileptons is closely related to the thermal radiation of the QGP. However, at higher center-of-mass energies this signal competes with significant contributions from open heavy-flavor decays such as $c\bar{c} \rightarrow e^+e^-X$, where such charm contributions may get modified by the medium. The prominent sources of dileptons in the low mass range ($M_{ll} < 1.1$ GeV/c^2), in addition to the Dalitz decays, are the direct leptonic decays of the $\rho(770)$, $\omega(782)$, and $\phi(1020)$ vector mesons. The in-medium modification of the spectral shape of these mesons may serve as signature of chiral symmetry restoration. The ρ meson is of special interest given that in thermal equilibrium its contribution to the low mass range is expected to dominate through its strong coupling to the $\pi\pi$ channel [2]. Moreover, its short lifetime of $\tau \sim 1.3$ fm/c, will make the spectral shape of this meson especially sensitive to in-medium modifications [3].

At SPS energies, the observed low-mass dilepton enhancement observed in both the CERES dielectron [3, 4] and NA60 dimuon data [5] could be explained in terms of in-medium effects on the spectral shape of the ρ meson. Moreover, the dimuon measurements by NA60 are found to favor significant broadening of the line shape over a mass-dropping model. At top RHIC energies, the PHENIX collaboration measured a significant enhancement in its dielectron measurements, with a strong p_T and centrality dependence [6]. However, models that have been able to describe the measurements at SPS energies are unable to consistently describe the PHENIX results in the most central collisions. The intermediate invariant mass range, between the ϕ and J/ψ mesons, opens an important window to thermal radiation from the QGP. However, contributions from the earlier mentioned semileptonic decays of open heavy-flavor hadrons are significant and need to be accounted for.

Direct photon measurements by the PHENIX collaboration in the range of $1 < p_T < 4$ GeV/c yielded a substantial elliptic flow, v_2, comparable to the v_2 of hadrons [7]. Model calculations which include a dominant QGP thermal emission source significantly underpredicted the observed v_2. However, incorporating a more detailed evolution of the hadronic phase, which involves the hadronic flow fields at chemical and kinetic freeze-out, appeared to improve the description of the direct photon v_2 reasonably well, while setting further constraints on the initial QGP temperatures [8]. Dilepton elliptic flow measurements as a function of p_T have been proposed as an independent measure to study the medium properties [9]. The combination of certain invariant mass and transverse momentum ranges allows for different observational windows on specific stages of the expansion. Dileptons can be used to further probe the early stages after a collision and possibly constrain the QGP equation of state.

The installation of the Time-of-Flight (TOF) detector [10] and the upgrade of its data acquisition (DAQ) system [11], allowed the STAR experiment to extend its large-acceptance particle identification capabilities and increase its DAQ rate. The TOF detector not only extends the reach of hadron identification to higher momenta, but also significantly improves the electron identification capability in the low momentum range. This, combined with the RHIC Beam Energy Scan (BES) program in 2010 and 2011, has put STAR in a unique position to measure dielectron spectra in the low and intermediate mass ranges from top RHIC beam energies down to SPS center-of-mass energies. In this paper, the preliminary dielectron results from the STAR experiment at top RHIC energy, $\sqrt{s_{NN}} = 200$ GeV, are discussed as well as the preliminary results at several BES energies. The low invariant mass measurements are compared with recent model calculations. This paper concludes with an outlook on the future of the STAR dilepton program.

2. Electron Identification and Background Reconstruction

The electron identification for the results reported in the next sections involves the STAR Time Projection Chamber (TPC) and the TOF detector. The TPC detector is the central tracking device of the STAR experiment and it provides charged particle tracking and momentum measurement. The energy-loss measurements, dE/dx, in the TPC are used for particle identification. The TOF detector, with full azimuthal coverage at mid-rapidity, extends the particle identification range to higher momenta. The combination of the TOF velocity information and the TPC energy loss allows for the removal of slower hadrons which contaminate the electron sample. The selection criteria for electron identification have been optimized for each RHIC beam energy. With an additional track momentum threshold of $p_T > 0.2$ GeV/c, the electron purity in the minimum bias Au+Au analysis is 95%.

The unlike-sign invariant mass distributions, which are reconstructed by combining electrons and positrons from the same event, contain both signal and background contributions. Especially in high-multiplicity events, the contribution of the combinatorial background is large, see Fig. 1, and two methods have been used to estimate the background.

The mixed-event method combines electrons and positrons from different events with similar total particle multiplicity, vertex position along the beam line, and event-plane angle. While the statistical accuracy of the background description can be arbitrarily improved by involving more events, the mixed-event method fails to reconstruct correlated background sources. At the lower invariant masses, such correlated pairs arise from jets, double Dalitz decays, Dalitz decays followed by a conversion of the decay photon, or two-photon decays followed by the conversion of both photons [12]. A background estimation based on the like-sign method can account for such correlated contributions. Its drawback, however, is that the statistical accuracy is only comparable to the unlike-sign, *i.e.* the original raw mass spectrum. Moreover, the like-sign method will need to consider detector acceptance differences, in contrast to the (unlike-sign) mixed-event method.

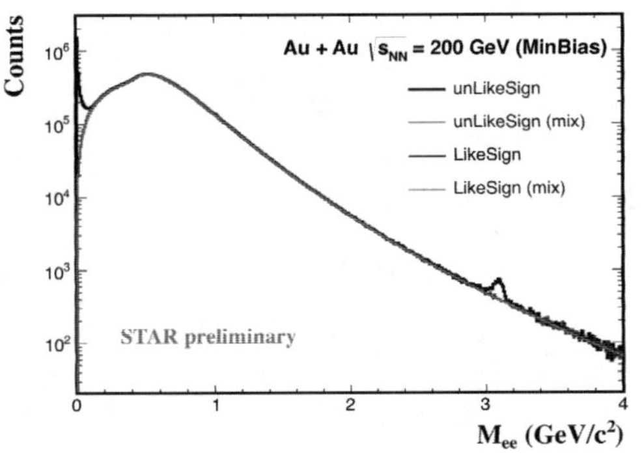

Figure 1. (Color online) Unlike-sign, like-sign, and mixed-event dielectron invariant mass distributions in Au+Au minimum bias collisions at $\sqrt{s_{NN}} = 200$ GeV.

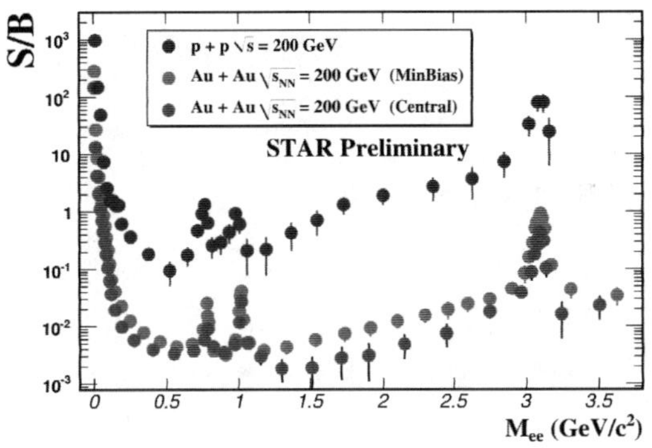

Figure 2. (Color online) Signal-to-background ratios in p+p, and Au+Au central, minimum bias events at $\sqrt{s_{NN}} = 200$ GeV [13].

The like-sign method is applied throughout the full mass range in Au+Au at $\sqrt{s_{NN}} =$ 19.6 GeV, in the low mass range for $M_{ee} < 750$ MeV/c^2 at $\sqrt{s_{NN}} = 200$ GeV, and for $M_{ee} < 900$ MeV/c^2 at 39 GeV and 62.4 GeV. Above these respective invariant mass thresholds, the mixed-event method is applied [13, 14]. The signal-to-background ratios for 200 GeV p+p and Au+Au are shown in Fig. 2.

3. Dielectron Measurements at $\sqrt{s_{NN}} = 200$ GeV

The STAR experiment has recently published its results on dielectron measurements in p+p at $\sqrt{s}$ =200 GeV [15]. These results were based on 107 million p+p events taken in the 2009 RHIC run, with only a partially installed TOF system. The agreement between the measured yields and the expected yields from a range of hadronic decays, heavy-flavor decays, and Drell-Yan production, provided an important verification of the analysis methods. More recently, the STAR experiment significantly improved its p+p data sample by about 700 million events with a fully installed TOF, effectively doubling its dielectron acceptance.

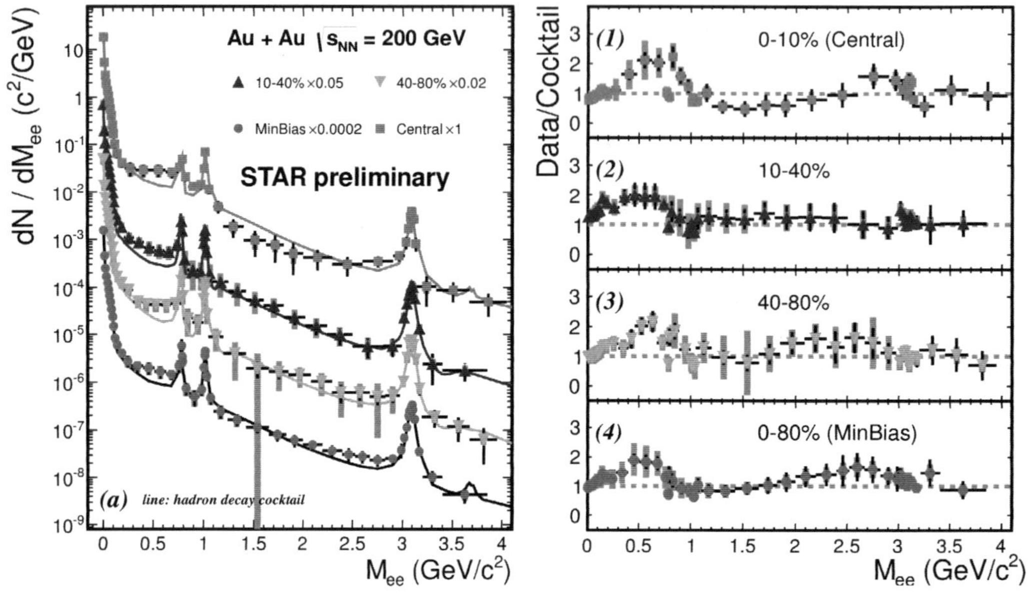

Figure 3. (Color online) Dielectron invariant mass spectra for Au+Au collisions at $\sqrt{s_{\rm NN}}$ =200 GeV for different centrality selections (left panel) and the ratio of data to cocktail (right panel). The systematical uncertainties are indicated by boxes.

The left panel of Fig. 3 presents for Au+Au collisions at $\sqrt{s_{\rm NN}}$=200 GeV the centrality dependence of the invariant mass spectra in the STAR acceptance ($|y_{ee}| < 1.0$, $|\eta_e| < 1$, and $p_{\rm T} > 0.2$ GeV/c). The measured yields are compared to a cocktail simulation of expected yields where the hadronic decays include the leptonic decay channels of the ω, ϕ, and J/ψ vector mesons, as well as the Dalitz decays of the π^0, η, η' mesons [13]. The input distributions to the simulations are based on Tsallis Blast-Wave function fits to the invariant yields of the measured mesons [12]. These functions serve as the input distributions for the GEANT detector simulation using the full STAR detector geometry. The ρ meson contributions have not been included in the cocktails as it may be sensitive to in-medium modifications which are expected to affect this meson's spectral line shape [16]. In the intermediate mass range, the $c\bar{c}$ cross section is based on PYTHIA simulations scaled by the number of nucleon-nucleon collisions [17]. The cocktail simulations are observed to overestimate the data in central collisions. This can indicate a modification of the charm contribution. However, the observed discrepancy is still consistent within the experimental uncertainties.

In the right panels of Fig. 3, the ratios of the data to cocktail yields have been depicted for different centrality selections. A clear enhancement in the low mass range can be observed. As the charm contributions scale with the number of binary collisions, the total cocktail yield increases with centrality, and only little centrality dependence can be observed in the ratio plots. On the other hand, as can be seen in Fig. 4, a comparison of the dilepton yield dN/dM_{ee} in the range of $150 < M_{ee} < 750$ MeV/c^2 scaled to the number of participants, $N_{\rm part}$, appears to indicate an increase of the low-mass-range enhancement with increasing centrality. Such an increase, albeit with large uncertainties, would agree with similar observations reported in [4].

Figure 5 shows the inclusive dielectron transverse mass slopes, in the intermediate mass range, measured in p+p and Au+Au at $\sqrt{s_{\rm NN}}$ =200 GeV. The $m_{\rm T}$ slopes in this plot have been determined for $1.1 < M_{ee} < 1.8$, $1.8 < M_{ee} < 2.8$ GeV/c^2 in Au+Au and $1.1 < M_{ee} < 1.6$, $1.6 < M_{ee} < 2.9$ GeV/c^2, respectively. These measurements are compared with the slope parameters of hadrons (circles) and charm/Drell-Yan subtracted dimuon measurements (open squares, [18]). While the p+p results (triangles) are consistent with PYTHIA calculations, the transverse mass slopes in Au+Au collisions (filled squares) are observed to be larger than those in p+p. This is indicative of a possible combination of both thermal dilepton production and charm modification. Future detector upgrades will allow STAR to further disentangle the potentially modified charm contributions and help provide improved measurements of the thermal QGP dilepton radiation.

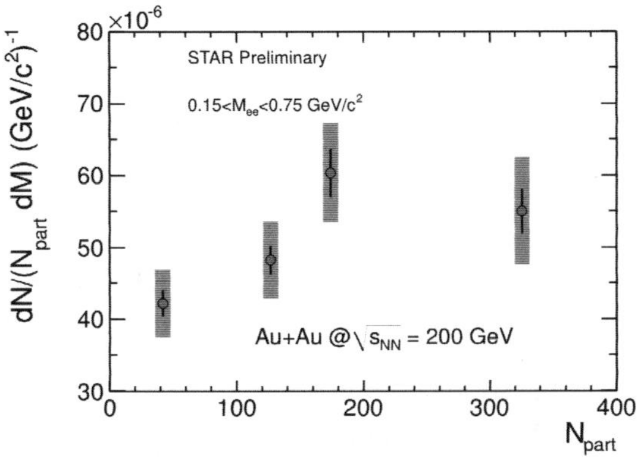

Figure 4. (Color online) Dielectron LMR yields scaled by $N_{\rm part}$ versus centrality ($N_{\rm part}$) for Au+Au collisions at $\sqrt{s_{\rm NN}}$=200 GeV [19]. The boxes indicate the systematical uncertainty.

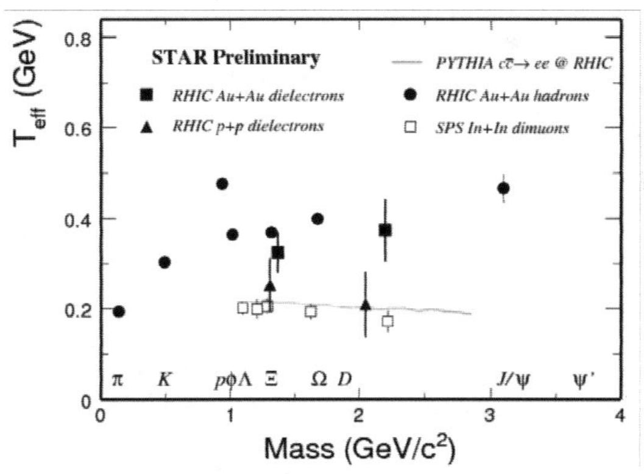

Figure 5. Transverse mass slope parameters from Au+Au at $\sqrt{s_{\rm NN}}$=200 GeV (RHIC) and In+In at $\sqrt{s_{\rm NN}}$=17.2 GeV (SPS) energies [13].

The STAR dielectron elliptic-flow measurements in Au+Au at $\sqrt{s_{\rm NN}}$ = 200 GeV as a function of the dielectron invariant mass are presented in Fig. 6. These preliminary results are based on 700 million minimum bias events involving an analysis that combined the data sets from the 2010 and 2011 RHIC Runs. The elliptic flow, v_2, is calculated using the event-plane method in which the event plane has been reconstructed from TPC tracks [20]. The "signal" elliptic flow, $v_2^{\rm signal}$, is calculated as follows [21]:

$$v_2^{\rm total}(M_{ee}) = v_2^{\rm signal}(M_{ee})\frac{r(M_{ee})}{1+r(M_{ee})} + v_2^{\rm bkgd}(M_{ee})\frac{1}{1+r(M_{ee})},$$

where $v_2^{\rm total}$ is the flow measurement of all dielectron candidates, $v_2^{\rm bkgd}$ the flow measurement of the dielectron background, and $r(M_{ee})$ the mass-dependent signal-to-background ratio. The expected v_2 from a cocktail simulation based on the contributions from π^0, η, ω, and ϕ mesons is within uncertainties consistent with the measurements.

Differential measurements have been done as a function of centrality (not shown here) and $p_{\rm T}$, as shown in Fig. 7. Both $v_2(p_{\rm T})$ for different dielectron mass windows, and its centrality dependence in the lower mass bin show a consistency between the measurements and the simulations. Work is underway to further extend these measurements into the intermediate

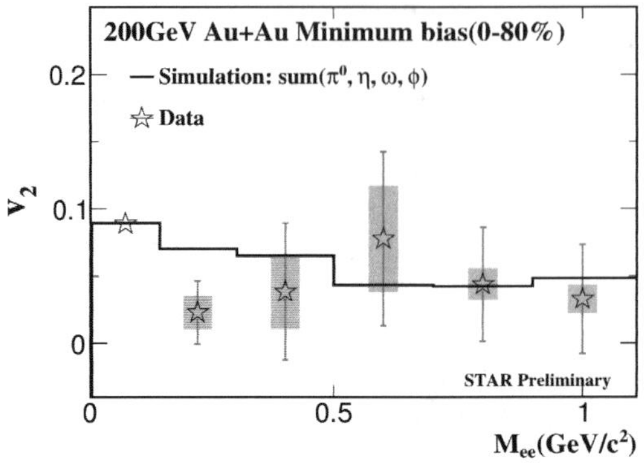

Figure 6. (Color online) Dielectron elliptic flow as a function of the invariant mass in minimum-bias Au+Au collisions at $\sqrt{s_{\rm NN}}$ =200 GeV. The boxes indicate the systematic uncertainty. The black line is the sum of a simulation which involved π^0, η, ω, and ϕ-mesons.

mass range. However, it will be important to disentangle the charm contributions at higher M_{ee}. The current experimental uncertainties on the STAR data points are still too large to allow for further constraints on the QGP equation of state, as is conjectured in *e.g.* [9].

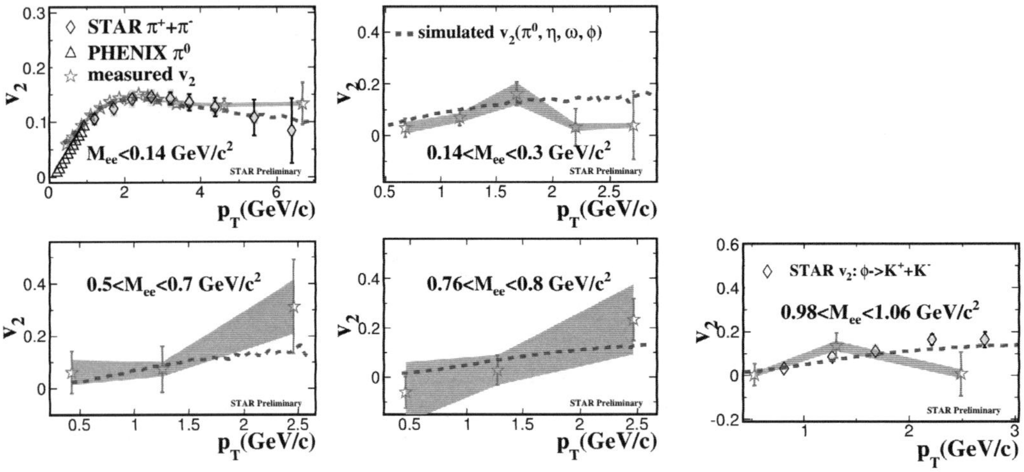

Figure 7. (Color online) Elliptic flow, v_2, as a function of $p_{\rm T}$ for various dielectron invariant mass ranges in minimum-bias collision Au+Au collisions at $\sqrt{s_{\rm NN}}$ =200 GeV (red stars). In addition, in the upper left panel the v_2 of π^0 [22] and $\pi^{\pm}$ mesons [23] are shown. The ϕ meson v_2 measurements in the lower right panel are from STAR. The blue dashed lines are the expected v_2 from cocktail simulations (see text). The grey bands indicate the systematic uncertainties.

4. Dielectron Measurements in the Beam Energy Scan Program

Measurements performed at top RHIC energies by both the STAR and PHENIX [6] collaborations indicate a significant enhancement in the low mass range when compared to a hadron cocktail simulation of the expected yields. Such an enhancement may point to in-medium modification effects that possibly result from chiral symmetry restoration. At the SPS, measurements performed by the NA60 [5] and CERES [3, 4] collaborations also show a low-mass enhancement which favors a broadening of the ρ meson spectral shape when compared to a dropping-mass scenario. The Beam Energy Scan (BES) program at RHIC allowed the STAR collaboration to systematically explore the dielectron production from $\sqrt{s_{NN}}$ = 200 GeV down to SPS beam energies.

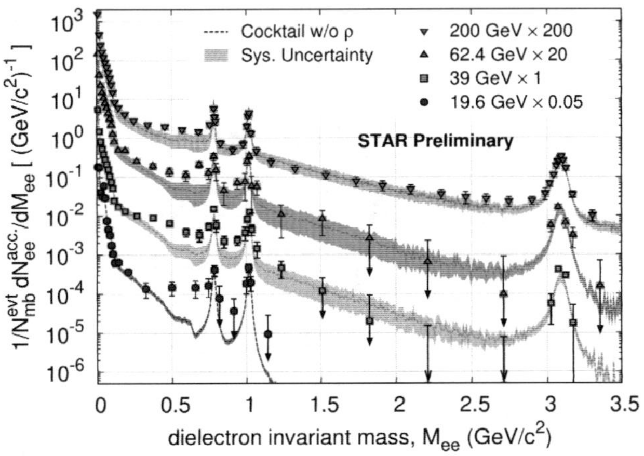

Figure 8. (Color online) Background-subtracted dielectron invariant-mass distributions from Au+Au collisions at $\sqrt{s_{NN}}$ = 19.6, 39, 62.4, and 200 GeV. The (colored) dotted lines show the hadron cocktails (excluding contributions from ρ mesons). The (color) shaded areas indicate the systematic uncertainties.

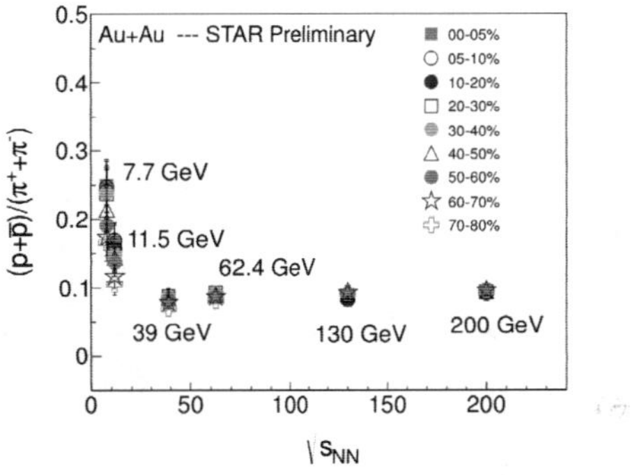

Figure 9. (Color online) $(p+\bar{p})/(\pi^++\pi^-)$ in Au+Au collisions as a function of the center-of-mass energy $\sqrt{s_{NN}}$.

In Fig. 8, the inclusive invariant mass spectra in Au+Au collisions for $\sqrt{s_{NN}}$ = 19.6, 39, 62.4, and 200 GeV are shown, together with the hadron cocktail simulations for each energy. The cocktail simulations exclude contributions from the ρ meson. For each of the energies a significant enhancement in the low mass range can be observed. While a quantitative discrepancy remains between the PHENIX and STAR measurements at $\sqrt{s_{NN}}$ = 200 GeV [6, 24], the preliminary STAR results are comparable with the CERES measurements in the Pb+Au system at $\sqrt{s_{NN}}$ =17.2 GeV albeit with differences in acceptance [3, 14]. Moreover, various models have shown good agreement with the low-mass spectrum measured by STAR in central Au+Au collisions at $\sqrt{s_{NN}}$ = 200 GeV [25, 26, 27]. The in-medium broadening of the ρ meson is expected to be driven by the strong coupling to baryons and thus the total baryon density since the vector mesons interact symmetrically with baryons and antibaryons [3, 27]. At SPS beam energies with substantial nuclear stopping, most baryons are participating nucleons. At RHIC top energies nuclear stopping and thus the net-baryon density will vanish and a significant baryon-antibaryon production is expected to compensate the total baryon density. As shown in Fig. 9, the total baryon density at freeze-out as a function of $\sqrt{s_{NN}}$ does not change significantly with center-of-mass energies down to 20 GeV. Accordingly, models that show good agreement at SPS and top-RHIC energies should be able to describe the low-mass enhancement throughout the BES center-of-mass energies.

In Fig. 10, the efficiency-corrected invariant mass spectra are shown for minimum bias Au+Au collisions at $\sqrt{s_{NN}}$ = 19.6, 62.4, and 200 GeV, respectively. In each of the three panels the hadron cocktail simulations include contributions from Dalitz decays, photon conversions (19.6 GeV only), and the dielectron decay of the ω and ϕ vector mesons. As is the case for the previously described 200 GeV cocktail simulations, contributions from ρ mesons have been excluded. Instead, the ρ meson is explicitly included in the model calculations by Rapp [2, 27] which involve in-medium modifications of the ρ meson spectral shape. In this model a complete evolution of the QGP and thermal dilepton rates in the QGP and hadron-gas (HG) phases are convoluted with an isentropic fireball evolution. In the HG phase the ρ "melts" when extrapolated close the conjectured phase transition boundary. Moreover, it has been noted that the top-down extrapolated QGP rates closely coincide with the bottom-up extrapolated hadronic rates [2]. The measured low-mass enhancement can consistently be described by these model

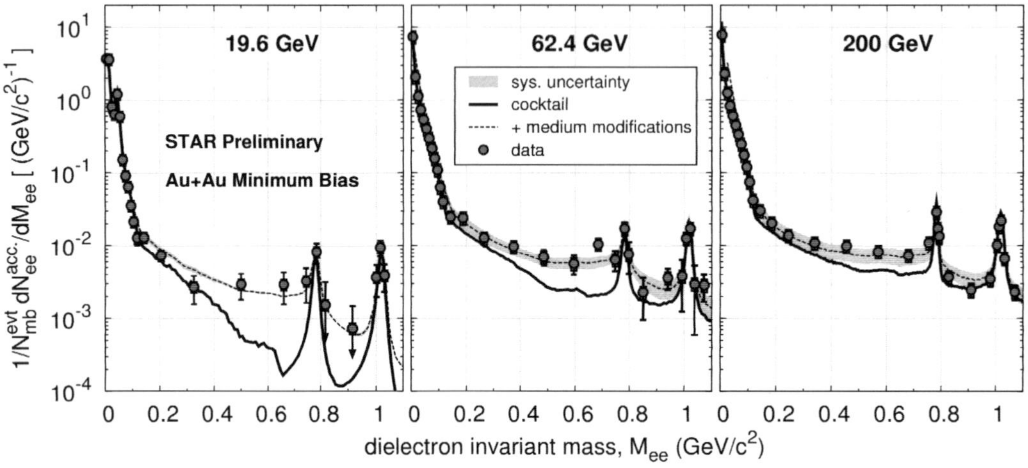

Figure 10. (Color online) BES dielectron invariant-mass distributions in the low invariant-mass range from Au+Au collisions at 19.6 GeV (left), 62.4 GeV (middle), and 200 GeV (right panel). The red data points include both statistical and systematic uncertainties (boxes). The black curve depicts the hadron cocktail, while the dashed line shows the sum of the cocktail and model calculations. The latter includes contributions from both the HG and the QGP phases. The systematic uncertainty on the former is shown by the light red band.

calculations and agrees with a scenario in which the in-medium modification of the ρ involves a broadening of its spectral function.

Differential studies of the low-mass dielectron distributions as a function of the center-of-mass energy, centrality, and p_T will allow for a more detailed comparison against these chiral hadronic models. Further comparisons between these models, lattice QCD calculations, and experimental data could help provide explicit evidence of chiral symmetry restoration in heavy-ion collisions.

5. Future Directions of the STAR Dilepton Program

Measuring dileptons is a challenging task where very small signals sit on top of large combinatorial and physics backgrounds. With the recent TOF upgrade, the STAR experiment is able to address some of the important physics questions that involve the low and intermediate dielectron mass ranges. In the low mass range, STAR's measurements agree with results at SPS energies in which the enhancement is attributed to the broadening of the ρ meson. Moreover, model calculations by Rapp [27] appear to be able to describe the BES data at the intermediate center-of-mass energies. In the future, additional verification of the robustness of such a description will be allowed by the inclusion of independent analyses of other BES data sets. Differential measurements will further verify the consistency of such models and advance the study of chiral symmetry restoration.

Journal of Physics: Conference Series **458** (2013) 012016 doi:10.1088/1742-6596/458/1/012016

As the typical production rates for dileptons are rather small, large event samples are needed to make significantly accurate measurements in both the low and intermediate mass ranges. Especially at the lower RHIC beam energies, the statistical uncertainties are very large (see Fig. 8). The $\sqrt{s_{\rm NN}}$=19.6 GeV measurements are based on 28 million Au+Au collisions. To achieve statistical uncertainties at a level of 10% and improve the understanding of the baryonic component to the in-medium effects on the vector mesons, a factor of 10 more statistics would be required. For such an increase in total luminosity at RHIC, requires the development and installation of electron cooling components.

Cocktail simulations indicate that at top RHIC energies the correlated charm contributions dominate the intermediate-mass range dielectrons. The PYTHIA simulations that are used for the cocktail simulations assume no medium effects and have been rescaled to match the charm cross sections to, *e.g.*, 0.96 mb in 200 GeV Au+Au collisions. Both assumptions result in large uncertainties, with some hints that point to a suppression of the charm contribution in central collisions when compared to the minimum-bias data (see Fig. 3). These uncertainties directly affect the extraction of any intermediate-mass dilepton signals related to the thermal QGP radiation. With the dominant source of intermediate-mass $e - \mu$ correlations originating from the $c\bar{c}$ correlated pair decays, measuring this correlation will be an essential tool for isolating the QGP thermal contribution in the intermediate mass range. Upcoming STAR detector upgrades [28] will, among many other key measurements, significantly improve our understanding of the charm contribution at intermediate dilepton masses. The Muon Telescope Detector (MTD) will provide a dedicated trigger for and the measurement of these electron-muon correlations, where both leptons are measured at mid-rapidity.

With the primary focus of the MTD physics program [29] on the quarkonia measurements at RHIC energies, the large-acceptance muon detector will also allow the STAR dilepton physics program to include the dimuon measurements of QGP thermal radiation, light vector mesons, and Drell-Yan production. The MTD is expected to be fully commissioned in 2014.

Acknowledgments
This work is supported by the US Department of Energy under grant DE-FG02-10ER41666.

References
[1] E. Shuryak, Phys. Lett. B78 (1978) 150.
[2] R. Rapp, J. Wambach, H. van Hees, arXiv:0901.3289v1 [hep-ph].
[3] D. Adamová *et al.*, Phys. Lett. B 666 (2008) 425.
[4] G. Agakichiev *et al.*, Eur. Phys. J. C 41 (2005) 475.
[5] R. Arnaldi *et al.*, Eur. Phys. J. 59 (2009) 607.
[6] A. Adare *et al.*, Phys. Rev. C 81 (2010) 034911.
[7] A. Adare *et al.*, arXiv:1105.4126.
[8] H. van Hees, C. Gale, and R. Rapp, Phys. Rev. C 84 (2011) 054906.
[9] R. Chatterjee *et al.*, Phys. Rev. C 75 (2007), 054909.
[10] W. Llope (for the STAR Collaboration), Nucl. Instr. and Meth. A 661 (2012), S110.
[11] J. Landgraf *et al.*, Nucl. Instr. and Meth. A 499 (2003), 762.
[12] L. Ruan (for the STAR Collaboration), Nucl. Phys. A 855 (2011), 269.
[13] J. Zhao (for the STAR Collaboration), J. Phys. G: Nucl. Part. Phys. 38 (2011), 124134.
[14] B. Huang, proceedings QM 2012, accepted by Nucl. Phys. A.
[15] L. Adamczyk *et al.*, Phys. Rev. C 86 (2012) 024906.
[16] J. Adams *et al.*, Phys. Rev. Lett. 92 (2004), 092301.
[17] J. Adams *et al.*, Phys. Rev. Lett. 94 (2005), 062301.
[18] R. Arnaldi *et al.*, Phys. Rev. Lett. 100 (2008), 022302.
[19] L. Ruan, *Rapporteur Talk: Heavy Flavor, Quarkonia and Electroweak Probes,* QuarkMatter 2012.
[20] B. Adelev *et al.*, Phys. Rev. C 77 (2008), 054901.
[21] B. Huang (for the STAR Collaboration), Acta Phys. Pol. B Proc. 5 (2012), 471.
[22] S. Afanasiev *et al.*, Phys. Rev. C 80 (2009), 054907.

[23] Y. Bai, Ph.D. Thesis, Universiteit Utrecht, The Netherlands (2007).
[24] F. Geurts (for the STAR Collaboration), arXiv:1208.3437 [nucl-ex]
[25] O. Linnyk *et al.*, Phys. Rev. C 85 (2012) 024910.
[26] H. Xu *et al.*, Phys. Rev. C 85 (2012) 024906.
[27] R. Rapp, Phys. Rev. C 63 (2001) 054907; R. Rapp, private communication.
[28] F. Videbaek, these proceedings.
[29] L. Ruan *et al.*, J. Phys. G: Nucl. Part. Phys. 36 (2009) 095001.

Journal of Physics: Conference Series **458** (2013) 012017 doi:10.1088/1742-6596/458/1/012017

Baryon femtoscopy in $\sqrt{s_{NN}} = 2.76$ TeV Pb–Pb collisions at ALICE

Jai Salzwedel

Department of Physics, The Ohio State University, 191 West Woodruff Ave, Columbus, OH 43210, USA

E-mail: `jai.salzwedel@cern.ch`

Abstract. We present femtoscopic results for proton and Λ correlation functions measured by ALICE in $\sqrt{s_{\rm NN}} = 2.76$ TeV Pb–Pb collisions. Femtoscopic radii are extracted from pp, $\bar{\rm p}\bar{\rm p}$, and p$\bar{\rm p}$ pairs. Comparisons of these radii with those from pion and kaon analyses reveal an approximate transverse mass scaling that is consistent with explanations of radial flow. Inelastic final state interactions are explored in baryon-antibaryon correlations functions to investigate their relationship with reduced proton yields at LHC energies.

1. Introduction

The study of two-particle correlations at low relative momentum is commonly known as femtoscopy. Femtoscopic analyses are capable of measuring spatio-temporal characteristics of heavy-ion collisions, with a particular emphasis made on estimating the homogeneity lengths (also called radii) of the particle-emitting source. These analyses traditionally study pions [1, 2] because of their ready availability and sensitivity to the spatial scale of the source. However, measurements of heavier particles such as kaons and baryons can serve to complement the pion results. One motivation for studying an assortment of heavier particles is to test the hydrodynamic prediction that radial flow should cause the source radii to scale with the transverse mass $m_{\rm T}$ of the particles [3, 4].

Baryon–(anti)baryon correlation functions are also useful in the study of final state interactions (FSI). Measurements of scattering lengths of pairs such as pΛ and $\Lambda\Lambda$ are of interest for rescattering calculations, and, in the latter case, for understanding the properties of neutron stars [5, 6]. Baryon-antibaryon correlations allow for the study of annihilation processes. It has been argued that annihilation in the hadronic rescattering phase should be taken into account when determining particle yields [7–9]. This annihilation should result in an anticorrelation in baryon-antibaryon correlation functions. Results from these femtoscopic analyses may be able to provide insight into the p/π^+ ratio measured in experiments at the LHC, which falls short of thermal model expectations [10].

In this study, we present femtoscopic correlation functions for pp, $\bar{\rm p}\bar{\rm p}$, and p$\bar{\rm p}$; p$\bar{\Lambda}$ and $\bar{\rm p}\Lambda$; and $\Lambda\bar{\Lambda}$ pairs. Source radii have been extracted for pp, $\bar{\rm p}\bar{\rm p}$, and p$\bar{\rm p}$ pairs, and these are compared with radii from kaon and pion analyses.

2. Data analysis

Femtoscopic analyses have been performed on approximately 40 million Pb–Pb events from ALICE at $\sqrt{s_{\rm NN}} = 2.76$ TeV. The Inner Tracking System (ITS), Time Projection Chamber (TPC), and Time-of-Flight detector (TOF) provided tracking for the particles [11]. Particle identification was performed using the TPC and TOF. Primary protons were selected based on the distance of closest approach (DCA) to the primary vertex. Λ ($\bar{\Lambda}$) candidates were selected by looking for daughter tracks that match the Λ's decay topology.

Two-particle correlation functions were constructed in terms of the one dimensional relative momentum $q_{\rm inv} = 2k^* = |P(m_1^2 - m_2^2)/P^2 - q|$, where P and q are the four-vector momentum sum and difference, respectively, and m_1 and m_2 are the masses of the two particles.

2.1. Analytical Model

The correlation functions can be calculated analytically [12] using

$$C(k^*) = 1 + \sum_S \rho_S \left[\frac{1}{2} \left| \frac{f^S(k^*)}{R} \right|^2 (1 - \frac{d_0^S}{2\sqrt{\pi}R}) + \frac{2\Re f^S(k^*)}{\sqrt{\pi}R} F_1(2k^*R) - \frac{\Im f^S(k^*)}{R} F_2(2k^*R) \right] \quad (1)$$

where $F_1(z) = \int_0^z \mathrm{d}x\, e^{x^2-z^2}/z$, $F_2(z) = (1 - e^{-z^2})/z$, and R is the source size. We assume the particles are produced unpolarized, with $\rho_{\rm S}$ being the fraction of pairs in each total spin state S. $f^S(k_*) = (1/f_0^S + \frac{1}{2}d_0^S k^* - ik^*)^{-1}$ is the spin-dependent scattering amplitude, though this analysis measures only spin-averaged values. Following the techniques of the STAR Collaboration in their pΛ analysis [13], the effective radius of interaction d_0^S is also neglected in order to simplify the fit parameters. The scattering amplitude is dependent upon the complex scattering length f_0^S. The real part of the scattering amplitude can contribute either a positive or a negative correlation, but either way the effect is relatively narrow in k^* (on the order of about one hundred MeV/c). The imaginary part of the scattering amplitude accounts for inelastic processes of baryon-antibaryon annihilation. Introducing a non-zero imaginary part to the scattering length produces a wide (hundreds of MeV/c) negative correlation. For identical and charged particles, additional terms [14] are necessary to account for quantum interference and the Coulomb interaction, respectively.

2.2. Experimental correlation functions

Coulomb and strong FSI dominate p$\bar{\rm p}$ correlations (Fig. 1), while p$\bar{\Lambda}$, $\bar{\rm p}\Lambda$, and $\Lambda\bar{\Lambda}$ correlations (Fig. 2) have only strong FSI. Nonetheless, one common feature is visible in the various correlation functions – a wide negative correlation in k^*. This shared feature demonstrates that the anticorrelation is not just a result of identical particle-antiparticle annihilation, but an overall expression of inelastic strong FSI resulting from the imaginary part of the scattering amplitude.

Fig. 3 shows correlation functions for pp and $\bar{\rm p}\bar{\rm p}$ pairs in three centrality ranges. These correlations are affected by Fermi–Dirac statistics, Coulomb interactions, and strong FSI. The pp and $\bar{\rm p}\bar{\rm p}$ correlations functions are observed to be consistent with each other within statistical uncertainties, as are the p$\bar{\Lambda}$ and $\bar{\rm p}\Lambda$ correlation functions. For all pairs analyzed, we observe that the strength of the correlation increases as a function of centrality (not shown for p$\bar{\Lambda}$ and $\bar{\rm p}\Lambda$). This behavior is consistent with a decrease of the system size.

Femtoscopic analyses are under way for pΛ and $\Lambda\Lambda$, as well as their respective antiparticle pairs.

2.3. Residual correlations

In pp ($\bar{\rm p}\bar{\rm p}$) correlations, a broad excess is observed in the 0.05 GeV/c < k^* < 0.1 GeV/c range that cannot be explained by correlations coming from the pp wave function. One potential explanation for this excess is the existence of residual pΛ correlations.

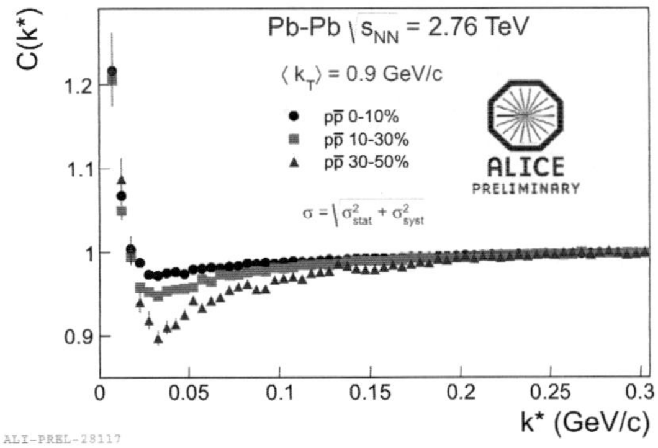

Figure 1. Correlation functions for $p\bar{p}$ in three centrality ranges.

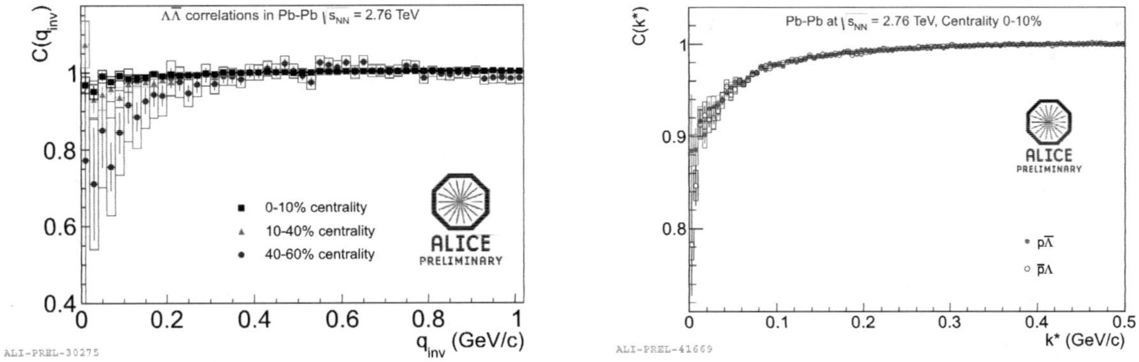

Figure 2. Correlation functions for $\Lambda\bar{\Lambda}$ (left panel) in three centrality ranges. $\mathrm{p}\bar{\Lambda}$ and $\bar{\mathrm{p}}\Lambda$ (right panel) are shown in the 0-10% centrality range.

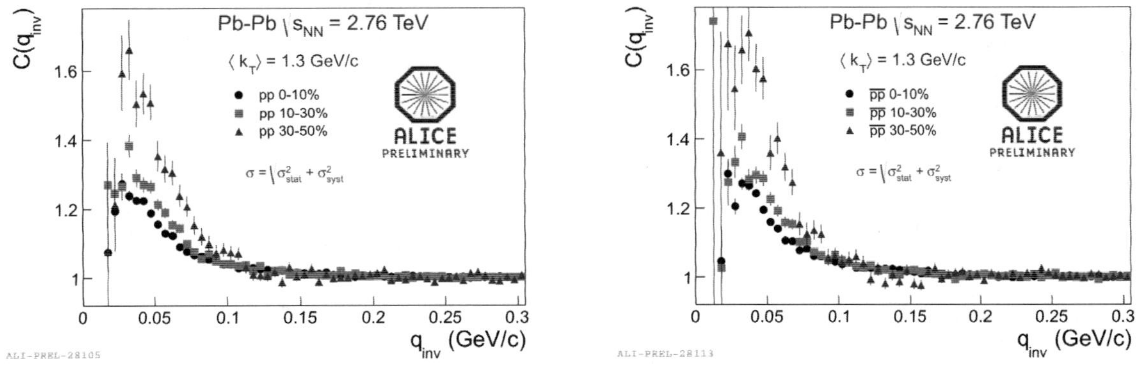

Figure 3. Correlation functions for pp (left panel) and $\bar{\mathrm{p}}\bar{\mathrm{p}}$ (right panel) in three centrality ranges.

Not every pp pair will contain two primary protons. Despite our DCA cuts to select primaries, contamination from secondary particles remains. For example, in a $\Lambda \rightarrow p + \pi^-$ decay, the proton carries most of the momentum of its parent. If that Λ is primary, its daughter proton will often pass the DCA cuts. In the case where a signal pair is comprised of one primary proton and one proton from a Λ decay, there cannot be any natural proton–proton correlation like quantum interference or FSI between the two particles. However, the primary proton and the Λ may be correlated. Taking into account momentum smearing due to the decay kinematics, that correlation could also be visible between the primary proton and the Λ's daughter proton. Attempts to measure a pp correlation will therefore see some amount of contamination from these residual pΛ correlations.

We attempt to quantify the residual contamination in our correlation functions by simultaneously fitting the data for both the primary correlation function and the residual correlation function. For pp correlations with pΛ feeddown, we fit our data using

$$C_{\mathrm{meas}}(k^*_{\mathrm{pp}}) = 1 + \lambda_{\mathrm{pp}}[C_{\mathrm{pp}}(k^*_{\mathrm{pp}}) - 1] + \lambda_{\mathrm{p}\Lambda}[C_{\mathrm{p}\Lambda}(k^*_{\mathrm{pp}}) - 1], \tag{2}$$

where

$$C_{\mathrm{p}\Lambda}(k^*_{\mathrm{pp}}) \equiv \sum_{k^*_{\mathrm{p}\Lambda}} C_{\mathrm{p}\Lambda}(k^*_{\mathrm{p}\Lambda}) T(k^*_{\mathrm{p}\Lambda}, k^*_{\mathrm{pp}}),$$

$C_{\mathrm{p}\Lambda}(k^*_{\mathrm{p}\Lambda})$ is the pΛ correlation function calculated from Eq. (1), and T is a THERMINATOR [15] transform matrix, which generates pairs of particles and determines kinematically how the k^* of the pair transforms when one of the particles decays.

Fig. 4 (left panel) shows ALICE data for a $\bar{\mathrm{p}}\bar{\mathrm{p}}$ correlation function that is fit using Eq. 2. The figure shows the full fit, as well as the individual $\bar{\mathrm{p}}\bar{\mathrm{p}}$ and $\bar{\mathrm{p}}\bar{\Lambda}$ portions. We see that the combined fit successfully reproduces both the peak at $k^* \approx .03$ GeV/c and the broad excess between 0.03 and 0.1 GeV/c. A similar procedure has been successful in fitting p$\bar{\mathrm{p}}$ correlations, as seen in the right panel of Fig. 4.

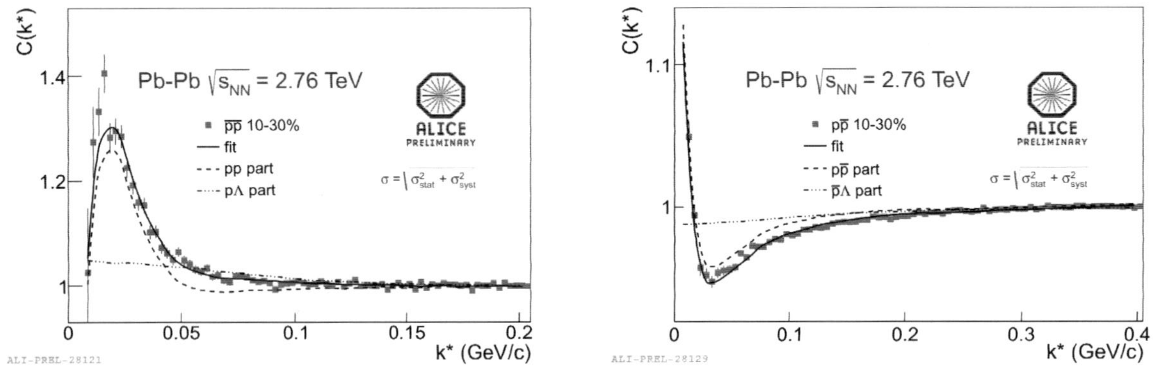

Figure 4. Fits of $\bar{\mathrm{p}}\bar{\mathrm{p}}$ (left panel) and p$\bar{\mathrm{p}}$ (right panel) with residual correlation included. The dashed line shows the primary correlation fit, the dash-dot line shows the residual correlation, and solid line—as the sum of the other two—is the total correlation function.

3. Fitting Results

Preliminary radii have been extracted from pp, $\bar{\mathrm{p}}\bar{\mathrm{p}}$, and p$\bar{\mathrm{p}}$ correlation functions using scattering lengths taken from [12]. These results are shown in Fig. 5 for three centrality bins and two transverse momentum ($k_{\mathrm{T}} = |\vec{p}_{\mathrm{T},1} + \vec{p}_{\mathrm{T},2}|/2$) bins. A tendency is observed for the radii to increase with multiplicity and decrease with k_{T}.

Fig. 6 shows extracted radii as a function of $m_{\mathrm{T}} = (k_{\mathrm{T}}^2 + m_{\mathrm{inv}}^2)^{-1/2}$ for pions, charged and neutral kaons, and protons. A kinematic scaling factor of $[(\sqrt{\gamma} + 2)/3]^{-1/2}$, with γ given by $m_{\mathrm{T}}/m_{\mathrm{inv}}$ of the pair, has been appended to the radii to approximate their size in the LCMS frame, where collective flow behavior is expected to be seen. We observe an approximate m_{T} scaling that is consistent with accounts of radial flow.

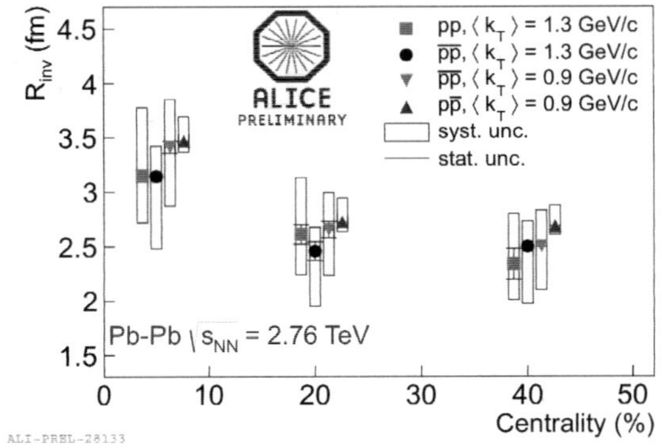

Figure 5. Radii extracted from fits to pp, $\bar{p}\bar{p}$, and $p\bar{p}$ correlation functions.

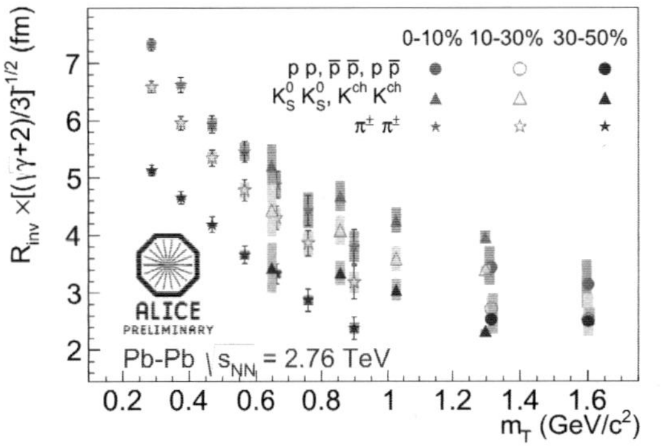

Figure 6. Transverse mass dependence of the kinematically scaled radius parameters for pion, kaon, and proton pairs.

4. Summary

Femtoscopic correlation functions have been constructed using protons and Λs. The correlation strength is observed to increase for more peripheral events. Emission source radii have been extracted from proton correlation functions, and they have been observed to decrease for higher centralities and pair transverse momenta. An approximate transverse mass scaling of pion, kaon, and proton radii is observed.

References

[1] Goldhaber G, Goldhaber S, Lee W Y and Pais A 1960 *Phys.Rev.* **120** 300–312

[2] Aamodt K *et al.* (ALICE Collaboration) 2011 *Phys.Lett.* **B696** 328–337 (*Preprint* 1012.4035)

[3] Csorgo T and Lorstad B 1996 *Phys.Rev.* **C54** 1390–1403 (*Preprint* hep-ph/9509213)

[4] Lisa M A, Pratt S, Soltz R and Wiedemann U 2005 *Ann.Rev.Nucl.Part.Sci.* **55** 357–402 (*Preprint* nucl-ex/0505014)

[5] Schaffner-Bielich J 2008 *Nucl.Phys.* **A804** 309–321 (*Preprint* 0801.3791)

[6] Wang Y and Shen H 2010 *Phys.Rev.* **C81** 025801 (*Preprint* 1002.0204)

[7] Werner K, Karpenko I, Bleicher M, Pierog T and Porteboeuf-Houssais S 2012 *Phys.Rev.* **C85** 064907 (*Preprint* 1203.5704)

[8] Karpenko I, Sinyukov Y and Werner K 2013 *Phys.Rev.* **C87** 024914 (*Preprint* 1204.5351)

[9] Steinheimer J, Aichelin J and Bleicher M 2013 *Phys.Rev.Lett.* **110** 042501 (*Preprint* 1203.5302)

[10] Preghenella R (ALICE Collaboration) 2012 (*Preprint* 1203.5904)

[11] Aamodt K *et al.* (ALICE Collaboration) 2008 *JINST* **3** S08002

[12] Lednicky R and Lyuboshitz V L 1982 *Sov. J. Nucl. Phys.* **35** 770

[13] Adams J *et al.* (STAR Collaboration) 2006 *Phys.Rev.* **C74** 064906 (*Preprint* nucl-ex/0511003)

[14] Aamodt K *et al.* (ALICE Collaboration) 2011 *Phys.Rev.* **D84** 112004 (*Preprint* 1101.3665)

[15] Chojnacki M, Kisiel A, Florkowski W and Broniowski W 2012 *Comput.Phys.Commun.* **183** 746–773 (*Preprint* 1102.0273)

Journal of Physics: Conference Series **458** (2013) 012018 doi:10.1088/1742-6596/458/1/012018

v-USPhydro: Bulk Viscosity Effects on Event-by-Event Relativistic Hydrodynamics

Jacquelyn Noronha-Hostler
Instituto de Física, Universidade de São Paulo, C.P. 66318, 05315-970 São Paulo, SP, Brazil
E-mail: jakinoronhahostler@gmail.com

Gabriel S. Denicol
Department of Physics, McGill University, 3600 University Street, Montreal, Quebec, H3A 2T8, Canada

Jorge Noronha
Instituto de Física, Universidade de São Paulo, C.P. 66318, 05315-970 São Paulo, SP, Brazil

Rone P. G. Andrade
Instituto de Física, Universidade de São Paulo, C.P. 66318, 05315-970 São Paulo, SP, Brazil

Frédérique Grassi
Instituto de Física, Universidade de São Paulo, C.P. 66318, 05315-970 São Paulo, SP, Brazil

Abstract.
v-USPhydro is a new relativistic 2+1 Lagrangian hydrodynamic code that incorporates the effects of bulk viscous hydrodynamics using Smoothed Particle Hydrodynamics (SPH) and is applicable to heavy ion collisions. Within this framework the bulk viscosity effects on collective flow harmonics i.e. $v_2 - v_5$ are studied on an event-by-event basis. We discuss which corrections to the Cooper Frye model are most appropriate when bulk viscosity is considered. An enhancement of all the Fourier harmonics is seen when bulk viscosity correction to the Cooper Frye is considered even when the bulk viscosity to entropy density ratio, ζ/s, is significantly smaller than $1/(4\pi)$.

1. Introduction
The Quark Gluon Plasma (QGP) found within heavy ion collisions is said to be a nearly perfect fluid. The large degree of collectivity evidenced by the Fourier harmonics of the azimuthal momentum distribution of charged hadrons are compatible [1] with viscous hydrodynamic calculations in which the shear viscosity to entropy density ratio, η/s, nears the uncertainty principle estimate $\sim 1/(4\pi)$ [2, 3]. Recent studies on η/s suggest that there may be a minimum close to the critical region [4, 5]. Then η/s is larger in the high temperature perturbative regime [6, 7], is low close to the critical temperature in the hadron gap phase due to Hagedorn States [8] , and increases again at low temperatures [9]. Thus far, the relativistic hydrodynamical models

used to study the collective flow effects have primarily only considered shear viscosity [10, 11] with the exception of a few that considered a nonzero ζ/s within averaged initial conditions [12, 13, 14, 15, 16, 17] while some aspects involving event-by-event simulations have been studied in [18]. There is no a priori reason to neglect bulk viscosity, especially on an event-by-event basis so the purpose of this proceedings is to explore the effects of bulk viscosity within the framework of a 2+1 relativisitic hydrodynamical code on an event-by-event basis. The hydrodynamical code is called viscous Ultrarelativistic Smoothed Particle hydrodynamics (v-USPhydro) [19], which is written in C++ using the Lagrangian method of Smoothed Particle Hydrodynamics (SPH).

2. Setup

Within v-USPhydro we assume a vanishing baryon chemical potential and the conservation of energy and momentum is given by

$$\frac{1}{\sqrt{-g}}\partial_\mu\left(\sqrt{-g}T^{\mu\nu}\right)+\Gamma^\nu_{\lambda\mu}T^{\lambda\mu}=0 \tag{1}$$

where $\sqrt{-g}=\tau$ and the Christoffel symbol is

$$\Gamma^\nu_{\lambda\mu}=\frac{1}{2}g^{\nu\sigma}\left(\partial_\mu g_{\sigma\lambda}+\partial_\lambda g_{\sigma\mu}-\partial_\sigma g_{\mu\lambda}\right). \tag{2}$$

The most general expression for the energy-momentum tensor (in the absence of shear viscosity effects) is

$$T^{\mu\nu}=\varepsilon u^\mu u^\nu-(p+\Pi)\,\Delta^{\mu\nu}, \tag{3}$$

where Π is the bulk viscous pressure, the spatial projector is $\Delta_{\mu\nu}=g_{\mu\nu}-u_\mu u_\nu$. We employ the simplest second order formulation of the fluid dynamical equations of motion that can be causal and stable [15]

$$\tau_\Pi\left(D\Pi+\Pi\theta\right)+\Pi+\zeta\theta=0, \tag{4}$$

where $D=u^\mu\partial_\mu$ is the comoving covariant derivative, $\theta=\tau^{-1}\partial_\mu\left(\tau u^\mu\right)$ is the fluid expansion rate, ζ is the bulk viscosity, and τ_Π is the relaxation time coefficient required to preserve causality [20].

This leaves us with two transport coefficients ζ/s and τ_π which are chosen in this work to be

$$\begin{aligned}\frac{\zeta}{s}&=\frac{1}{8\pi}\left(\frac{1}{3}-c_s^2\right)\\ \tau_\Pi&=9\,\frac{\zeta}{\varepsilon-3p}\end{aligned} \tag{5}$$

which are inspired by the calculations in [21] and [22], respectively. The coefficients are shown in Fig. 1. Here ε is the energy density and p is the pressure. We choose the time step $dt=0.1$ fm in our calculations. Note that, in order to solve the relaxation-type equations correctly, we must ensure that the relaxation time is larger than the time step used in the simulation. Since tau_{pi} goes to zero at large temperatures, this makes the simulation very time consuming. In our calculations, we fixed the time step to be $dt=0.1$ and use this as a lower bound for taupi, hence, the flattening in Fig. 1. Because the time step is small, such a lower bound does not affect the observables computed in this work. The initial conditions are taken from a Monte Carlo Glauber code [23]. Our initial time is $t_0=1fm/c$, our freeze-out temperature is taken to be $T=150$ MeV, and we use the Lattice based equation of state (S95n-v1) in [24]. At this point no decays are included and all graphs in the Results Section only consider direct π^+ results for the flow harmonics. In future work we will include hadronic decays and it would be especially interesting to include the effects of heavier resonances known as Hagedorn states [25].

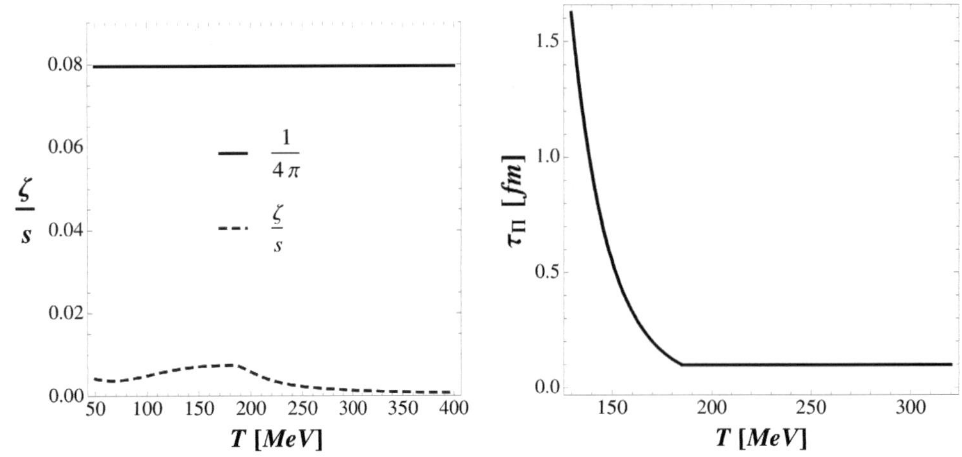

Figure 1. The temperature dependence of the transport coefficients ζ/s and τ_Π, see Eq. (5).

As stated previously, we employ SPH, which is a Lagrangian method for solving hydrodynamics. In the SPH approach one introduces a conserved reference density current $J^\mu = \sigma u^\mu$ where σ is the local density of a fluid element in its rest frame and h is known as the SPH length scale. As the fluid flows, the cell is deformed but its density obeys $D\sigma + \sigma\theta = 0$, which in hyperbolic coordinates is equivalent to $\partial_\mu(\tau\sigma u^\mu) = 0$. In terms of this reference density, the equations of motion defined above can be written as

$$\gamma\frac{d}{d\tau}\left[\frac{(\varepsilon+p+\Pi)}{\sigma}u^\mu\right] = \frac{1}{\sigma}\partial^\mu(p+\Pi) \tag{6}$$

$$\gamma\frac{d}{d\tau}\left(\frac{s}{\sigma}\right)+\left(\frac{\Pi}{\sigma}\right)\frac{\theta}{T} = 0 \tag{7}$$

$$\tau_\Pi\gamma\frac{d}{d\tau}\left(\frac{\Pi}{\sigma}\right)+\frac{\Pi}{\sigma}+\left(\frac{\zeta}{\sigma}\right)\theta = 0\,. \tag{8}$$

These equations are completely equivalent to the above equations of motion but they are more suitable for the Lagrangian implementation via SPH.

The effects of the bulk viscosity must also be applied to the Cooper Frye method of freezeout. The distribution function can be split into an ideal term with a bulk correction for π^+

$$f_{\mathbf{k}}^{(\pi)} = f_{0\mathbf{k}}^{(\pi)} + \delta f_{\mathbf{k}}^{(\pi)} \tag{9}$$

where the ideal term $f_{0\mathbf{k}}^{(\pi)}$ is described by

$$f_{0\mathbf{k}}^{(\pi)} = \left(e^{E_{(\pi)}/T} + a_{(\pi)}\right)^{-1} \tag{10}$$

where $E_{(\pi)}$ is the energy of the pion, T is the freeze out temperature, and $a_{(\pi)}$ is -1 since π^+ is a boson. Using the method of moments described in Refs. [26, 27], we compute the non-equilibrium contribution $\delta f_{\mathbf{k}}^{(\pi)}$ associated with bulk viscosity effects to the momentum distribution function of a hadronic mixture. We obtain the following δf description

$$\delta f_{\mathbf{k}}^{(\pi)} = f_{0\mathbf{k}}^{\pi}\,\Pi\left[B_0^{(\pi)} + D_0^{(\pi)} u\cdot k_\pi + E_0^{(\pi)}\left(u\cdot k_\pi\right)^2\right]. \tag{11}$$

where the coefficients B_0, D_0, and E_0 for pions with freeze-out temperature $T_{FO} = 150$ MeV are

$$\begin{aligned} B_0^{(\pi)} &= -65.85\ fm^4\,, \\ D_0^{(\pi)} &= 171.27\ fm^4/GeV\,, \\ E_0^{(\pi)} &= -63.05\ fm^4/GeV^2\,, \end{aligned} \tag{12}$$

see details in [19]. For freeze-out temperatures lower than 150 MeV, the δf contribution to the distribution function can become comparable to the ideal distribution f_0, which makes a perturbative analysis of the viscous effects at freeze-out unreliable. Thus, we used $T_{FO} = 150$ MeV for the calculations in this paper. We also consider the two other derivations of δf described in [12] (MH) and in [17] (DS). A comparison of the three methods can be seen in Fig. 2 in which only the moments method has a well-behaved description all the way up to $p_T = 3$ GeV for the elliptic flow and, thus, will be used for the rest of this paper. The results are for mid-rapidity RHIC's $\sqrt{200}$ GeV $20-30\%$ most central collisions in the case where the initial condition corresponds to a single average Glauber initial condition averaged over 150 events. The ideal case is the solid black line, our result for v_2 computed using the δf obtained via the Moments method is the long dashed black line, results for the δf described in [17] is the short dashed red curve, while the short and long dashed brown curve is the result computed using the δf described in [12]. One can see also for the spectra that the results from [12] and [17] are significantly steeper.

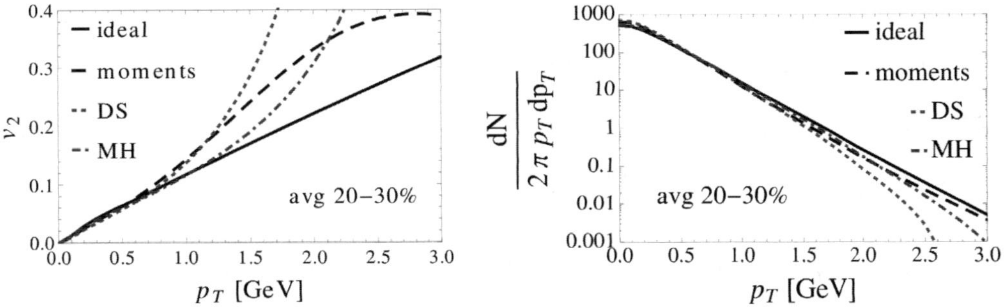

Figure 2. Dependence of the direct π^+ differential elliptic flow (left) and spectrum (right) on the specific formula for the viscous δf contribution from the bulk viscosity that enters in the Cooper-Frye freeze-out. The results are for RHIC's $\sqrt{200}$ GeV $20-30\%$ most central collisions in the case where the initial condition corresponds to a single average Glauber initial condition. The ideal case is the solid black line, our result for v_2 computed using the δf obtained via the Moments method is the long dashed black line, results for the δf described in [17] is the short dashed red curve, while the short and long dashed brown curve is the result computed using the δf described in [12].

Note that our moments method takes certain assumptions into account. We use only 100 of the known hadrons (up to mass $M = 1.2$ GeV) due to the complexity of the calculations involved when further hadrons are included, we assume a constant cross-section, and we assume Navier Stokes scaling, which implies that τ_π is small at freeze out [19]. All of these assumptions we plan to further test in an upcoming paper.

3. Results

All results below are shown for direct, thermal π^+ on an event-by-event basis (for 150 events) using the event plane method [28]. We consider only $p_T = 0-3$ GeV for RHIC's $\sqrt{s} = 200$ GeV most central collisions ($0-5\%$) and peripheral collisions ($20-30\%$). The direct, thermal π^+ spectra, $dN/(dyp_Tdp_T)$, are shown for $0-5\%$ and $20-30\%$ centrality classes in Fig. 3. Because we fit the integrated π^+ yields to 123 pions for most central collisions (roughly 41% of pions are direct pions at our freeze out temperature of $T = 150$ MeV and RHIC measures 300 total π^+ for most central collisions [29]) then the only difference between curves is the slope of the graphs themselves. The effect of bulk viscosity steepens the curve compared to the ideal fluid. However, the only noticeable differences comes from the correction to the Cooper Frye method, not from the hydrodynamical evolution itself. The difference is consistent across both centrality classes. As we will see in the following graphs the smaller number of high p_T pions contribute to an enhancement in the flow harmonics at high p_T.

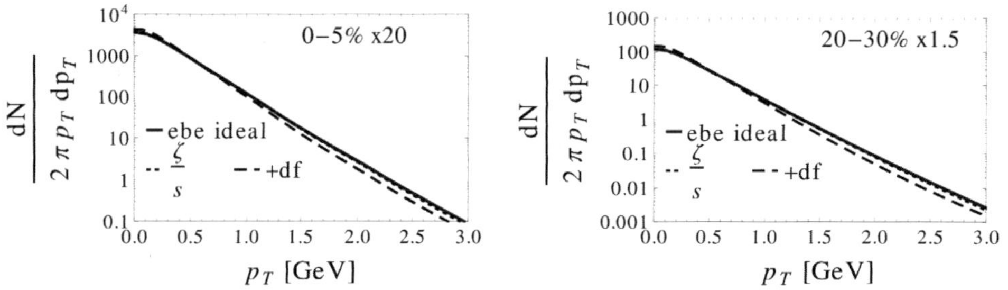

Figure 3. The π^+ spectra $dN/(2\pi p_T dp_T)$ for $0-5\%$ and $20-30\%$ centrality classes. The ideal fluid case is shown in a solid blue line, the result in the case where effects of bulk viscosity are included only on the hydrodynamical evolution but not on the freeze-out is shown by the short dashed black line while the long dashed black curve includes bulk effects on both the hydro evolution and freeze-out.

Results for the flow harmonics v_2 to v_5 for both most central collisions ($0-5\%$) and peripheral collisions ($20-30\%$) are shown in Fig. 4 for event-by-event Glauber initial conditions. The black solid lines indicate the ideal fluid whereas the short dashed lines include bulk viscosity corrections within the hydrodynamical evolution but at freezeout and the long dashed lines include bulk viscosity corrections in both the hydrodynamical evolution and freezeout. One can clearly see that the effects from the bulk viscosity solely in the hydrodynamical evolution are minimal. However, once the corrections to the freezeout are included the bulk viscosity is enhanced in the region of $p_T = 1-3$ GeV. Both centrality classes experience this enhancement and it occurs for all v_n's. When one compares these results to previous studies with shear viscosity, one is not surprised that the effects come primarily through the Cooper Frye correction term because the same effect was observed with shear viscosity. However, while shear viscosity depresses the flow harmonics for the higher p_T region, bulk viscosity has the opposite effect. One could expect that in the case where both shear and bulk viscosity are included in event by event simulations there could be some competition between the two effects.

4. Conclusions

Within the framework of our 2+1 viscous, Lagrangian relativistic hydrodynamical code we were able to study the effects of bulk viscosity on an event-by-event basis and compare them directly to an ideal fluid. We found that all flow harmonics $v_2(p_T)$-$v_5(p_T)$ there was an enhancement

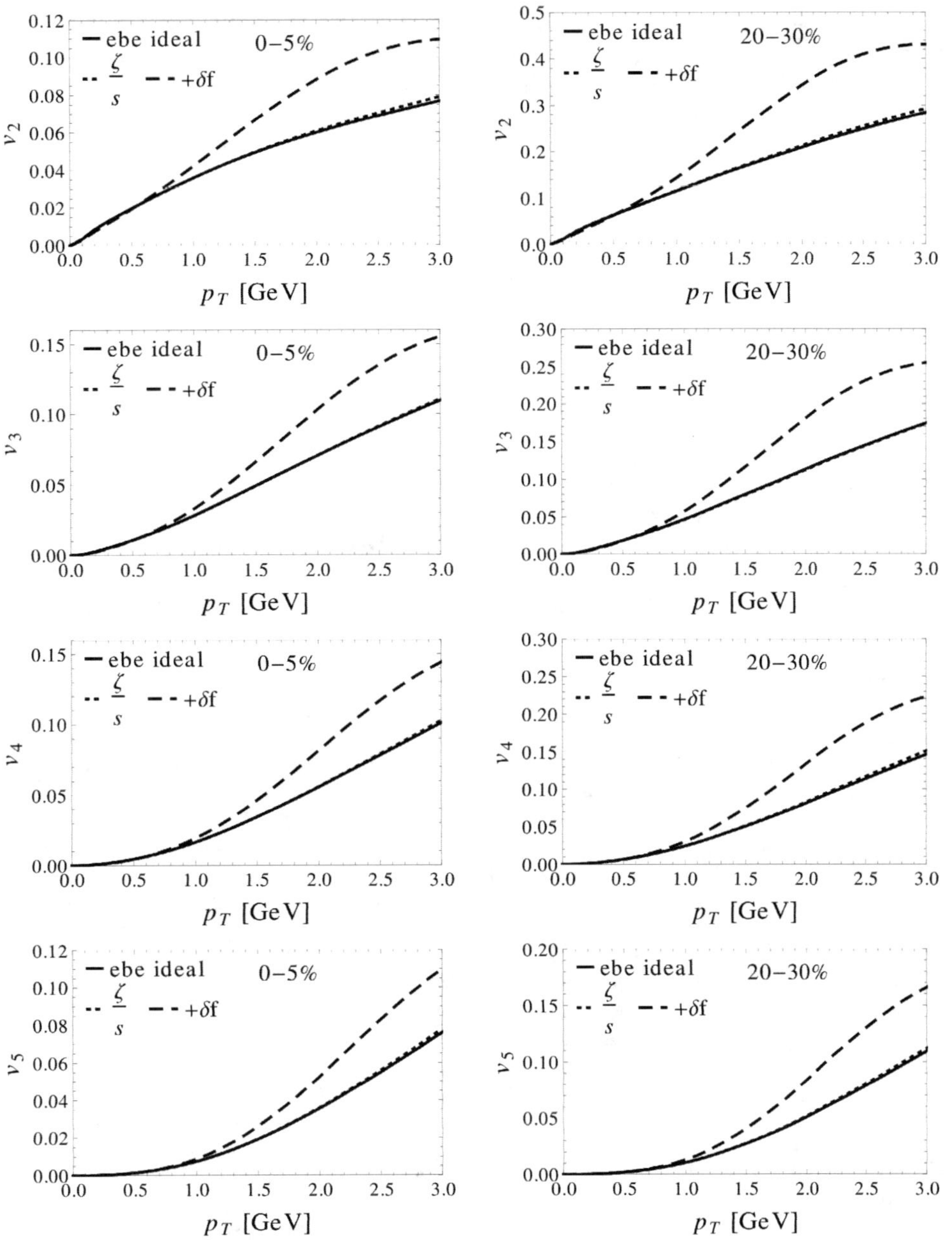

Figure 4. Results for the bulk viscosity ζ/s shown for most central collisions $(0-5\%)$ and non-central collisions $(20-30\%)$ computed using event-by-event simulations. The solid lines corresponds to the ideal fluid result, the short dashed lines include bulk viscosity only on the hydrodynamical evolution but not at freeze-out while the long dashed lines include bulk viscosity effects both on the hydro evolution and at freeze-out.

due to bulk viscosity in the momentum range above $p_T > 1$ GeV for both central and non-central collisions. Even though our value of ζ/s was almost an order of magnitude smaller than the commonly used value of shear viscosity to entropy density ratio $\eta/s = 1/(4\pi)$, we found significant deviations in our calculations compared to the ideal fluid flow harmonics.

Furthermore, this deviation shows up almost entirely through the non-equillibrium viscous correction to the Cooper Frye freeze out and not through the hydrodynamical process itself, as was expected from previous studies with shear viscosity.

Incidentally, the bulk viscosity has the opposite effect of the shear viscosity [11] when it comes to the p_T dependence of the flow harmonics. Previous studies have found that shear viscosity universally depresses the flow harmonics (for higher p_T's in the range of $p_T = 1-3$ GeV) whereas we found that the bulk viscosity does the opposite and enhances the flow harmonics in the same range. Because of this, it could be that much of the suppression of the flow harmonics due to shear viscosity is counteracted by the bulk viscosity. Thus, it it is vital that studies that include shear viscosity also include the effect of bulk viscosity on an event-by-event basis in order to get a more accurate description and take this compensation into account. Additionally, because the effect is primarily seen through the non-equilibrium correction to the Cooper Frye it is vital that we have an accurate description and continue to invest in more accurate models.

4.1. Acknowledgments

J. Noronha-Hostler, R. P. G. Andrade, J. Noronha, and F. Grassi acknowledge Fundação de Amparo à Pesquisa do Estado de São Paulo (FAPESP) and Conselho Nacional de Desenvolvimento Científico e Tecnológico (CNPq) for financial support. G. S. Denicol is supported by the Natural Sciences and Engineering Research Council of Canada.

5. References

[1] For a recent review, see U. W Heinz, R. Snellings, arXiv:1301.2826 [nucl-th].
[2] P. Danielewicz, M. Gyulassy, Phys. Rev. D **31**, 53 (1985).
[3] P. Kovtun, D. T. Son and A. O. Starinets, Phys. Rev. Lett. **94**, 111601 (2005) [hep-th/0405231].
[4] T. Hirano and M. Gyulassy, Nucl. Phys. A **769**, 71 (2006) [nucl-th/0506049].
[5] L. P. Csernai, J. .I. Kapusta, L. D. McLerran, Phys. Rev. Lett. **97**, 152303 (2006) [nucl-th/0604032].
[6] P. B. Arnold, G. D. Moore, L. G. Yaffe and , JHEP **0011**, 001 (2000) [hep-ph/0010177].
[7] P. B. Arnold, G. D. Moore, L. G. Yaffe and , JHEP **0305**, 051 (2003) [hep-ph/0302165].
[8] J. Noronha-Hostler, J. Noronha and C. Greiner, Phys. Rev. Lett. **103**, 172302 (2009) [arXiv:0811.1571 [nucl-th]].
[9] M. Prakash, M. Prakash, R. Venugopalan and G. Welke, Phys. Rep. **227**, 321 (1993).
[10] B. Schenke, S. Jeon, C. Gale, Phys. Rev. Lett. **106**, 042301 (2011) [arXiv:1009.3244 [hep-ph]].
[11] B. Schenke, S. Jeon, C. Gale, Phys. Rev. C **85**, 024901 (2012) [arXiv:1109.6289 [hep-ph]].
[12] A. Monnai, T. Hirano, Phys. Rev. C **80**, 054906 (2009) [arXiv:0903.4436 [nucl-th]].
[13] H. Song, U. W Heinz, Phys. Rev. C **81**, 024905 (2010) [arXiv:0909.1549 [nucl-th]].
[14] P. Bozek, Phys. Rev. C **81**, 034909 (2010) [arXiv:0911.2397 [nucl-th]].
[15] G. S. Denicol, T. Kodama, T. Koide and P. .Mota, Phys. Rev. C **80**, 064901 (2009) [arXiv:0903.3595 [hep-ph]]; G. S. Denicol, T. Kodama and T. Koide, J. Phys. G **37**, 094040 (2010) [arXiv:1002.2394 [nucl-th]].
[16] V. Roy, A. K. Chaudhuri, Phys. Rev. C **85**, 024909 (2012) [Erratum-ibid. C **85**, 049902 (2012)] [arXiv:1109.1630 [nucl-th]].
[17] K. Dusling, T. Schfer, Phys. Rev. C **85**, 044909 (2012) [arXiv:1109.5181 [hep-ph]].
[18] P. Bozek and W. Broniowski, Phys. Rev. C **85**, 044910 (2012) [arXiv:1203.1810 [nucl-th]].
[19] J. Noronha-Hostler, J. Noronha, G. S. Denicol, R. P. G. Andrade, F. Grassi and C. Greiner, "v-USPhydro: Bulk Viscosity Effects in Event-by-Event Hydrodynamics" To appear soon.
[20] G. S. Denicol, T. Kodama, T. Koide and P. .Mota, J. Phys. G **35**, 115102 (2008) [arXiv:0807.3120 [hep-ph]].
[21] A. Buchel, Phys. Lett. B **663**, 286 (2008) [arXiv:0708.3459 [hep-th]].
[22] X. -G. Huang, T. Kodama, T. Koide and D. H. Rischke, Phys. Rev. C **83**, 024906 (2011) [arXiv:1010.4359 [nucl-th]].
[23] H. -J. Drescher and Y. Nara, Phys. Rev. C **75**, 034905 (2007); Phys. Rev. C **76**, 041903 (2007).
[24] P. Huovinen and P. Petreczky, Nucl. Phys. A **837**, 26 (2010) [arXiv:0912.2541 [hep-ph]].
[25] J. Noronha-Hostler, J. Noronha, G. S. Denicol, R. P. G. Andrade, F. Grassi, C. Greiner, arXiv:1302.7038 [nucl-th].
[26] G. S. Denicol, H. Niemi, E. Molnar and D. H. Rischke, Phys. Rev. D **85**, 114047 (2012) [arXiv:1202.4551 [nucl-th]].
[27] G. S. Denicol, H. Niemi, arXiv:1212.1473 [nucl-th].

[28] A. M. Poskanzer and S. A. Voloshin, Phys. Rev. C **58**, 1671 (1998) [nucl-ex/9805001].
[29] I. Arsene *et al.* [BRAHMS Collaboration], Nucl. Phys. A **757**, 1 (2005) [nucl-ex/0410020].

29th Winter Workshop on Nuclear Dynamics (WWND2013) IOP Publishing
Journal of Physics: Conference Series **458** (2013) 012019 doi:10.1088/1742-6596/458/1/012019

Parton-hadron matter in- and out-off equilibrium

E L Bratkovskaya[1,2], V Ozvenchuk[2], W Cassing[2], V P Konchakovski[2] and O Linnyk[2]

[1] Institute for Theoretical Physics, Frankfurt University, Frankfurt-am-Main, Germany
[2] Frankfurt Institut for Advanced Studies, Frankfurt University, Frankfurt-am-Main, Germany
[2] Institute for Theoretical Physics, University of Giessen, Giessen, Germany

E-mail: Elena.Bratkovkaya@th-physik.uni-frankfurt.de

Abstract. We study the shear and bulk viscosities of partonic and hadronic matter - as well as the electric conductivity - as functions of temperature T within the Parton-Hadron-String Dynamics (PHSD) off-shell transport approach. Dynamical hadronic and partonic systems in equilibrium are studied by the PHSD simulations in a finite box with periodic boundary conditions. The ratio of the shear viscosity to entropy density $\eta(T)/s(T)$ from PHSD shows a minimum (with a value of about 0.1) close to the critical temperature T_c, while it approaches the perturbative QCD (pQCD) limit at higher temperatures in line with lattice QCD results. For $T < T_c$, i.e. in the hadronic phase, the ratio η/s rises fast with decreasing temperature due to a lower interaction rate of the hadronic system and a significantly smaller number of degrees-of-freedom. The bulk viscosity $\zeta(T)$ – evaluated in the relaxation time approach – is found to strongly depend on the effects of mean fields (or potentials) in the partonic phase. We find a significant rise of the ratio $\zeta(T)/s(T)$ in the vicinity of the critical temperature T_c, when consistently including the scalar mean-field from PHSD, which is also in agreement with that from lQCD calculations. Furthermore, we present the results for the ratio $(\eta + 3\zeta/4)/s$, which is found to depend non-trivially on temperature and to generally agree with the lQCD calculations as well. Within the PHSD calculations, the strong maximum of $\zeta(T)/\eta(T)$ close to T_c has to be attributed to mean-fields (or potential) effects that in PHSD are encoded in the temperature dependence of the quasiparticle masses, which is related to the infrared enhancement of the resummed (effective) coupling $g(T)$. We also find that the dimensionless ratio of the electric conductivity over temperature σ_0/T rises above T_c approximately linearly with T up to $T = 2.5T_c$, but approaches a constant above $5T_c$, as expected qualitatively from perturbative QCD (pQCD).

1. Introduction

High energy heavy-ion reactions are studied experimentally and theoretically to obtain information about the properties of nuclear matter under the extreme conditions of high density and/or temperature. Ultra-relativistic heavy ion collisions at the BNL Relativistic Heavy Ion Collider (RHIC) and the Large Hadron Collider (LHC) at CERN have produced a new state of matter, the quark gluon plasma (QGP). The produced quark gluon plasma behaves as a strongly-interacting fluid unlike a weakly-interacting gas [1, 2]. Large values of the azimuthal asymmetry of charged particles in momentum space, in particular the elliptic flow v_2, observed in these experiments [3, 4, 5, 6, 7] could quantitatively be well described by means of ideal hydrodynamics up to transverse momenta on the order of 1.5 GeV/c [8, 9, 10, 11, 12]. An ideal fluid has been defined as having a zero shear viscosity η; yet semiclassical arguments have been

Journal of Physics: Conference Series **458** (2013) 012019 doi:10.1088/1742-6596/458/1/012019

given suggesting that the shear viscosity cannot be arbitrarily small [13]. Indeed, the lower bound for the shear viscosity to entropy density ratio $\eta/s \geq 1/4\pi$ was obtained by Kovtun-Son-Starinets (KSS) [14] for infinitely coupled supersymmetric Yang-Mills gauge theory based on the AdS/CFT duality conjecture. Recent relativistic viscous hydrodynamic calculations using the Israel-Stewart framework require a very small η/s of $0.08 - 0.24$ in order to reproduce the RHIC elliptic flow v_2 data [15, 16, 17, 18]. The main uncertainty in these estimates results from the equation of state and the initial conditions employed.

There is strong evidence from atomic and molecular systems that η/s should have a minimum in the vicinity of the phase transition (or rapid crossover) between the hadronic matter and the quark-gluon plasma [19], and that the ratio of bulk viscosity to entropy density ζ/s should be maximum or even diverge at a second-order phase transition [20, 21, 22, 23, 24].

The shear and bulk viscosities of strongly-interacting systems have been calculated within different approaches. Calculations have been performed at extremely high temperatures, where perturbation theory can be applied [25, 26], as well as at extremely low temperatures [26, 27, 28]. First results for shear and bulk viscosities obtained within lattice QCD simulations just above the critical temperature of pure gluon/glueball matter have been published in [29, 30, 31, 32]. In the literature there are several methods for the calculation of the shear and bulk viscosities, such as the Relaxation Time Approximation (RTA) [34], the Chapmann-Enskog (CE) method [35] and the Green-Kubo method [36, 37].

In this contribution we extract the shear and bulk viscosities from the 'infinite' parton-hadron matter employing different methods within the Parton-Hadron-String Dynamics (PHSD) transport approach [38, 39], which is based on generalized transport equations on the basis of the off-shell Kadanoff-Baym equations [40, 41] for Green's functions in phase-space representation (in first order gradient expansion beyond the quasiparticle approximation). The approach consistently describes the full evolution of a relativistic heavy-ion collision from the initial hard scatterings and string formation through the dynamical deconfinement phase transition to the strongly-interacting quark-gluon plasma (sQGP) as well as hadronization and the subsequent interactions in the expanding hadronic phase. In the hadronic sector PHSD is equivalent to the Hadron-String-Dynamics (HSD) transport approach [42, 43] that has been used for the description of pA and AA collisions from SIS to RHIC energies in the past and the partonic dynamics is based on the Dynamical QuasiParticle Model (DQPM) [44, 45], which describes QCD properties in terms of single-particle Green's functions (in the sense of a two-particle irreducible (2 PI) approach) and reproduces lattice QCD results – including the partonic equation of state – in thermodynamic equilibrium. The PHSD approach [38, 39, 46] has been applied to pp, $p + A$ and $A + A$ collisions from $\sqrt{s_{NN}} = 5$ GeV up to 2.76 TeV providing a consistent description of hadronic single-particle spectra, collective flow coefficients of hadrons as well as dilepton radiation from the partonic and hadronic phase. Since transport coefficients are no inherent parameters of PHSD we may ask the question: what are the transport coefficients from PHSD in equilibrium - in particular as a function of temperature - that are compatible with the experimental observations? Here we summarize our findings from Refs. [47, 48, 49].

2. Shear viscosity coefficient: the Kubo formalism

In this Section we concentrate on the extraction of the shear viscosity from the 'infinite' parton-hadron matter employing the Kubo formalism. We simulate the 'infinite' matter within a cubic box with periodic boundary conditions at various values for the quark density (or chemical potential) and energy density. The size of the box is fixed to 9^3 fm^3. The initialization is done by populating the box with light (u, d) and strange (s) quarks, antiquarks and gluons. The system is initialized out of equilibrium and approaches kinetic and chemical equilibrium during it's evolution by PHSD. If the energy density in the system is below the critical energy density ($\varepsilon_c \approx 0.5$ GeV/fm^3), the evolution proceeds through the dynamical phase transition (as

Journal of Physics: Conference Series **458** (2013) 012019 doi:10.1088/1742-6596/458/1/012019

described in Ref. [47]) and ends up in an ensemble of interacting hadrons. For more details we refer the reader to Ref. [47].

The Kubo formalism relates linear transport coefficients such as heat conductivity, shear and bulk viscosities to non-equilibrium correlations of the corresponding dissipative fluxes, and treats dissipative fluxes as perturbations to local thermal equilibrium [36, 37]. The Green-Kubo formula for the shear viscosity η is given by [50]:

$$\eta = \frac{1}{T}\int d^3r \int_0^\infty dt \langle \pi^{xy}(\mathbf{0},0)\pi^{xy}(\mathbf{r},t)\rangle, \tag{1}$$

where T is the temperature of the system, t refers to the time after the system equilibrates, which is set at $t = 0$, while $\langle ... \rangle$ denotes the ensemble average in thermal equilibrium and π^{xy} is the shear component (traceless part) of the energy momentum tensor $\pi^{\mu\nu}$:

$$\pi^{xy}(\mathbf{x},t) \equiv T^{xy}(\mathbf{x},t) = \int \frac{d^3p}{(2\pi)^3}\frac{p^x p^y}{E} f(\mathbf{x},\mathbf{p};t), \tag{2}$$

where the mean-field U_s enters in the energy $E = \sqrt{\mathbf{p}^2 + U_s}$.

In our numerical simulation the volume averaged shear component of the energy momentum tensor can be written as

$$\pi^{xy}(t) = \frac{1}{V}\sum_{i=1}^{N}\frac{p_i^x p_i^y}{E_i}, \tag{3}$$

where V is the volume of the system and the sum is over all particles in the box at time t. Note that the scalar mean-field contribution U_s only enters via the energy E. Taking into account that point particles are uniformly distributed in our box (implying $\pi^{xy}(\mathbf{r},t) = \pi^{xy}(t)$), we can simplify the Kubo formula for the shear viscosity to

$$\eta = \frac{V}{T}\int_0^\infty dt \langle \pi^{xy}(0)\pi^{xy}(t)\rangle. \tag{4}$$

The correlation functions $\langle \pi^{xy}(0)\pi^{xy}(t)\rangle$ are empirically found to decay exponentially in time,

$$\langle \pi^{xy}(0)\pi^{xy}(t)\rangle = \langle \pi^{xy}(0)\pi^{xy}(0)\rangle \; e^{-t/\tau} \;, \tag{5}$$

as shown in Fig. 1 (lhs), where τ is the relaxation time. Finally, we end up with the Green-Kubo formula for the shear viscosity,

$$\eta = \frac{V}{T}\langle \pi^{xy}(0)^2\rangle\tau, \tag{6}$$

which we use to extract the shear viscosity from the PHSD simulations in the box.

3. The relaxation time approximation

3.1. Calculation of the shear and bulk viscosities

The starting hypothesis of the relaxation time approximation (RTA) is that the collision integral can be approximated by

$$C[f] = -\frac{f - f^{eq}}{\tau}, \tag{7}$$

where τ is the relaxation time. In this approach it has been shown that the shear and bulk viscosities (without mean-field or potential effects) can be written as [51, 52, 53]:

$$\eta = \frac{1}{15T}\sum_a \int \frac{d^3p}{(2\pi)^3}\frac{|\mathbf{p}|^4}{E_a^2}\tau_a(E_a) f_a^{eq}(E_a/T), \tag{8}$$

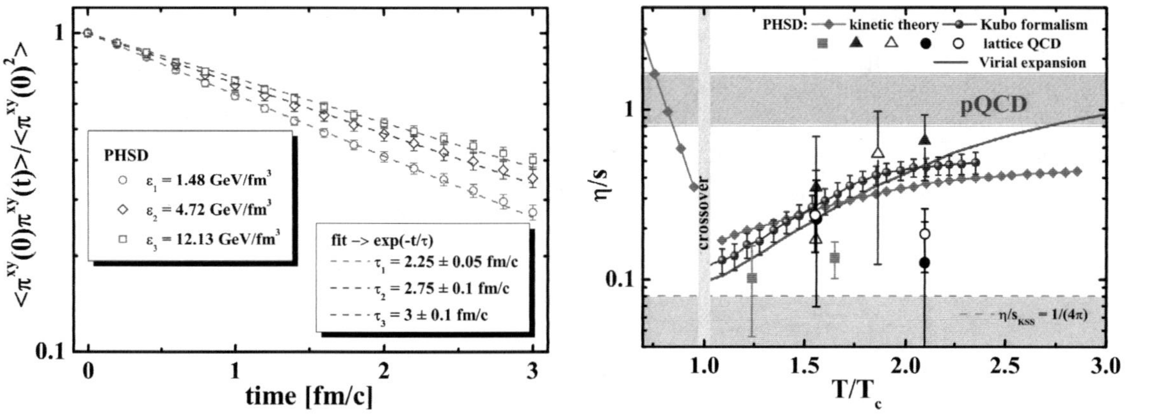

Figure 1. (lhs) The correlation functions $\langle\pi^{xy}(0)\pi^{xy}(t)\rangle$, which are normalized by $\langle\pi^{xy}(0)^2\rangle$, as a function of time obtained by the PHSD simulations in the box (open symbols) for systems at different energy densities and corresponding exponential fits (dashed lines) with extracted relaxation times.
(rhs) The shear viscosity to entropy density ratio η/s as a function of temperature of the system obtained by the PHSD simulations using different methods: the relaxation time approximation (red line+diamonds) and the Kubo formalism (blue line+dots). The others symbols denote lattice QCD data for pure $SU_c(3)$ gauge theory from [29] (magenta squares), from [31] (wine open and full triangles), and from [32] (black open and full circles). The orange dashed line demonstrates the Kovtun-Son-Starinets bound [14] $(\eta/s)_{KSS} = 1/(4\pi)$. For comparison, the virial expansion approach (green line) [55] is shown as a function of temperature, too.

$$\zeta = \frac{1}{9T}\sum_a \int \frac{d^3p}{(2\pi)^3}\frac{\tau_a(E_a)}{E_a^2}[(1-3v_s^2)E_a^2 - m_a^2]^2 \; f_a^{eq}(E_a/T), \tag{9}$$

where the sum is over particles of different type a (in our case, $a = q, \bar{q}, g$). In the PHSD transport approach the relaxation time can be expressed in the following way:

$$\tau_a = \Gamma_a^{-1}, \tag{10}$$

where Γ_a is the width of particles of type $a = q, \bar{q}, g$, which is defined within the DQPM. In our numerical simulation the volume averaged shear and bulk viscosities assume the following expressions:

$$\eta = \frac{1}{15TV}\sum_{i=1}^{N}\frac{|\mathbf{p}_i|^4}{E_i^2}\Gamma_i^{-1}, \tag{11}$$

$$\zeta = \frac{1}{9TV}\sum_{i=1}^{N}\frac{\Gamma_i^{-1}}{E_i^2}[(1-3v_s^2)E_i^2 - m_i^2]^2, \tag{12}$$

where the speed of sound $v_s = v_s(T)$ is taken from [54] or the DQPM, alternatively. Note that $v_s(T)$ from both approaches is practically identical since it is governed by the DQPM, which reproduces the lattice QCD results.

In Fig. 1 (rhs) we present the shear viscosity to entropy density ratio as a function of temperature of the system extracted from the PHSD simulations in the box employing different methods: the relaxation time approximation (red line+diamonds) and the Kubo formalism (blue line+dots). For comparison, the virial expansion approach (green line) [55] and lattice QCD data for pure $SU_c(3)$ gauge theory are shown as a function of temperature, too.

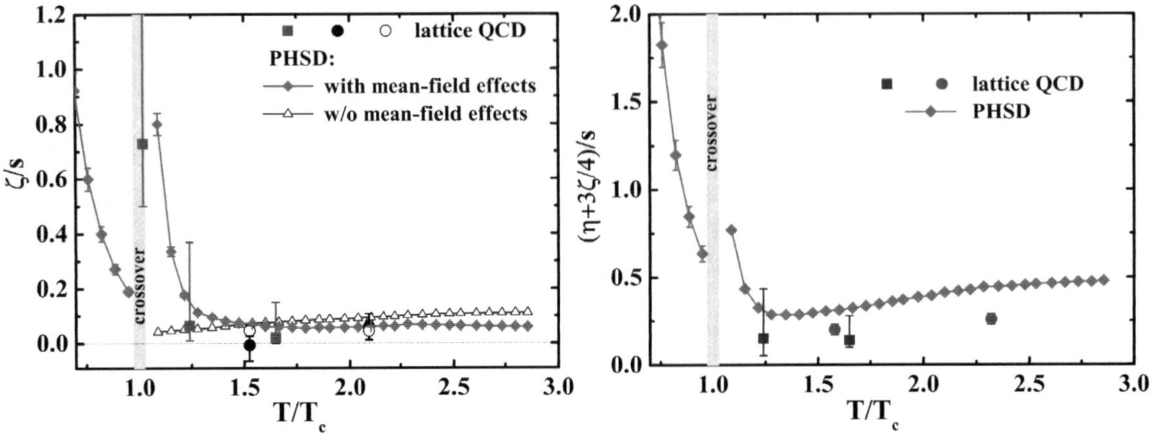

Figure 2. (lhs) The bulk viscosity to entropy density ratio as a function of temperature of the system extracted from the PHSD simulations in the box using the relaxation time approximation including mean-field effects (red line+diamonds) and without potential effects (blue line+open triangles). The available lattice QCD data from [30] (green squares) and from [32] (black open and full circles) are shown, too.
(rhs) The specific sound channel $(\eta+3\zeta/4)/s$ as a function of temperature of the system obtained by the PHSD simulations in the box using the relaxation time approximation including mean-field effects (red line+diamonds). It is compared with lattice QCD data from [33] (green circles) and from combining results of [30] and [32] (blue squares).

3.2. Mean-field or potential effects

In the absence of a chemical potential for quarks there should be no sizeable vector or tensor fields, only scalar fields. This affects the bulk viscosity, but not the shear viscosity. The expression for the bulk viscosity with potential effects is given by [53]

$$\begin{aligned} \zeta &= \frac{1}{T}\sum_a \int \frac{d^3p}{(2\pi)^3}\frac{\tau_a(E_a)}{E_a^2} f_a^{eq}(E_a/T) \\ &\times \left[\left(\frac{1}{3}-v_s^2\right)|\mathbf{p}|^2 - v_s^2\left(m_a^2 - T^2\frac{dm_a^2}{dT^2}\right)\right]^2. \end{aligned} \quad (13)$$

In the numerical simulation the volume averaged bulk viscosity with mean-field effects is used

$$\zeta = \frac{1}{TV}\sum_{i=1}^{N}\frac{\Gamma_i^{-1}}{E_i^2}\left[\left(\frac{1}{3}-v_s^2\right)|\mathbf{p}|^2 - v_s^2\left(m_i^2 - T^2\frac{dm_i^2}{dT^2}\right)\right]^2. \quad (14)$$

Using the DQPM expressions for masses of quarks and gluons (for $\mu_q = 0$)

$$m_q^2(T) = \frac{1}{3}g^2(T)T^2, \qquad m_g^2(T) = \frac{3}{4}g^2(T)T^2$$

we can calculate the derivative dm^2/dT^2 as well as v_s^2.

In Fig. 2 (lhs) we show the bulk viscosity to entropy density ratio as a function of temperature of the system obtained by the PHSD simulations in the box employing the relaxation time approximation including mean-field (or potential) effects (red line+diamonds) and without potential effects (blue line+open triangles) as well as the available lattice QCD data from [30, 32].

Journal of Physics: Conference Series **458** (2013) 012019 doi:10.1088/1742-6596/458/1/012019

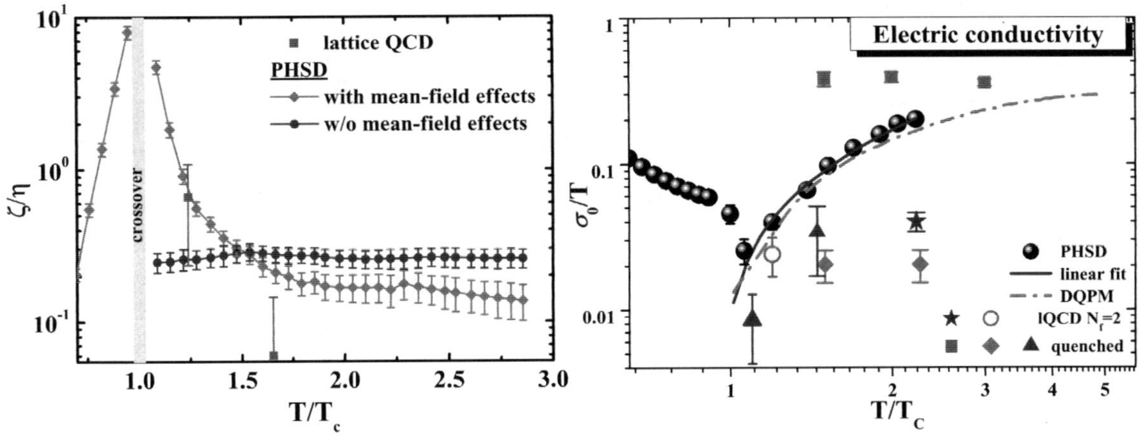

Figure 3. (lhs) The bulk to shear viscosity ratio ζ/η as a function of temperature of the system obtained by the PHSD simulations in the box employing the relaxation time approximation including mean-field effects (red line+diamonds) and without potential effects (blue line+circles).
(rhs) The ratio of the electric conductivity over temperature σ_0/T as a function of the scaled temperature T/T_c (T_c = 158 MeV). The full round symbols show the PHSD results, the solid blue line is a linear fit to the PHSD results (above T_c), while the dash-dotted red line gives the corresponding ratio in the relaxation-time approach (employing the DQPM parameters). The scattered symbols with error bars represent the results from lattice QCD calculations: triangles – Refs. [56], diamonds – Ref. [57], squares – Ref. [58], star – Ref. [59], open circle – Ref. [60].

A sound wave propagation in the z-direction with wavelength $\lambda = 2\pi/k$ is damped according to

$$T_{03}(t,k) \propto \exp\left[-\frac{(\frac{4}{3}\eta + \zeta)k^2 t}{2(\varepsilon + p)}\right], \tag{15}$$

where T_{03} is the momentum density in the z-direction, ε is the energy density and p is the pressure. Thus both the shear η and bulk ζ viscosities contribute to the damping of sound waves in the medium. In Fig. 2 (rhs) we present the specific sound channel $(\eta + 3\zeta/4)/s$ as a function of temperature of the system obtained by the PHSD simulations in the box using the relaxation time approximation including mean-field effects (red line+diamonds). It is compared with lQCD data from [33] (green circles) and from combining results of [30] and [32] (blue squares).

Furthermore, in Fig. 3 (lhs), we show the bulk to shear viscosity ratio ζ/η as a function of temperature of the system extracted from the PHSD simulations in the box using the relaxation time approximation including mean-field (or potential) effects (red line+diamonds) and without potential effects (blue line+circles). This ratio shows a pronounced maximum close to T_c when including the expected mean-field effects.

3.3. Electric conductivity

A further quantity of interest is the electric conductivity in the partonic/hadronic matter which controls the electromagnetic emissivity of the system in equilibrium. Here we briefly present the results from Ref. [49] which are displayed in Fig. 3 (rhs) for the dimensionless ratio of the electric conductivity over temperature σ_0/T which shows a pronounced minimum close to T_c and approaches a constant at higher temperature.

4. Summary and conclusions

We have employed the off-shell Parton-Hadron-String Dynamics (PHSD) approach in a finite box with periodic boundary conditions for the study of the shear and bulk viscosities - as well as the electric conductivity - as a function of temperature (or energy density) for dynamical infinite partonic and hadronic systems in equilibrium. The PHSD transport model is based on a lQCD equation of state [69] and well describes the entropy density $s(T)$, the energy density $\varepsilon(T)$ as well as the pressure $p(T)$ in thermodynamic equilibrium in comparison to the lQCD results. We have employed the Kubo formalism as well as the relaxationtime approximation to calculate the shear viscosity $\eta(T)$. We find that both methods provide very similar results for the ratio η/s with a minimum close to the critical temperature T_c while approaching the perturbative QCD (pQCD) limit at higher temperatures. For $T < T_c$, i.e. in the hadronic phase, the ratio η/s rises fast with decreasing temperature due to a lower interaction rate of the hadronic system and a significantly smaller number of degrees-of-freedom (or entropy density). Our results are, furthermore, also in almost quantitative agreement with the ratio $\eta(T)/s(T)$ from the virial expansion approach in Ref. [55] as well as with lQCD data for the pure gauge sector.

We have, furthermore, evaluated the bulk viscosity $\zeta(T)$ in the relaxation time approach and focused on the effects of mean fields (or potentials) in the partonic phase. Here we find a significant rise of the ratio $\zeta(T)/s(T)$ in the vicinity of the critical temperature T_c due to the scalar mean-fields from PHSD. The result for this ratio is in line with that from lQCD calculations. Additionally, the specific sound channel $(\eta + 3\zeta/4)/s(T)$ has been calculated and presents a non-trivial temperature dependence; the absolute value for this combination of the shear and bulk viscosities is in an approximate agreement with the lattice gauge theory. Furthermore, the ratio $\zeta(T)/\eta(T)$ within the PHSD calculations shows a strong maximum close to T_c, which has to be attributed to mean-field (or potential) effects that in PHSD are encoded in the infrared enhancement of the resummed coupling $g(T)$ (from the DQPM).

We also find that the dimensionless ratio of the electric conductivity over temperature σ_0/T rises above T_c approximately linearly with T up to $T = 2.5T_c$, but approaches a constant above $5T_c$, as expected from pQCD. This finding is naturally explained within the relaxation-time approach using the DQPM spectral functions. Below T_c the ratio σ_0/T rises with decreasing temperature because the system merges to a moderately interacting gas of pions with a larger charge (squared) to mass ratio than in the partonic phase and a longer relaxation time.

Since the PHSD calculations have proven to describe single-particle as well as collective observables and also dilepton data from relativistic nucleus-nucleus collisions from lower SPS to top RHIC energies [61, 62], the extracted transport coefficients $\eta(T)$, $\zeta(T)$ and $\sigma_0(T)$ are compatible with experimental observations in a wide energy density (temperature) range. Furthermore, the qualitative and partly quantitative agreement with lQCD results is striking.

Acknowledgement

This work in part has been supported by DFG as well as by the LOEWE center HIC for FAIR.

References

[1] Gyulassy M and McLerran L D 2005 *Nucl. Phys.* A **750** 30
[2] Shuryak E V 2005 *Nucl. Phys.* A **750** 64
[3] Adams J *et al.* (STAR Collaboration) 2005 *Nucl. Phys.* A **757** 102
[4] Adcox K *et al.* (PHENIX Collaboration) 2005 *Nucl. Phys.* A **757** 184
[5] Arsene I *et al.* (BRAHMS Collaboration) 2005 *Nucl. Phys.* A **757** 1
[6] Back B B *et al.* (PHOBOS Collaboration) 2005 *Nucl. Phys.* A **757** 28
[7] Aamodt K *et al.* (ALICE Collaboration) 2010 *Phys. Rev. Lett.* **105** 252302
[8] Huovinen P Kolb P F Heinz U W Ruuskanen P V and Voloshin S A 2001 *Phys. Lett.* B **503** 58
[9] Kolb P F Huovinen P Heinz U and Heiselberg H 2001 *Phys. Lett.* B **500** 232
[10] Teaney D Lauret J and Shuryak E V 2001 *Phys. Rev. Lett.* **86** 4783
[11] Hirano T and Tsuda K 2002 *Phys. Rev.* C **66** 054905
[12] Kolb P F and Rapp R 2003 *Phys. Rev.* C **67** 044903

[13] Danielewicz P and Gyulassy M 1985 *Phys. Rev.* D **31** 53
[14] Policastro P Son D T and Starinets A O 2001 *Phys. Rev. Lett.* **87** 081601;
Kovtun P K Son D T and Starinets A O 2005 *Phys. Rev. Lett.* **94** 111601
[15] Romatschke P and Romatschke U 2007 *Phys. Rev. Lett.* **99** 172301
[16] Song H and Heinz U W 2008 *Phys. Rev.* C **77** 064901
[17] Luzum M and Romatschke P 2008 *Phys. Rev.* C **78** 034915
[18] Schenke B Jeon S and Gale C 2010 *Phys. Rev.* C **82** 014903
[19] Csernai L P Kapusta J I and McLerran L D 2006 *Phys. Rev. Lett.* **97** 152303
[20] Kharzeev D and Tuchin K 2008 *JHEP* **09** 093
[21] Karsch F Kharzeev D and Tuchin K 2008 *Phys. Lett.* B **663** 217
[22] Romatschke P and Son D T 2009 *Phys. Rev.* D **80** 065021
[23] Moore G D and Saremi O 2008 *JHEP* **09** 015
[24] Sasaki C and Redlich K 2009 *Phys. Rev.* C **79** 055207; 2010 *Nucl. Phys.* A **832** 62
[25] Arnold P B Moore G D and Yaffe L G 2010 *JHEP* **11** 001
[26] Arnold P B Dogan C and Moore G D 2006 *Phys. Rev.* D **74** 085021
[27] Prakash M *et al.* 1993 *Phys. Rep.* **227** 321
[28] Chen J W and Wang J 2009 *Phys. Rev.* C **79** 044913
[29] Meyer H B 2007 *Phys. Rev.* D **76** 101701
[30] Meyer H B 2008 *Phys. Rev. Lett.* **100** 162001
[31] Nakamura A and Sakai S 2005 *Phys. Rev. Lett.* **94** 072305
[32] Sakai S and Nakamura A 2007 *Pos* **LAT2007** 221
[33] Meyer H B 2009 *Nucl. Phys. A* **830** 641c
[34] Reif F 1965 *Fundamentals of Statistical and Thermal Physics*, McGraw-Hill Book Company, New York
[35] de Groot S R van Leeuwen W A and Weert C 1980 *Relativistic Kinetic Theory, Principles and Applications*, North-Holland Company, Amsterdam
[36] Green M S 1954 *J. of Chem. Phys.* **22** 398
[37] Kubo R 1957 *J. Phys. Soc. Japan* **12** 570; 1966 *Rep. Prog. Phys.* **29** 255
[38] Cassing W and Bratkovskaya E L 2009 *Nucl. Phys.* A **831** 215
[39] Cassing W and Bratkovskaya E L 2008 *Phys. Rev.* C **78** 034919
[40] Kadanoff L P and Baym G 1962 *Quantum Statistical Mechanics*, Benjamin, New York
[41] Juchem S Cassing W and Greiner C 2004 *Phys. Rev.* D **69** 025006; 2004 *Nucl. Phys.* A **743** 92
[42] Cassing W and Bratkovskaya E L 1999 *Phys. Rept.* **308** 65
[43] Bratkovskaya E L *et al.* 2004 *Phys. Rev. Lett.* **92** 032302
[44] Cassing W 2007 *Nucl. Phys.* A **795** 70
[45] Cassing W 2007 *Nucl. Phys.* A **791** 365
[46] Bratkovskaya E L *et al.* 2011 *Nucl. Phys.* A **856** 162
[47] Ozvenchuk V *et al.* 2013 *Phys. Rev.* C **87** 024901
[48] Ozvenchuk V *et al.* 2012 *arXiv:1212.5393 [hep-ph]*
[49] Cassing W *et al.* 2013 *arXiv:1302.0906 [hep-ph] Phys. Rev. Lett.* in press
[50] Zubarev R and Morozov O 1996 *Statistical Mechanics of Nonequilibrium Processes Volume 2: Relaxation and Hydrodynamic Processes* Akademie Verlag GmbH
[51] Hosoya A and Kajantie K 1985 *Nucl. Phys.* B **250** 666
[52] Gavin S 1985 *Nucl. Phys.* A **435** 826
[53] Chakraborty P and Kapusta J I 2011 *Phys. Rev.* C **83** 014906
[54] Borsanyi S *et al.*, 2010 *hep-lat/10072580*
[55] Mattiello S and Cassing W 2010 *Eur. Phys. J.* C **70** 243
[56] Ding H T *et al.* 2011 *Phys. Rev.* D **83** 034504; 2013 *arXiv:1301.7436 [hep-lat]*
[57] Aarts G *et al.* 2007 *Phys. Rev. Lett.* **99** 022002
[58] Gupta S 2004 *Phys. Lett.* B **597** 57
[59] Buividovich P V *et al.* 2010 *Phys. Rev. Lett.* **105** 132001
[60] Brandt B B Francis A Meyer H B and Wittig H 2013 *arXiv:1302.0675 [hep-lat]*
[61] Konchakovski V P *et al.* 2012 *J. Phys. Conf. Ser.* **389** 012015
[62] Bratkovskaya E L *et al.* 2012 *J. Phys. Conf. Ser.* **389** 012016

29th Winter Workshop on Nuclear Dynamics (WWND2013) IOP Publishing
Journal of Physics: Conference Series **458** (2013) 012020 doi:10.1088/1742-6596/458/1/012020

New developments in EPOS: Parton saturation

K. Werner[(a)], **B. Guiot**[(a)], **Iu. Karpenko**[(b,c)], **T. Pierog**[(d)]

[(a)] SUBATECH, University of Nantes – IN2P3/CNRS– EMN, Nantes, France
[(b)] Bogolyubov Institute for Theoretical Physics, Kiev 143, 03680, Ukraine
[(c)] FIAS, Johann Wolfgang Goethe Universitaet, Frankfurt am Main, Germany
[(d)]Karlsruhe Inst. of Technology, KIT, Campus North, Inst. f. Kernphysik, Germany

Abstract. We discuss latest developments in EPOS concerning parton saturation and its relation with energy conservation in the multiple scattering framework.

Inclusive cross sections are particularly simple, quantum interference helps to provide simple formulas referred to a "factorization". Although factorization is widely used, strict mathematical proofs exist only in very special cases, and certainly not for hadron production in *pp* scattering. To go beyond factorization and to formulate a consistent multiple scattering theory is difficult. A possible solution is Gribov's Pomeron calculus, which can be adapted to our language by identifying Pomeron and parton ladder. Multiple scattering means that one has contributions with several parton ladders in parallel.

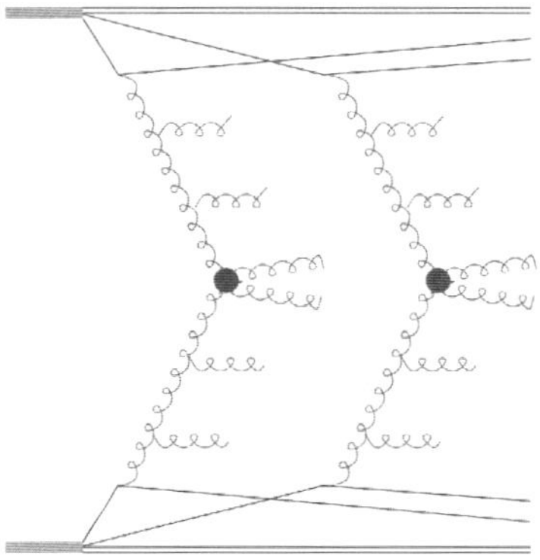

Figure 1. Multiple scattering

We indicated several years ago inconsistencies in this approach, proposing an "energy conserving multiple scattering treatment". The main idea is simple: in case of multiple scattering, when it comes to calculating partial cross sections for double, triple ... scattering, one has to explicitly care about the fact the partons ladders which happen to be parallel to each other (see figure 1) share the collision energy. This energy sharing has been implemented in EPOS, which is a multiple scattering approach corresponding to a marriage of Gribov-Regge theory and perturbative QCD (for details see [1]). An elementary scattering corresponds to a parton ladder, containing a hard scattering calculable based on pQCD, including initial and final state radiation.

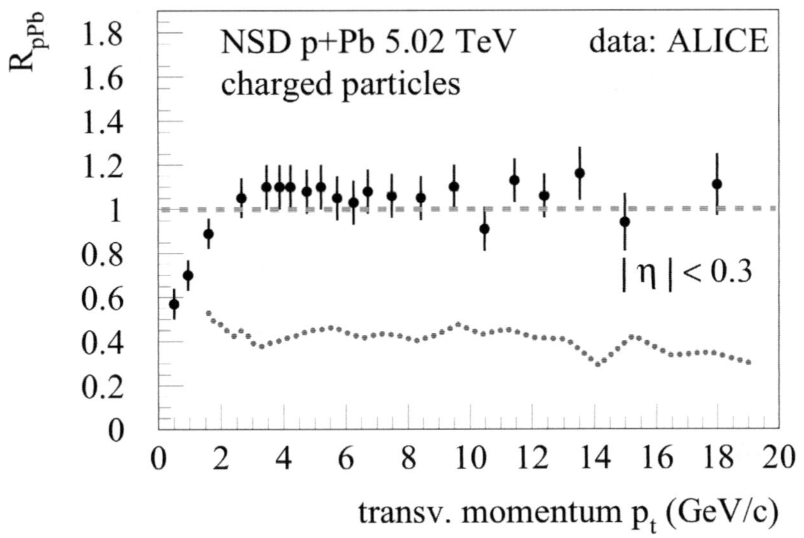

Figure 2. Nuclear modification factor in pPb at 5 TeV/c.

The energy sharing scheme generalizes straightforwardly to proton-nucleus collisions. Here one expects (and observes) the so-called binary scaling, which means that the scaled AA cross section (cross section divided by the number of binary collisions) is equal to the proton-proton one,

$$\frac{1}{N_{\rm coll}}\left.\frac{dn}{dp_t}\right|_{AA}=\left.\frac{dn}{dp_t}\right|_{pp},$$

or in other words, the nuclear modification factor (ratio l.h.s. to r.h.s.) should be unity.

In figure 2, we show the results for the nuclear modification factor in pPb at 5 TeV. The data are represented as black points, the calculations as red dotted line. Whereas the data are close to unity above two GeV/c, the simulations are clearly below. In fig. 3, we plot the scaled parton p_t distributions for different centralities in pPb at 5 TeV/c. In case of binary scaling all these curves should coincide, which is clearly not the case. So the simulations show a strong violation of this scaling, due to the imposed energy conservation.

After many attempts, we finally found a solution to the problem: The usual constant "soft scale" Q_0^2 (the lower cutoff of virtualities in the parton ladder) has to be replaced by a centrality dependent saturation scale [3, 4, 5, 6, 7, 8],

$$Q_s^2=Q_0^2\left(1+B_{\rm satur}N_{\rm part}(i,j)\right),$$

where $N_{\rm part}(i,j)$ is the number of participating nucleons connected to a given parton ladder between projectile nucleon i and target nucleon j. So each parton ladder has "its own" saturation scale! This new procedure (implemented in EPOS3, with a single free parameter $B_{\rm satur}$) allows to completely recover binary scaling, see figure 4.

Also in PbPb collisions we obtain binary scaling at large p_t. So for the first time energy sharing is compatible with binary scaling in a multiple scattering scheme.

I is no surprise that the nuclear modification factor is now close to unity, as seen in fig. 6.

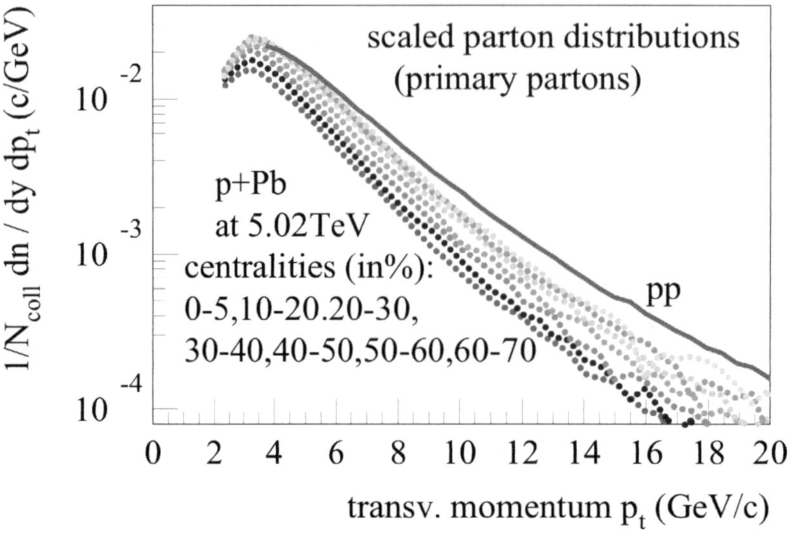

Figure 3. Scaled parton p_t distributions for different centralities in pPb at 5 TeV/c, using a constant soft scale Q_0.

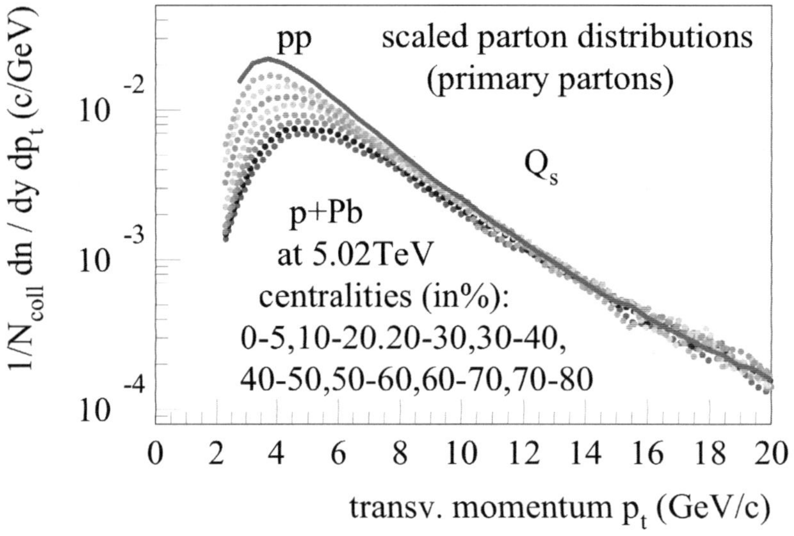

Figure 4. Scaled parton p_t distributions for different centralities in pPb at 5 TeV/c, using a saturation scale Q_s.

[1] H. J. Drescher, M. Hladik, S. Ostapchenko, T. Pierog and K. Werner, Phys. Rept. 350, 93, 2001
[2] K.Werner, Iu.Karpenko, M. Bleicher, T.Pierog, S. Porteboeuf-Houssais, arXiv:1203.5704, accepted Phys. Rev. C.

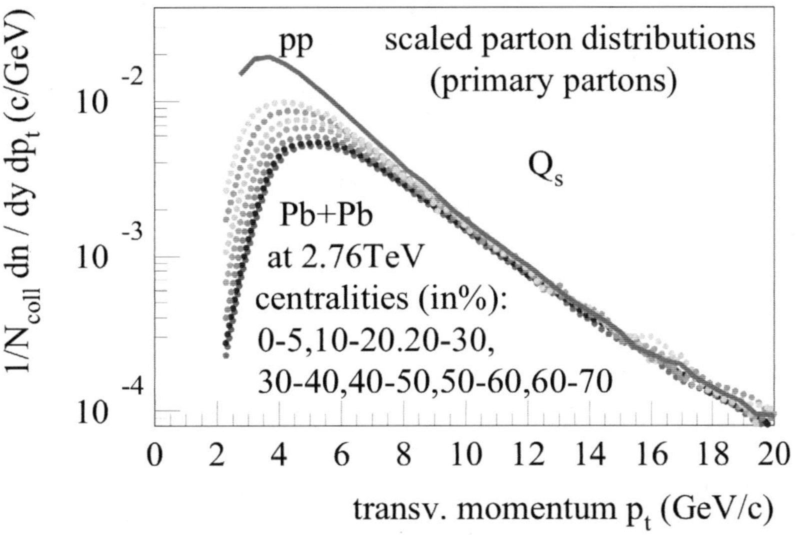

Figure 5. Scaled parton p_t distributions for different centralities in PbPb at 2.76 TeV/c, using a saturation scale Q_s.

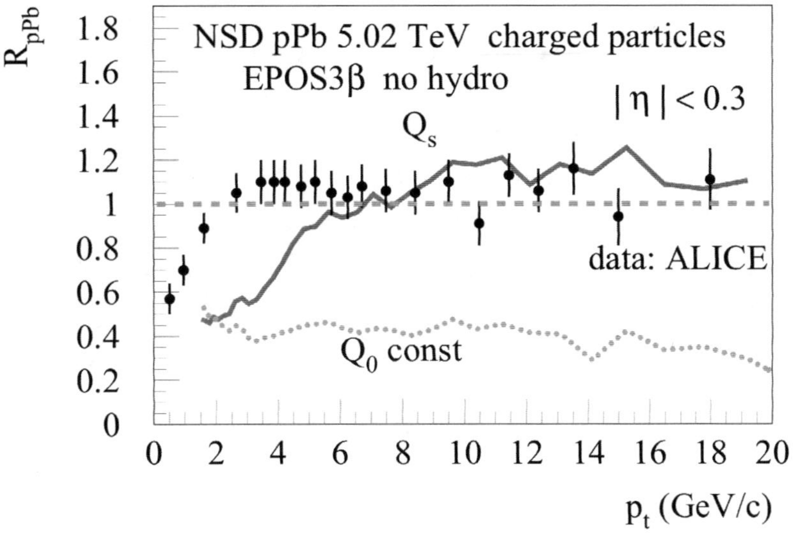

Figure 6. Nuclear modification factor in pPb at 5 TeV/c.

[3] L. McLerran, R. Venugopalan, Phys. Rev. D 49 (1994) 2233; L. McLerran, R. Venugopalan, Phys. Rev. D 49 (1994) 3352; L. McLerran, R. Venugopalan, Phys. Rev. D 50 (1994) 2225.
[4] I. Balitsky, Nucl. Phys. B 463 (1996) 99
[5] Yu. Kovchegov, Phys. Rev. D 60 (2000) 034008
[6] J. Jalilian-Marian, A. Kovner, A. Leonidov, H. Weigert, Phys. Rev. D 59 (1999) 014014, hep-ph/9706377; J.

Jalilian-Marian, A. Kovner, A. Leonidov, H. Weigert, Nucl. Phys. B 504 (1997) 415, hep-ph/9701284
[7] E. Iancu, A. Leonidov, L.D. McLerran, Phys. Lett. B 510 (2001) 133, hep-ph/0102009; E. Iancu, A. Leonidov, L.D. McLerran, Nucl. Phys. A 692 (2001) 583, hep-ph/0011241
[8] H. Weigert, Nucl. Phys. A 703 (2002) 823, hep-ph/0004044

Journal of Physics: Conference Series **458** (2013) 012021 doi:10.1088/1742-6596/458/1/012021

Early thermalization of quark-gluon matter initially created in high-energy nucleus-nucleus collisions

Xiao-Ming Xu

Department of Physics, Shanghai University, Baoshan, Shanghai 200444, China

E-mail: xmxu@mail.shu.edu.cn

Abstract. Elastic parton-parton-parton scattering is briefly reviewed and is included in transport equations of quark-gluon matter. We solve the transport equations and get thermal states from initially produced quark-gluon matter. Both gluon matter and quark matter take early thermalization, but gloun matter has a shorter thermalization time than quark matter.

1. Introduction

A large amount of particles move randomly and momentum distribution functions in different directions are identical. The distribution depends only on the absolute value of the three-dimensional momentum. The dependence indicates the existence of a thermal state and exposes the basic quantity temperature, which value is acquired from the analytic expression of the distribution (e.g., temperature is in the Boltzmann distribution). Unfortunately, the system of gluons, quarks and antiquarks produced in initial nucleus-nucleus collisions at the Relativistic Heavy Ion Collider (RHIC) energies does not possess temperature. This is because the distribution in the incoming nucleus beam direction is much larger than the distribution in the direction perpendicular to the beam direction [1]. However, the system evolves rapidly into a thermal state (a quark-gluon plasma) as concluded from the elliptic flow data of hadrons [2] and the corresponding explanation of hydrodynamic calculations which assume early thermalization and ideal relativistic fluid flow [3, 4]. The early thermalization or the rapid creation of temperature is generally interesting. Parton-parton scattering was employed to study thermalization of initially produced quark-gluon matter, but is not enough to explain the early thermalization [5, 6, 7, 8]. Elastic gluon-gluon-gluon scattering was proposed to explain the early thermalization of gluon matter [9]. Nevertheless, elastic quark-quark scattering and elastic quark-quark-quark scattering give a long thermalization time of quark matter [10]. Recently we have obtained that quark matter takes the early thermalization due to all types of elastic scattering that involves quarks [11]. In the next section I briefly introduce the elastic scattering of quark-quark-quark [10], quark-quark-antiquark [12], quark-antiquark-antiquark, antiquark-antiquark-antiquark, gluon-quark-quark [13], gluon-quark-antiquark [13], gluon-antiquark-antiquark, gluon-gluon-quark [11], gluon-gluon-antiquark or gluon-gluon-gluon [9]. In section 3 we present transport equations that include all types of elastic parton-parton-parton scattering and numerical results of the equations. Conclusions are in the last section.

Table 1. The number of diagrams for elastic parton-parton-parton scattering, the number of diagrams in a class, the number of the triple-gluon vertex and the number of the four-gluon vertex of a diagram.

elastic scattering	number of diagrams for a type	class	number of diagrams in a class	number of the 3-gluon vertex	number of the 4-gluon vertex
quark-quark-quark 1	42	1	36	0	0
		2	6	1	0
quark-quark-quark 2	14	1	12	0	0
		2	2	1	0
quark-quark-quark 3	7	1	6	0	0
		2	1	1	0
quark-quark-antiquark 1	58	1	52	0	0
		2	6	1	0
quark-quark-antiquark 2	14	1	12	0	0
		2	2	1	0
quark-quark-antiquark 3	29	1	26	0	0
		2	3	1	0
quark-quark-antiquark 4	7	1	6	0	0
		2	1	1	0
gluon-quark-quark 1	72	1	40	0	0
		2	24	1	0
		3	8	2	0
or				0	1
gluon-quark-quark 2	36	1	20	0	0
		2	12	1	0
		3	4	2	0
or				0	1
gluon-quark-antiquark 1	76	1	40	0	0
		2	28	1	0
		3	8	2	0
or				0	1
gluon-quark-antiquark 2	36	1	20	0	0
		2	12	1	0
		3	4	2	0
or				0	1
gluon-gluon-quark	123	1	24	0	0
		2	36	1	0
		3	38	2	0
or				0	1
		4	25	3	0
or				1	1
gluon-gluon-gluon	220	1	105	4	0
		2	96	2	1
		3	19	0	2

Journal of Physics: Conference Series **458** (2013) 012021 doi:10.1088/1742-6596/458/1/012021

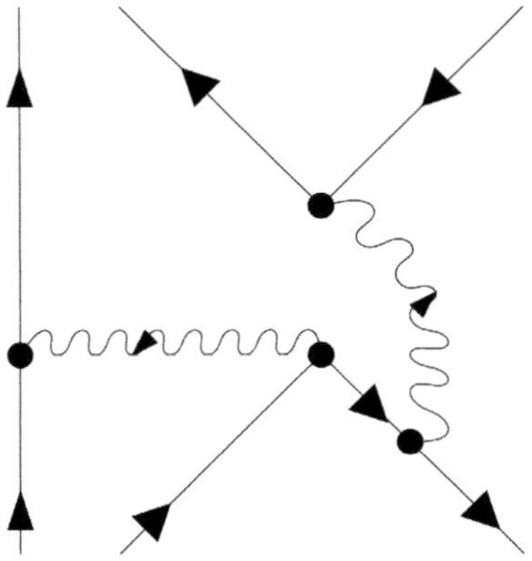

$F_{\bar{Q}DE}$

Figure 1. Elastic gluon-quark-antiquark scattering. The wiggly (solid) lines stand for gluons (quarks and antiquarks).

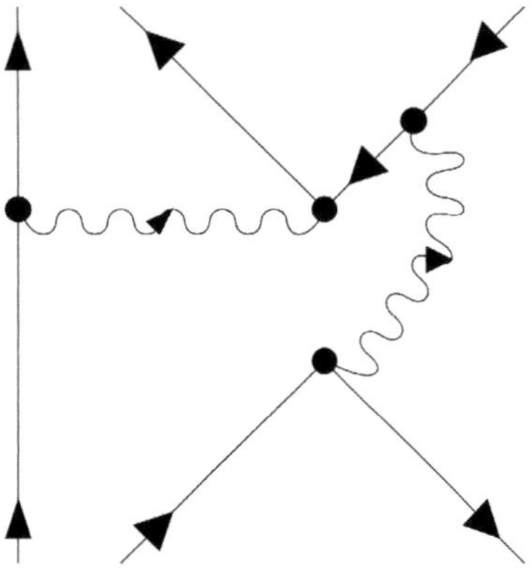

$F_{\bar{Q}UE}$

Figure 2. Same as figure 1.

2. Elastic parton-parton-parton scattering

Each type of elastic 3-to-3 scattering is divided into 2, 3 or 4 classes according to the number of the triple-gluon vertex and the number of the four-gluon vertex. In Table 1 we list the number of diagrams for each type or in a class, the number of the triple-gluon vertex and the number of the four-gluon vertex of a diagram. The number following the entry quark-quark-quark has the meanings: the three quarks are identical at 1; only two quarks are identical at 2; the three quarks are different at 3. The number following the entry quark-quark-antiquark has the meanings: the two quarks and the antiquark have the same flavor at 1; only the two quarks have the same flavor at 2; only one quark has the same flavor as the antiquark at 3; the three flavors of the two quarks and the antiquark are not identical at 4. The number following the entry gluon-quark-quark (gluon-quark-antiquark) has the meanings: the two quarks (the quark and the antiquark) have the same flavor at 1; the two quarks (the quark and the antiquark) take different flavors at 2.

Feynman diagrams for the elastic quark-antiquark-antiquark scattering (antiquark-antiquark-antiquark, gluon-antiquark-antiquark, gluon-gluon-antiquark) are obtained from the Feynman diagrams for the elastic quark-quark-antiquark (quark-quark-quark, gluon-quark-quark, gluon-gluon-quark) via the replacement of quark by antiquark and vice versa. To illustrate elastic 3-to-3 scattering, we show four Feynman diagrams in figures 1-4. In figures 1 and 2 quark-antiquark annihilation and creation occur. In both figure 3 and figure 4 two triple-gluon vertices and one four-gluon vertex are involved.

The elastic gluon-gluon-gluon scattering is complicated but rich in the process as shown by the 220 Feynman diagrams at the tree level [9]. To better understand the scattering, we sort out Feynman diagrams in every class. We select three triple-gluon vertices of a diagram to subclassify the diagrams in the first class and six subclasses are obtained according to the initial gluons and the final gluons possessed by the selected vertices. The subclasses are listed in Table 2. Every diagram in the second class has a four-gluon vertex and two triple-gluon vertices. We select the four-gluon vertex and a triple-gluon vertex to subclassify the diagrams in the second class and nine subclasses are obtained according to the initial gluons and the final gluons possessed by the four-gluon vertex and the selected triple-gluon vertex. The subclasses are listed in Table 3. Diagrams in the third class can not be subclassified.

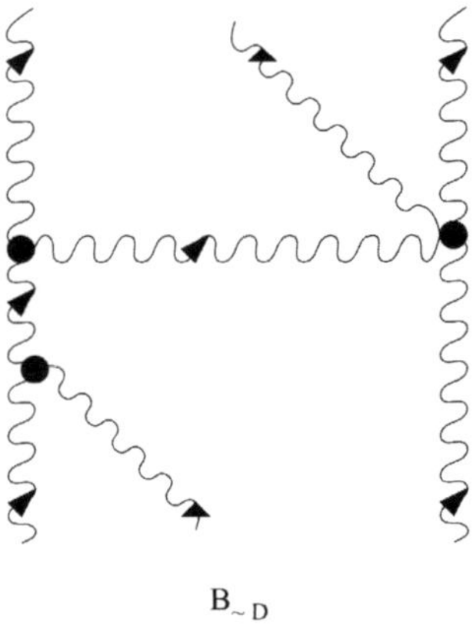

Figure 3. Elastic gluon-gluon-gluon scattering. The wiggly lines stand for gluons.

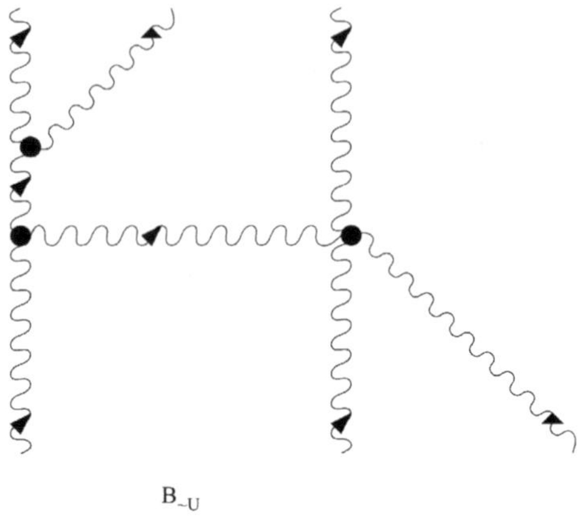

Figure 4. Same as figure 3.

Table 2. Number of diagrams in a subclass identified by the three selected triple-gluon vertices. n_{1i} and n_{1f} (n_{2i} and n_{2f}, n_{3i} and n_{3f}) are numbers of initial and final gluons possessed by the first (second, third) triple-gluon vertex, respectively.

subclass	number of diagrams	n_{1i}	n_{1f}	n_{2i}	n_{2f}	n_{3i}	n_{3f}
1	36	1	1	1	1	1	0
2	18	0	2	1	1	1	0
3	18	2	0	1	1	0	1
4	18	2	0	0	2	1	0
5	6	1	1	1	1	0	0
6	9	2	0	0	2	0	0

Table 3. Number of diagrams in a subclass identified by the four-gluon vertex and the selected triple-gluon vertex. n_{i4} and n_{f4} (n_{i3} and n_{f3}) are numbers of initial and final gluons possessed by the 4-gluon vertex (the 3-gluon vertex), respectively.

subclass	number of diagrams	n_{i4}	n_{f4}	n_{i3}	n_{f3}
1	18	1	2	1	1
2	18	2	1	1	1
3	9	1	2	2	0
4	9	2	1	0	2
5	3	0	3	2	0
6	3	3	0	0	2
7	18	1	1	1	1
8	9	0	2	2	0
9	9	2	0	0	2

3. Transport equations and thermal states

We assume that quark-gluon matter consists of gluons, quarks and antiquarks with up and down flavors and the two flavors (the quark and the antiquark as well) have the same amount. Let the distribution functions for the gluon, the up-quark, the down-quark, the up-antiquark and the down-antiquark be f_{gi}, f_{ui}, f_{di}, $f_{\bar{u}i}$ and $f_{\bar{d}i}$, respectively, where i labels the ith parton in scattering, and $f_{ui} = f_{di} = f_{\bar{u}i} = f_{\bar{d}i} = f_{qi}$. The transport equation for gluon matter is

$$\begin{aligned}
\frac{\partial f_{g1}}{\partial t} + \vec{v}_1 \cdot \vec{\nabla}_{\vec{r}} f_{g1} = & -\frac{1}{2E_1}\int \frac{d^3p_2}{(2\pi)^3 2E_2}\frac{d^3p_3}{(2\pi)^3 2E_3}\frac{d^3p_4}{(2\pi)^3 2E_4}(2\pi)^4\delta^4(p_1+p_2-p_3-p_4) \\
& \times\left\{\frac{g_G}{2} \mid \mathcal{M}_{gg\to gg} \mid^2 [f_{g1}f_{g2}(1+f_{g3})(1+f_{g4}) - f_{g3}f_{g4}(1+f_{g1})(1+f_{g2})]\right. \\
& + g_Q(\mid \mathcal{M}_{gu\to gu} \mid^2 + \mid \mathcal{M}_{gd\to gd} \mid^2 + \mid \mathcal{M}_{g\bar{u}\to g\bar{u}} \mid^2 + \mid \mathcal{M}_{g\bar{d}\to g\bar{d}} \mid^2) \\
& \left.\times[f_{g1}f_{q2}(1+f_{g3})(1-f_{q4}) - f_{g3}f_{q4}(1+f_{g1})(1-f_{q2})]\right\} \\
& -\frac{1}{2E_1}\int \frac{d^3p_2}{(2\pi)^3 2E_2}\frac{d^3p_3}{(2\pi)^3 2E_3}\frac{d^3p_4}{(2\pi)^3 2E_4}\frac{d^3p_5}{(2\pi)^3 2E_5}\frac{d^3p_6}{(2\pi)^3 2E_6} \\
& \times(2\pi)^4\delta^4(p_1+p_2+p_3-p_4-p_5-p_6)\left\{\frac{g_G^2}{12} \mid \mathcal{M}_{ggg\to ggg} \mid^2\right. \\
& \times[f_{g1}f_{g2}f_{g3}(1+f_{g4})(1+f_{g5})(1+f_{g6}) - f_{g4}f_{g5}f_{g6}(1+f_{g1})(1+f_{g2})(1+f_{g3})] \\
& + \frac{g_G g_Q}{2}(\mid \mathcal{M}_{ggu\to ggu} \mid^2 + \mid \mathcal{M}_{ggd\to ggd} \mid^2 + \mid \mathcal{M}_{gg\bar{u}\to gg\bar{u}} \mid^2 + \mid \mathcal{M}_{gg\bar{d}\to gg\bar{d}} \mid^2) \\
& \times[f_{g1}f_{g2}f_{q3}(1+f_{g4})(1+f_{g5})(1-f_{q6}) - f_{g4}f_{g5}f_{q6}(1+f_{g1})(1+f_{g2})(1-f_{q3})] \\
& + g_Q^2[\frac{1}{4} \mid \mathcal{M}_{guu\to guu} \mid^2 + \frac{1}{2}(\mid \mathcal{M}_{gud\to gud} \mid^2 + \mid \mathcal{M}_{gdu\to gdu} \mid^2) + \frac{1}{4} \mid \mathcal{M}_{gdd\to gdd} \mid^2 \\
& + \mid \mathcal{M}_{gu\bar{u}\to gu\bar{u}} \mid^2 + \mid \mathcal{M}_{gu\bar{d}\to gu\bar{d}} \mid^2 + \mid \mathcal{M}_{gd\bar{u}\to gd\bar{u}} \mid^2 + \mid \mathcal{M}_{gd\bar{d}\to gd\bar{d}} \mid^2 \\
& + \frac{1}{4} \mid \mathcal{M}_{g\bar{u}\bar{u}\to g\bar{u}\bar{u}} \mid^2 + \frac{1}{2}(\mid \mathcal{M}_{g\bar{u}\bar{d}\to g\bar{u}\bar{d}} \mid^2 + \mid \mathcal{M}_{g\bar{d}\bar{u}\to g\bar{d}\bar{u}} \mid^2) + \frac{1}{4} \mid \mathcal{M}_{g\bar{d}\bar{d}\to g\bar{d}\bar{d}} \mid^2] \\
& \left.\times[f_{g1}f_{q2}f_{q3}(1+f_{g4})(1-f_{q5})(1-f_{q6}) - f_{g4}f_{q5}f_{q6}(1+f_{g1})(1-f_{q2})(1-f_{q3})]\right\},
\end{aligned} \tag{1}$$

and the transport equation for up-quark matter is

$$\begin{aligned}
\frac{\partial f_{q1}}{\partial t} + \vec{v}_1 \cdot \vec{\nabla}_{\vec{r}} f_{q1} = & -\frac{1}{2E_1}\int \frac{d^3p_2}{(2\pi)^3 2E_2}\frac{d^3p_3}{(2\pi)^3 2E_3}\frac{d^3p_4}{(2\pi)^3 2E_4}(2\pi)^4\delta^4(p_1+p_2-p_3-p_4) \\
& \times\left\{g_G \mid \mathcal{M}_{ug\to ug} \mid^2 [f_{q1}f_{g2}(1-f_{q3})(1+f_{g4}) - f_{q3}f_{g4}(1-f_{q1})(1+f_{g2})]\right. \\
& + g_Q(\frac{1}{2} \mid \mathcal{M}_{uu\to uu} \mid^2 + \mid \mathcal{M}_{ud\to ud} \mid^2 + \mid \mathcal{M}_{u\bar{u}\to u\bar{u}} \mid^2 + \mid \mathcal{M}_{u\bar{d}\to u\bar{d}} \mid^2) \\
& \left.\times [f_{q1}f_{q2}(1-f_{q3})(1-f_{q4}) - f_{q3}f_{q4}(1-f_{q1})(1-f_{q2})]\right\} \\
& -\frac{1}{2E_1}\int \frac{d^3p_2}{(2\pi)^3 2E_2}\frac{d^3p_3}{(2\pi)^3 2E_3}\frac{d^3p_4}{(2\pi)^3 2E_4}\frac{d^3p_5}{(2\pi)^3 2E_5}\frac{d^3p_6}{(2\pi)^3 2E_6} \\
& \times(2\pi)^4\delta^4(p_1+p_2+p_3-p_4-p_5-p_6)\left\{\frac{g_G^2}{4} \mid \mathcal{M}_{ugg\to ugg} \mid^2\right. \\
& \times[f_{q1}f_{g2}f_{g3}(1-f_{q4})(1+f_{g5})(1+f_{g6}) - f_{q4}f_{g5}f_{g6}(1-f_{q1})(1+f_{g2})(1+f_{g3})] \\
& + g_Q g_G(\frac{1}{2} \mid \mathcal{M}_{uug\to uug} \mid^2 + \mid \mathcal{M}_{udg\to udg} \mid^2 + \mid \mathcal{M}_{u\bar{u}g\to u\bar{u}g} \mid^2 + \mid \mathcal{M}_{u\bar{d}g\to u\bar{d}g} \mid^2) \\
& \times[f_{q1}f_{q2}f_{g3}(1-f_{q4})(1-f_{q5})(1+f_{g6}) - f_{q4}f_{q5}f_{g6}(1-f_{q1})(1-f_{q2})(1+f_{g3})]
\end{aligned}$$

$$
\begin{aligned}
&+g_Q^2[\frac{1}{12} \mid \mathcal{M}_{uuu\to uuu} \mid^2 + \frac{1}{4}(\mid \mathcal{M}_{uud\to uud} \mid^2 + \mid \mathcal{M}_{udu\to udu} \mid^2) + \frac{1}{4} \mid \mathcal{M}_{udd\to udd} \mid^2 \\
&+\frac{1}{2} \mid \mathcal{M}_{uu\bar{u}\to uu\bar{u}} \mid^2 + \frac{1}{2} \mid \mathcal{M}_{uu\bar{d}\to uu\bar{d}} \mid^2 + \mid \mathcal{M}_{ud\bar{u}\to ud\bar{u}} \mid^2 + \mid \mathcal{M}_{ud\bar{d}\to ud\bar{d}} \mid^2 \\
&+\frac{1}{4} \mid \mathcal{M}_{u\bar{u}\bar{u}\to u\bar{u}\bar{u}} \mid^2 + \frac{1}{2}(\mid \mathcal{M}_{u\bar{u}\bar{d}\to u\bar{u}\bar{d}} \mid^2 + \mid \mathcal{M}_{u\bar{d}\bar{u}\to u\bar{d}\bar{u}} \mid^2) + \frac{1}{4} \mid \mathcal{M}_{u\bar{d}\bar{d}\to u\bar{d}\bar{d}} \mid^2] \\
&\times [f_{q1}f_{q2}f_{q3}(1-f_{q4})(1-f_{q5})(1-f_{q6}) - f_{q4}f_{q5}f_{q6}(1-f_{q1})(1-f_{q2})(1-f_{q3})]\},
\end{aligned}
\tag{2}
$$

where $\vec{v}_1$ is the parton velocity; g_G (g_Q) is the gluon (quark) color-spin degeneracy factor; $p_i (i = 1, \cdots, 6)$ denote the four-momenta of initial and final partons; E_i is the energy component of p_i. In equation (1) $\mathcal{M}_{gd\bar{d}\to gd\bar{d}}$ is the amplitude for the elastic scattering of a gluon, a down-quark and a down-antiquark; in equation (2) $\mathcal{M}_{u\bar{d}\bar{u}\to u\bar{d}\bar{u}}$ is the amplitude for the elastic scattering of an up-quark, a down-antiquark and an up-antiquark. Other notations can be similarly understood. The transport equations for down-quark matter, up-antiquark matter and down-antiquark matter can be established just as the equation for up-quark matter.

We take the following replacement in the transport equations

$$f_{g1}f_{g2}(1+f_{g3})(1+f_{g4}) - f_{g3}f_{g4}(1+f_{g1})(1+f_{g2}) \to f_{g1}f_{g2} - f_{g3}f_{g4},$$

$$f_{g1}f_{q2}(1+f_{g3})(1-f_{q4}) - f_{g3}f_{q4}(1+f_{g1})(1-f_{q2}) \to f_{g1}f_{q2} - f_{g3}f_{q4},$$

$$f_{g1}f_{g2}f_{g3}(1+f_{g4})(1+f_{g5})(1+f_{g6}) - f_{g4}f_{g5}f_{g6}(1+f_{g1})(1+f_{g2})(1+f_{g3}) \to f_{g1}f_{g2}f_{g3} - f_{g4}f_{g5}f_{g6},$$

$$f_{g1}f_{g2}f_{q3}(1+f_{g4})(1+f_{g5})(1-f_{q6}) - f_{g4}f_{g5}f_{q6}(1+f_{g1})(1+f_{g2})(1-f_{q3}) \to f_{g1}f_{g2}f_{q3} - f_{g4}f_{g5}f_{q6},$$

$$f_{g1}f_{q2}f_{q3}(1+f_{g4})(1-f_{q5})(1-f_{q6}) - f_{g4}f_{q5}f_{q6}(1+f_{g1})(1-f_{q2})(1-f_{q3}) \to f_{g1}f_{q2}f_{q3} - f_{g4}f_{q5}f_{q6},$$

$$f_{q1}f_{g2}(1-f_{q3})(1+f_{g4}) - f_{q3}f_{g4}(1-f_{q1})(1+f_{g2}) \to f_{q1}f_{g2} - f_{q3}f_{g4},$$

$$f_{q1}f_{q2}(1-f_{q3})(1-f_{q4}) - f_{q3}f_{q4}(1-f_{q1})(1-f_{q2}) \to f_{q1}f_{q2} - f_{q3}f_{q4},$$

$$f_{q1}f_{g2}f_{g3}(1-f_{q4})(1+f_{g5})(1+f_{g6}) - f_{q4}f_{g5}f_{g6}(1-f_{q1})(1+f_{g2})(1+f_{g3}) \to f_{q1}f_{g2}f_{g3} - f_{q4}f_{g5}f_{g6},$$

$$f_{q1}f_{q2}f_{g3}(1-f_{q4})(1-f_{q5})(1+f_{g6}) - f_{q4}f_{q5}f_{g6}(1-f_{q1})(1-f_{q2})(1+f_{g3}) \to f_{q1}f_{q2}f_{g3} - f_{q4}f_{q5}f_{g6},$$

$$f_{q1}f_{q2}f_{q3}(1-f_{q4})(1-f_{q5})(1-f_{q6}) - f_{q4}f_{q5}f_{q6}(1-f_{q1})(1-f_{q2})(1-f_{q3}) \to f_{q1}f_{q2}f_{q3} - f_{q4}f_{q5}f_{q6}.$$

Then, any Monte Carlo method can be applied to solve the equations. We generate 1500 gluons, 250 up-quarks, 250 down-quarks, 250 up-antiquarks and 250 down-antiquarks within $-0.3 < z < 0.3$ fm in the longitudinal direction and $r < 6.4$ fm in the transverse direction from HIJING for central Au-Au collisions at $\sqrt{s_{NN}} = 200$ GeV. The number of partons moving in the longitudinal direction is much larger than the number of partons moving in the transverse direction. The anisotropically distributed partons undergo elastic 2-to-2 scattering [14, 15] and elastic 3-to-3 scattering to eventually become isotropic in momentum space (i.e. become a thermal state) at a time. But from gluon matter to quark matter the time is different, i.e. the thermalization time of gluon matter differs from the thermalization time of quark matter. At $t = 0.52$ fm/c the gluon momentum distribution meets the Jüttner distribution

$$f_g(\vec{p}) = \frac{\lambda_g}{e^{|\vec{p}|/T} - \lambda_g}, \tag{3}$$

with the temperature $T = 0.52$ GeV and the gluon fugacity $\lambda_g = 0.328$ that label the thermal state of gluon matter. The corresponding thermalization time is 0.32 fm/c. At $t = 0.86$ fm/c the quark momentum distribution meets the Jüttner distribution

$$f_q(\vec{p}) = \frac{\lambda_q}{e^{|\vec{p}|/T} + \lambda_q}, \tag{4}$$

with $T = 0.46$ GeV and the quark fugacity $\lambda_q = 0.143$ that label the thermal state of quark matter. The corresponding thermalization time is 0.66 fm/c. The thermalization time of quark matter is larger than the one of gluon matter.

4. Conclusions

The elastic 2-to-2 scattering and the elastic 3-to-3 scattering lead to the rapid creation of temperature or the early thermalization of quark-gluon matter. Solutions of the transport equations offer the thermalization times and the thermal states of gluon matter and quark matter. Gluon matter thermalizes faster than quark matter.

Acknowledgments
This work was supported by the National Natural Science Foundation of China under grant no 11175111.

References

[1] Xu X-M 2006 *Eur. Phys. J.* A **29** 39
[2] Ackermann K H *et al* (STAR Collaboration) 2001 *Phys. Rev. Lett.* **86** 402
[3] Kolb P F, Huovinen P, Heinz U and Heiselberg H 2001 *Phys. Lett.* B **500** 232
[4] Teaney D, Lauret J and Shuryak E V 2001 A hydrodynamic description of heavy ion collisions at the SPS and RHIC *Preprint* nucl-th/0110037
[5] Shuryak E 1992 *Phys. Rev. Lett.* **68** 3270
[6] Wong S M H 1996 *Phys. Rev.* C **54** 2588
[7] Shin G R and Müller B 2003 *J. Phys. G: Nucl. Part. Phys.* **29** 2485
[8] Blaizot J-P, Gelis F, Liao J, McLerran L and Venugopalan R 2012 Thermalization and Bose-Einstein condensation in overpopulated plasmas *Preprint* arXiv:1210.6838
[9] Xu X-M, Sun Y, Chen A-Q and Zheng L 2004 *Nucl. Phys.* A **744** 347
[10] Xu X-M, Peng R and Weber H J 2005 *Phys. Lett.* B **629** 68
[11] Xu X-M, Shen Z-Y, Ye Z-C and Xu W-J 2013 *Phys. Rev.* C in press
[12] Xu X-M, Ma C-C, Chen A-Q and Weber H J 2007 *Phys. Lett.* B **645** 146
[13] Xu X-M and Xu L-S 2010 *J. Phys. G: Nucl. Part. Phys.* **37** 115003
[14] Cutler R and Sivers D 1978 *Phys. Rev.* D **17** 196
[15] Combridge B L, Kripfganz J and Ranft J 1977 *Phys. Lett.* B **70** 234

Journal of Physics: Conference Series **458** (2013) 012022 doi:10.1088/1742-6596/458/1/012022

Cold nuclear matter effect measured with high p_{T} hadrons and jets in 200GeV d+Au collisions in PHENIX

Takao Sakaguchi, for the PHENIX collaboration
Physics Department, Brookhaven National Laboratory, Upton, NY 11973, USA
E-mail: takao@bnl.gov

Abstract. High p_{T} π^0 and η as well as jets are measured using high statistics d+Au collision data collected in RHIC Year-2008 run. Both $R_{d\mathrm{A}}$ and R_{cp} for three observables are found to be very consistent each other within quoted systematic and statistical uncertainties. It was found that the $R_{d\mathrm{A}}$ is strongly centrality dependent as opposed to the expectations from theoretical models. An explanation of the centrality dependence from the experimental point of view is presented.

1. Introduction
It has been of interest that whether or not the initial hard scattering process is modified in relativistic heavy ion collisions, after the suppression of high transverse momentum (p_{T}) hadron yields was discovered in Au+Au collisions at $\sqrt{s_{NN}}$=130 GeV in RHIC Year-2000 run [1]. The first d+Au collision experiment was carried out in RHIC Year-2003 run and confirmed that the suppression attributes to a final state interaction of hard scattered partons with a medium created in Au+Au collisions [2]. After these results were published, many theoretical works have been dedicated to precisely determine the initial state effect that affects to the evaluation of the final state effect such as parton energy loss. Figure 1 shows a nuclear modification (R_i^A(x)) to a free proton PDF from the CTEQ6.1M set [3] in the $\overline{MS}$ scheme, where i denotes the parton flavor [4]. The hard scattering process is factorized into a fragmentation function, a reaction cross-section, and a parton distribution function (PDF). Since the d+Au collisions will not produce a bulk hot dense medium, the yield difference of the hard scattering probe in d+Au and $p+p$ collisions are predominantly from the modification of PDF. Therefore, the nuclear modification factor $R_{d\mathrm{A}}$ ($\equiv (dN_{d\mathrm{A}}/dydp_T)/(\langle T_{d\mathrm{A}}\rangle d\sigma_{pp}/dydp_T)$) is expected to reflect the modification of PDF. Eskola and companies tried to construct the centrality dependent nuclear PDF using the previously published π^0 $R_{d\mathrm{A}}$ result as shown in Figure 2 [4]. Obviously, the statistical uncertainty is too large to observe the centrality dependence of the nuclear PDF. We analyzed another d+Au collision data set collected in RHIC Year-2008 in order to tackle this interesting subject.

2. Measurements of π^0, η, and jets in PHENIX
The schematic view of the PHENIX detector [5] in RHIC Year-2008 is shown in Figure 3. PHENIX sampled an integrated luminosity of 80 nb^{-1} during that run, a factor of 30 increase over

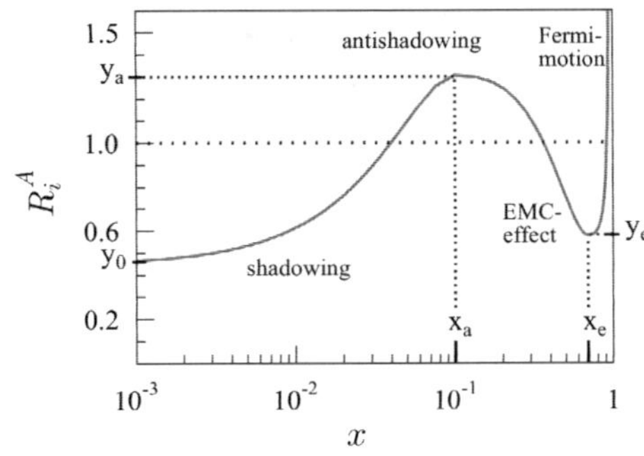

Figure 1. Nuclear modification ($R_i^A(x)$) to a free proton PDF in a nucleus.

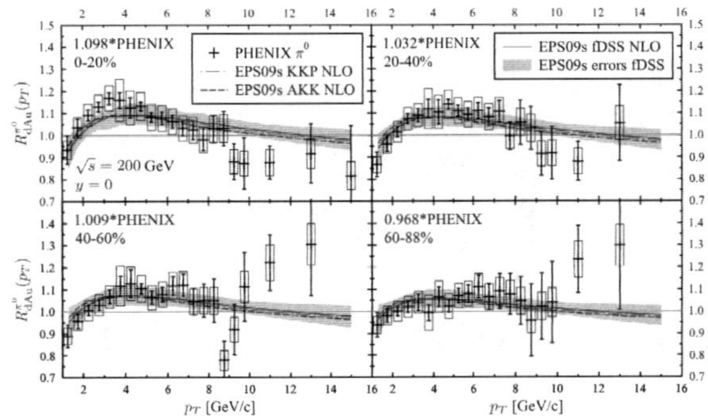

Figure 2. Comparison of measured π^0 R_{dA} and a theoretical model.

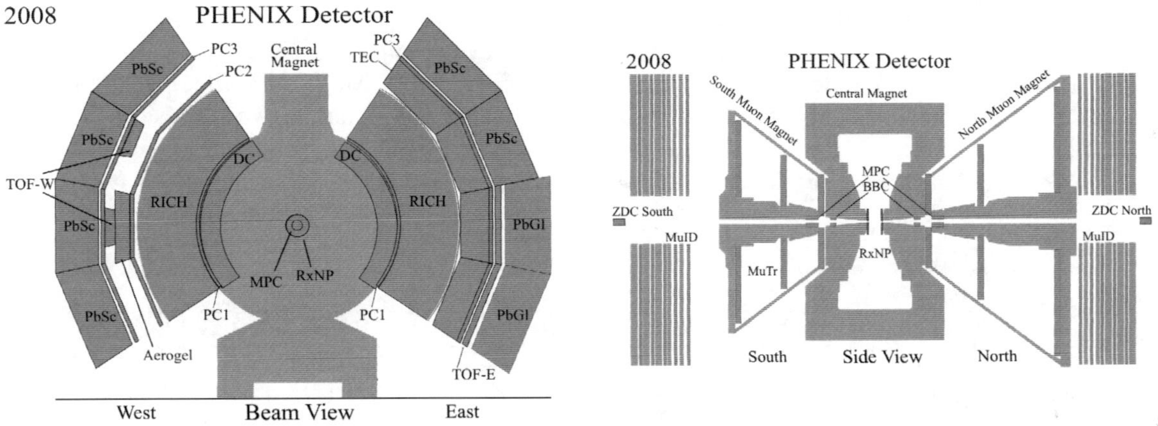

Figure 3. PHENIX detector in RHIC Year-2008 run.

the RHIC Year-2003 d+Au data set. The minimum bias events are triggered by a coincidence of the signals from the beam-beam counters (BBC) located at the south and north side of the detector system ($3.1 < |\eta| < 3.9$). We define the forward (positive) rapidity as the Au-going side (pointing to the south BBC). The collision centrality is determined using the charge signal from the south BBC. The underlying average number of binary nucleon-nucleon collisions (N_{coll}) and the number of participant nucleons (N_{part}) are estimated with a Glauber model Monte Carlo simulation. Then the result is combined with a negative binomial distribution (NBD) with parameters (μ, k) to model the charge distribution from the south BBC. Figure 4 shows the charge distribution in the south BBC (left) and the decomposed N_{coll} distributions for each centrality (right), respectively. Due to the auto-correlation between having a high p_T particle in the central PHENIX spectrometer and the charge multiplicity in the BBC, an additional correction factor is calculated. This is done with the same Glauber Monte Carlo simulation used for the calculation of N_{coll}.

π^0's and η's are reconstructed through 2γ's measured with eight electromagnetic calorimeters (EMCal) [6, 7]. The EMCal are located most outside of the central arm detector system, which covers an acceptance of $|\eta| < 0.35$ and a half azimuth. The contribution of random $\gamma\gamma$ combination is estimated using the mixed event technique, and subtracted from the measured foreground distribution. The drift chamber (DC) and pad chamber (PC) are also used to veto

Journal of Physics: Conference Series **458** (2013) 012022 doi:10.1088/1742-6596/458/1/012022

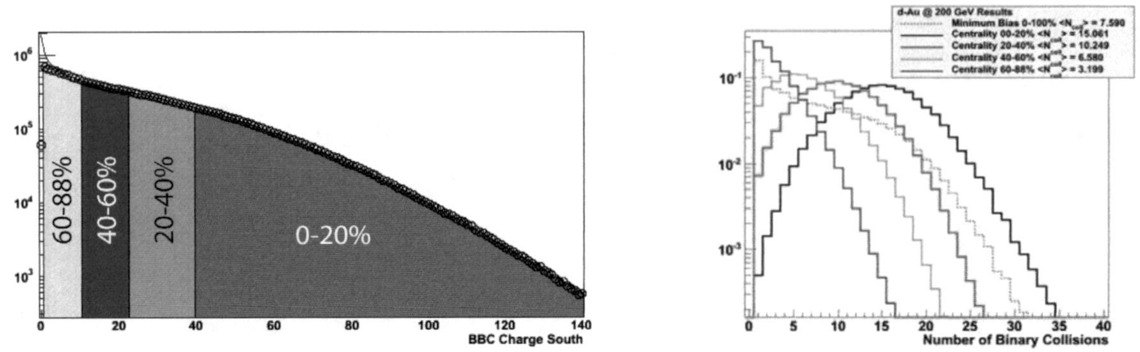

Figure 4. BBC south charge (left) and N_{coll} (right) distributions for each centrality obtained from decomposition of the BBC charge distribution.

charged particles including charged hadrons and electrons. The jets are reconstructed using neutral electromagnetic clusters measured in EMCal and charged particles measured in DC and PC. Those clusters/particles with reconstructed p_{T} of $p_{\mathrm{T}} > 400\,\mathrm{MeV}/c$ are included in the jet reconstruction procedure. We employed Gaussian filtering algorithm with a filter width of σ=0.3. This algorithm was also used in $p+p$ and Cu+Cu collision analysis at PHENIX, and cross-checked with anti-k_{T} algorithm. A mild underlying events in d+Au collisions shifts the

Table 1. Systematic uncertainties for π^0, η, and jet p_{T} measurement.

	source	type	5 GeV/c	10 GeV/c	15 GeV/c	20 GeV/c
π^0	Peak extraction	B	2 %	2 %	2 %	2 %
	Acceptance	C	2.5 %	2.5 %	2.5 %	2.5 %
	PID efficiency	B	7 %	8 %	8.5 %	9 %
	Energy scale	B	7.5 %	8 %	8 %	8 %
	Photon conversion	C	2 %	2 %	2 %	2 %
	Cluster merging	B	0 %	0 %	8 %	18 %
	source	type	5 GeV/c	10 GeV/c	15 GeV/c	20 GeV/c
η	Peak extraction	B	2 %	2 %	2 %	-
	Acceptance	C	2.5 %	2.5 %	2.5 %	-
	PID efficiency	B	7 %	8 %	8.5 %	-
	Energy scale	B	11 %	12 %	12 %	-
	Photon conversion	C	2 %	2 %	2 %	-
	source	type	9-12 GeV/c	12-15 GeV/c	15-20 GeV/c	20+ GeV/c
jets	Trigger efficiency	B	5-8 %	5-8 %	5 %	5 %
	Unfolding	B	4-10 %	4-10 %	4-6 %	4-6 %
	Residual fake rate	B	5 %	5 %	5 %	5 %
	Acceptance	C	3 %	3 %	3 %	3 %
	Vertex dependence	C	1 %	1 %	1 %	1 %

measured p_T to a higher value than the true one. This effect is evaluated and corrected for by embedding jets into real events. A small residual fake rate (<5 %) was observed for p_{T} >9 GeV/c and was accounted in a systematic uncertainty. The systematic uncertainties for π^0, η, and jets p_T measurements are summarized in Table 1. The type-B errors are the errors correlated from

p_T to p_T. The type-C errors are the errors of overall normalizations.

3. Results

The p_{T} spectra of π^0 and η are shown in Figures 5 and 6, respectively. With the better

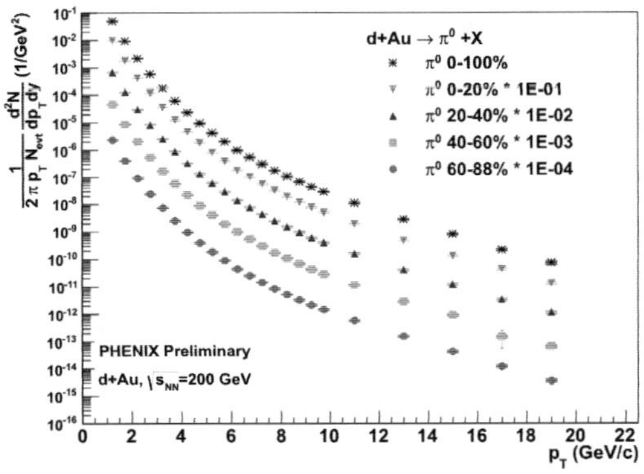

Figure 5. p_{T} spectra of π^0 measured in d+Au collisions.

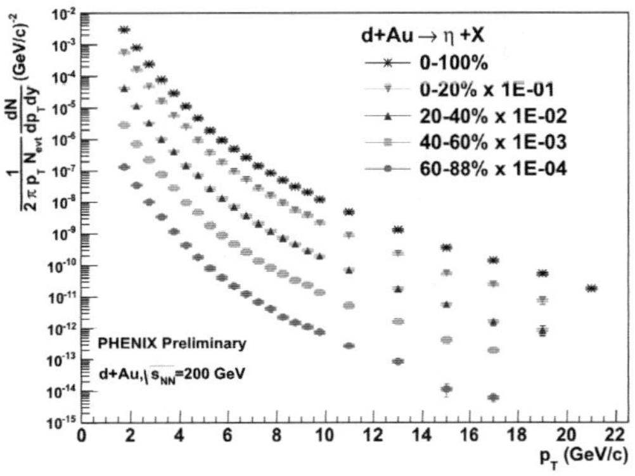

Figure 6. p_{T} spectra of η measured in d+Au collisions.

statistics in RHIC Year-2008 run, we extended p_{T} reach by more than 5 GeV/c. As going to higher p_{T} ($p_{\mathrm{T}} >\sim$ 12 GeV/c), the opening angle of $\pi^0 \rightarrow \gamma\gamma$ becomes so small that they can not be resolved in EMCal because of its limited granularity. It is called as the cluster merging effect. An η has a four times larger opening angle ($\theta_{\gamma\gamma} \propto M/p_{\mathrm{T}}$) and thus the merging of two γ's does not happen until ~50 GeV/c. Due to this effect, the ratio of raw η to π^0 yield increases at higher p_{T} [7]. The p_{T} spectra of jets is shown in Figure 7. Jets are reconstructed

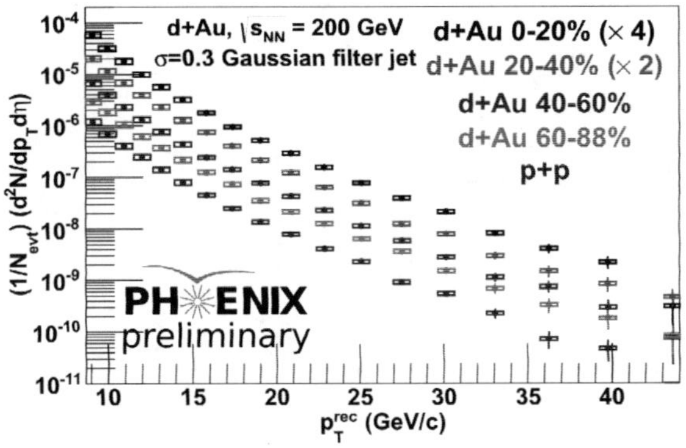

Figure 7. p_{T} spectra of jets measured in d+Au collisions.

from multi-particles, resulting in reaching to much higher p_{T}, as high as 45 GeV/c, with the same integrated luminosity. We computed the $R_{d\mathrm{A}}$ for π^0, η, and jets as shown in Figure 8.

Although jets and single identified hadrons can not be directly compared at the same p_{T}, they are agreeing each other very well within quoted uncertainties. For reader's sake, it may be useful to note that the mean $z(\equiv p_{\mathrm{T}}^{had}/p_{\mathrm{T}}^{jet})$ for π^0 and η is ~0.7. It was found that the $R_{d\mathrm{A}}$ is strongly centrality dependent; the most central collisions show a small suppression, while the most peripheral collisions show a sizable enhancement, compared to the expectation from $p+p$ collisions. It should also be noted that the enhancement in the peripheral collisions increases as

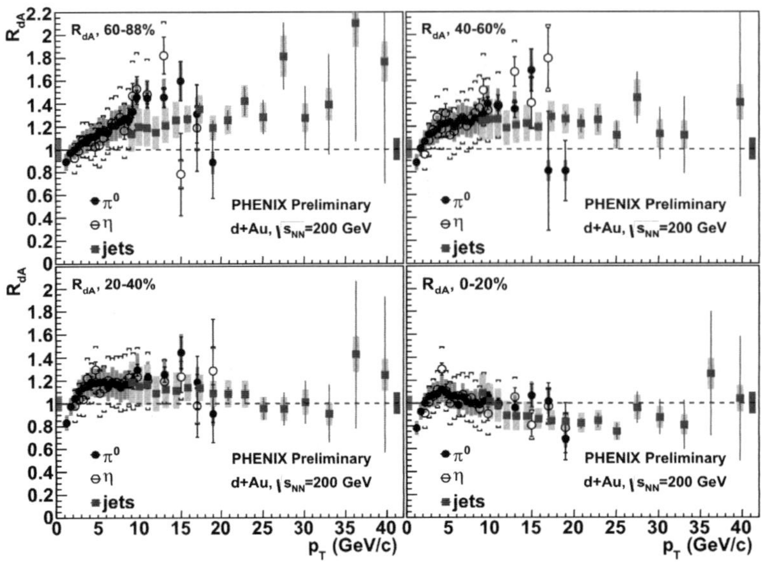

Figure 8. R_{dA} for π^0, η and jets in d+Au collisions for various centralities.

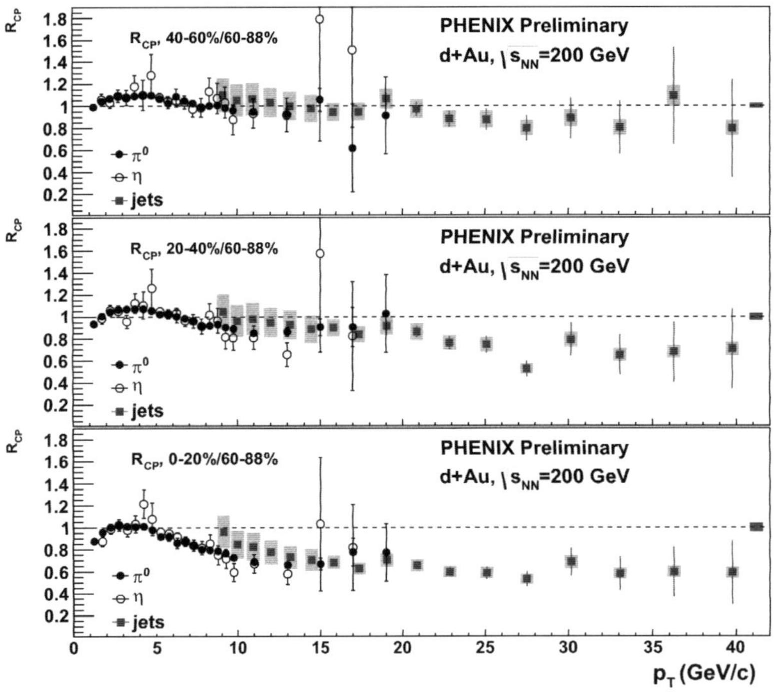

Figure 9. R_{cp} for π^0, η and jets in d+Au collisions for various centralities. The 60-88 % centrality is taken as the denominator.

going to higher p_T. This trend shows that the nuclear PDF may be strongly dependent of impact parameters as opposed to what was predicted by theoretical models. In order to investigate the suppression/enhancement trend with reduced systematic uncertainties, the ratio of the yield in central to peripheral collisions (R_{cp}) is computed and shown in Figure 9. In this ratio, the error from $p+p$ baseline measurement will be excluded. It was found that the π^0, η, and jet R_{cp} show the same trend and are consistent each other within even smaller uncertainties. The decreasing trend as going to higher p_T is consistent with the fact that the R_{dA} is enhanced in peripheral collisions and suppressed in central collisions.

4. Global feature in d+Au collisions

Before one concludes that the nuclear PDF is strongly impact parameter dependent, one may want to revisit some global features in d+Au collisions such as triggering or centrality definition. In any collisions, one has to generate a trigger using signals from detectors whose efficiencies are not 100 % in reality. The efficiency of a trigger is often particle multiplicity dependent. Therefore, when triggering minimum bias events in small collision system such like d+Au, the peripheral collisions are most biased. In Figure 10, we tried to sketch the collision dynamics for events with low and high p_T observables in peripheral collisions. As explained before, we trigger

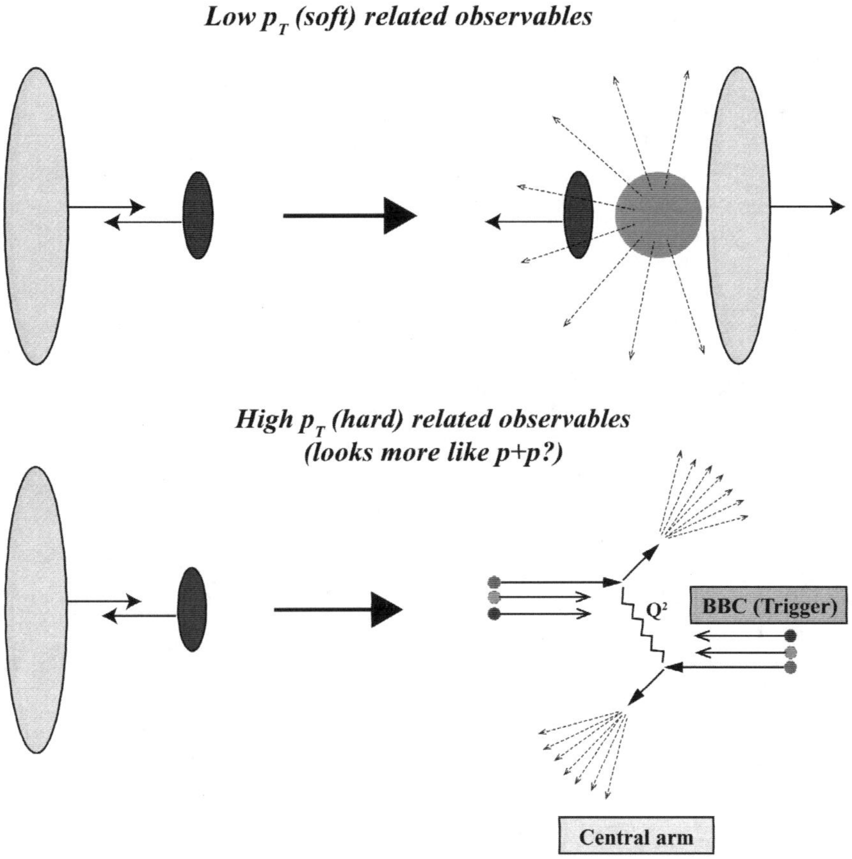

Figure 10. Collision dynamics for events with low p_T (top) and high p_T (bottom) observables.

minimum bias events and define their centrality classes based on the charge signals from the south side BBC (Au-going direction). As seen in Figure 10, the soft-process-dominated events emit

Journal of Physics: Conference Series **458** (2013) 012022 doi:10.1088/1742-6596/458/1/012022

particles isotropically. Therefore, the BBC charge is proportional to the centrality. However, in the case that hardest jets are produced, less energy will be available for soft production at high η, e.g., at the BBC location. This leads to a possible p_T dependence of the trigger efficiency in most peripheral collisions. We performed a simulation study for the simplest case, $p + p$ collisions, using PYTHIA [8] and AMPT [9] event generators to see this effect. The left side of the Figure 11 shows the number of BBC hits as a function of jet E_T obtained from a PYTHIA simulation. Since the collision system is symmetric, we took the BBC hit distributions from

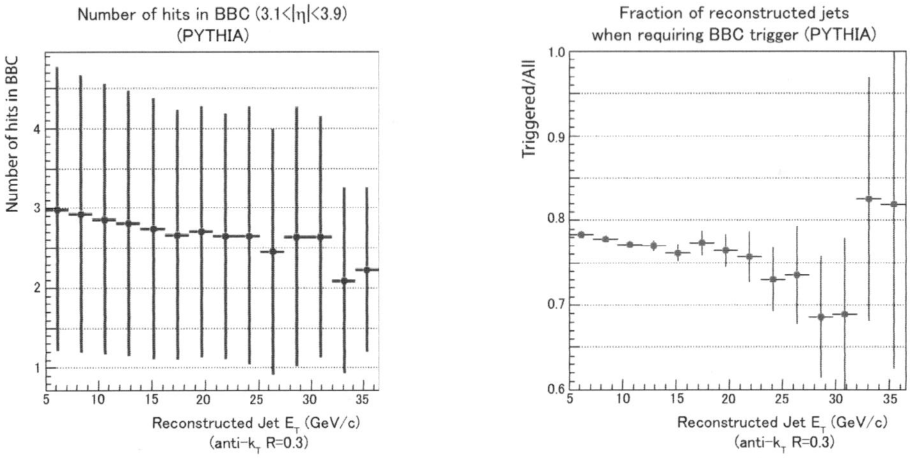

Figure 11. PYTHIA simulation of the number of BBC hits (left) and the fraction of jets seen in central arm when requiring BBC hits (right).

both south and north sides. The error bars show RMS of the distributions. As going to higher E_T, the number of hits in BBC decreases, and the trigger efficiency of events is reduced as shown in the right side of the same figure. At $\sim$30 GeV/c the trigger efficiency becomes 70 %. Figure 12 shows a AMPT simulation of $p+p$ minimum bias collisions. In this figure, the number

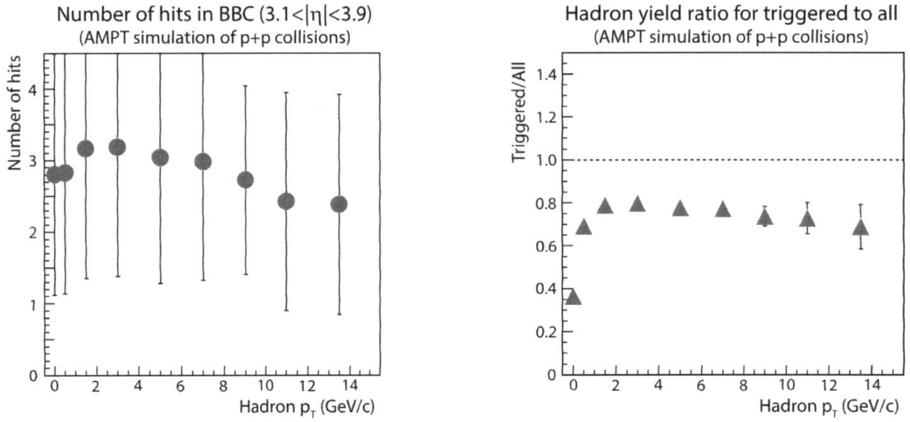

Figure 12. AMPT simulation of the number of BBC hits (left) and the fraction of π^0's seen in central arm when requiring BBC hits (right).

of BBC hits and the trigger efficiency are plotted against the p_T of π^0's. The yield of π^0's starts decreasing at ~4 GeV/c, and reaches to ~30 % deficit at 14 GeV/c. The trend and magnitude are found to be consistent with what PYTHIA simulation showed, by taking the difference of the energy scale between jets and π^0's into account. This implies a trigger bias is expected from a simple hard scattering kinematics, and manifests both in jet E_T and single hadron p_T. We are now looking for a same effect in the d+Au collision case using the simulations.

5. Summary and Outlook

High p_T π^0 and η as well as jets are measured using high statistics d+Au collision data collected in RHIC Year-2008 run. Both $R_{d\mathrm{A}}$ and R_cp for three observables are found to be very consistent each other within quoted systematic and statistical uncertainties. It was found that the $R_{d\mathrm{A}}$ is strongly centrality dependent as opposed to the expectations from theoretical models. A possible explanation is given based on the kinematics of hard scattering process. A simulation study in $p+p$ collision case using PYTHIA and AMPT event generators is presented for supporting the explanation. The study for d+Au collisions is on-going.

Recently, a dedicated workshop on d+A and p+A collisions at RHIC and LHC was organized [10]. In the workshop, it was pointed out that the collision dynamics in d+Au collisions is a sequence of 2→2 hard processes, therefore, the statistical treatment of collisions, such like using N_{coll} may lead a problem. We expect a rich physics in this small collision system.

References

[1] K. Adcox *et al.* [PHENIX Collaboration], Phys. Rev. Lett. **88**, 022301 (2002) [nucl-ex/0109003].
[2] S. S. Adler *et al.* [PHENIX Collaboration], Phys. Rev. Lett. **91**, 072303 (2003) [nucl-ex/0306021].
[3] D. Stump, J. Huston, J. Pumplin, et al., JHEP **0310**, 046 (2003) [hep-ph/0303013].
[4] K. J. Eskola, H. Paukkunen and C. A. Salgado, JHEP **0904**, 065 (2009) [arXiv:0902.4154 [hep-ph]].
[5] K. Adcox *et al.* [PHENIX Collaboration], Nucl. Instrum. Meth. A **499**, 469 (2003).
[6] A. Adare *et al.* [PHENIX Collaboration], Phys. Rev. C **87**, 034911 (2013) [arXiv:1208.2254 [nucl-ex]].
[7] A. Adare *et al.* [PHENIX Collaboration], Phys. Rev. C **82**, 011902 (2010) [arXiv:1005.4916 [nucl-ex]].
[8] T. Sjostrand, S. Mrenna and P. Z. Skands, JHEP **0605**, 026 (2006) [hep-ph/0603175].
[9] Z. -W. Lin, C. M. Ko, B. -A. Li, B. Zhang and S. Pal, Phys. Rev. C **72**, 064901 (2005) [nucl-th/0411110].
[10] http://www.bnl.gov/jqr2013/index.php

Journal of Physics: Conference Series **458** (2013) 012023 doi:10.1088/1742-6596/458/1/012023

Long-range correlations in proton-lead collisions at $\sqrt{s_{\mathrm{NN}}} = 5.02$ TeV from ATLAS

Jiangyong Jia on behalf of the ATLAS Collaboration
Chemistry Department, Stony Brook University, Stony Brook, NY 11794, USA
Physics Department, Brookhaven National Laboratory, Upton, NY 11796, USA

E-mail: `jjia@bnl.gov`

Abstract. Two-particle correlations in relative azimuth $\Delta\phi$ and relative pseudorapidity $\Delta\eta$ are studied in p+Pb collisions at $\sqrt{s_{\mathrm{NN}}} = 5.02$ TeV with the ATLAS detector at LHC. The correlations are studied as a function of charged particle p_{T} and the collision $\Sigma E_{\mathrm{T}}^{\mathrm{Pb}}$ summed over $3.1 < \eta < 4.9$ in the direction of the Pb beam. After subtracting the known sources of correlations such as dijets, resonances and momentum conservation, estimated using events with low $\Sigma E_{\mathrm{T}}^{\mathrm{Pb}}$, the resulting correlations exhibit a $\Delta\phi$ modulation that is flat in $\Delta\eta$ out to $|\Delta\eta| = 5$. The modulation is comparable in magnitude to similar modulations observed in heavy ion collisions, and can be described by a $1 + 2c_2 \cos 2\Delta\phi + 2c_3 \cos 3\Delta\phi$ function over $0.5 < p_{\mathrm{T}} < 7$ GeV in broad ranges of $\Sigma E_{\mathrm{T}}^{\mathrm{Pb}}$. The correlation analysis is repeated for event classes defined by the number of reconstructed charged particles $N_{\mathrm{ch}}^{\mathrm{rec}}$. This analysis gives nearly the same result as the analysis based on $\Sigma E_{\mathrm{T}}^{\mathrm{Pb}}$ for the long-range correlation at the near-side ($\Delta\phi \sim 0$), but leads to biases in the long-range correlations at the away-side ($\Delta\phi \sim \pi$). HIJING simulation suggests that this bias is mainly associated with the contributions from dijets which are correlated strongly with the $N_{\mathrm{ch}}^{\mathrm{rec}}$.

1. Introduction

Recent studies of two-particle angular correlations (2PC) in relative azimuthal angle, $\Delta\phi = \phi_a - \phi_b$, and relative pseudorapidity, $\Delta\eta = \eta_a - \eta_b$, in $p+p$, p+A and A+A collisions have generated considerable interest in both high energy and heavy-ion physics communities. In events with very large multiplicity, an enhanced production of particle pairs is observed at small relative azimuthal angle $\Delta\phi \sim 0$, and this excess extends over a long-range in $\Delta\eta$, i.e. extends to at least $|\Delta\eta| = 5$. This so called "ridge" phenomena was first observed in Au+Au or Pb+Pb collisions at RHIC and LHC [1–6], but was found to also exist in high-multiplicity $p+p$ collisions [7] and p+Pb collisions [8–10] (see Fig. 1). The "ridge" is a reflection of QCD dynamics in high-density system, where many hadrons are produced in a small volume. However, the effective mechanism may depend on the system size. In heavy-ion collisions, the ridge is commonly believed to be the result of collective flow of the produced matter in the final state. That flow has been successfully described by the relativistic hydrodynamic calculations [11]. In high-multiplicity $p+p$ and p+Pb collisions, however, the mechanism responsible for the ridge is not yet clear. Some models interpret the ridge as resulting from finial-state effects via the hydrodynamic picture [12–15], while others argue that the ridge is due to initial-state gluon saturation effects [16, 17].

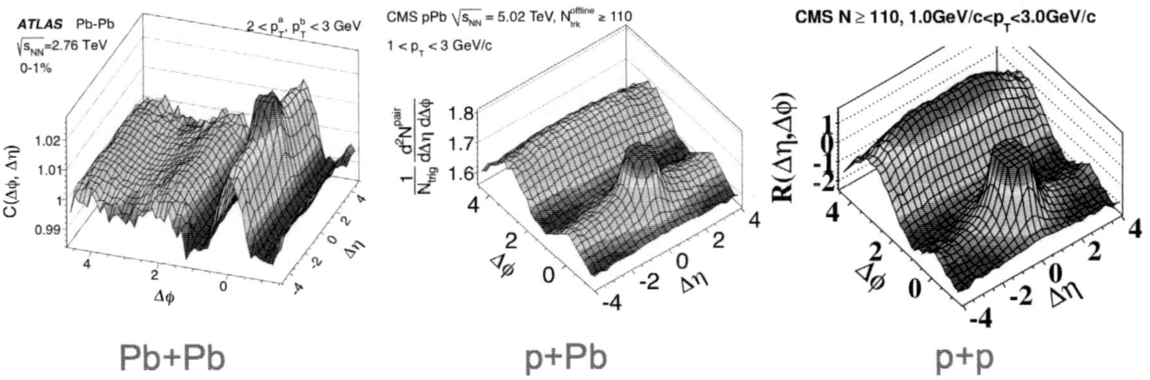

Figure 1. Examples of two-particle correlations in $\Delta\phi$ and $\Delta\eta$ in Pb+Pb [4], p+Pb [8] and $p+p$ [7] collisions.

One striking feature of the long-range correlation in Au+Au/Pb+Pb collisions is that the ridge is not only restricted to $\Delta\phi \sim 0$ (near-side), but also appears around $\Delta\phi \sim \pi$ (away-side). The detailed shape of the ridge in $\Delta\phi$ depends on the p_{T} selection of particles in the pair, and has been described in the framework of harmonic flow [4–6]. If the ridge in high-multiplicity $p+p$ or p+Pb collisions is due to a similar mechanism as that responsible for the ridge in large systems such as Pb+Pb, one would also expect the presence of a ridge at the away-side. The extraction of the away-side ridge in $p+p$ or p+Pb collisions, however, is complicated by the correlations arising from known sources, such as dijets, resonance and momentum conservation, collectively referred to as "recoil" [4]. In a recent publication [10], ATLAS has developed a data-driven procedure to separate the contribution from away-side recoil and the away-side long-range correlations in p+Pb collisions. This subtraction permits a detailed study of the long-range correlation over the full $\Delta\phi$ range, as well as the extraction of the associated harmonic coefficients as a function of p_{T}, η, and event multiplicity or activity.

This proceedings provides details of the ATLAS p+Pb ridge results at Ref. [10] with an extended set of plots given at Ref. [18]. We first describe how ATLAS classifies the activity of the events, and then discuss the recoil-removal procedure and explain why it can be used to expose the away-side ridge in p+Pb collisions. We review the properties of the ridge as a function of event multiplicity, p_{T}, $\Delta\eta$ and charge combination. We present the multiplicity and p_{T} dependence of the harmonic coefficients obtained via a Fourier analysis of the $\Delta\phi$ distribution. Finally, this proceedings comments on potential auto-correlation biases associated with different definitions of event classes.

2. Event and track selections

The result is based on p+Pb collisions corresponding to 1 μb^{-1}, or about 2 million events, at $\sqrt{s_{\mathrm{NN}}} = 5.02$ TeV recorded during a short run in September 2012. The 2PC analysis is based on charged particle tracks with $p_{\mathrm{T}} > 0.4$ GeV and $|\eta| < 2.5$, reconstructed by the ATLAS inner detector [19] with an overall efficiency of about 70%. The events are divided into 11 narrow activity classes based on the total transverse energy, $\Sigma E_{\mathrm{T}}^{\mathrm{Pb}}$, measured by the forward calorimeter over $3.1 < \eta < 4.9$ in the Pb-going direction (see Fig. 2). Four larger intervals, $\Sigma E_{\mathrm{T}}^{\mathrm{Pb}} > 80$ GeV, $55 < \Sigma E_{\mathrm{T}}^{\mathrm{Pb}} < 80$ GeV, $25 < \Sigma E_{\mathrm{T}}^{\mathrm{Pb}} < 55$ GeV and $\Sigma E_{\mathrm{T}}^{\mathrm{Pb}} < 20$ GeV, are used for detailed studies of the 2PC as a function of p_{T}. Event activity classes are also defined by selecting on the charged particle track multiplicity N_{ch} in the detector. However, this approach is found to introduce auto-correlation bias to the 2PC measurements (see Sec. 6). On the other hand, the average

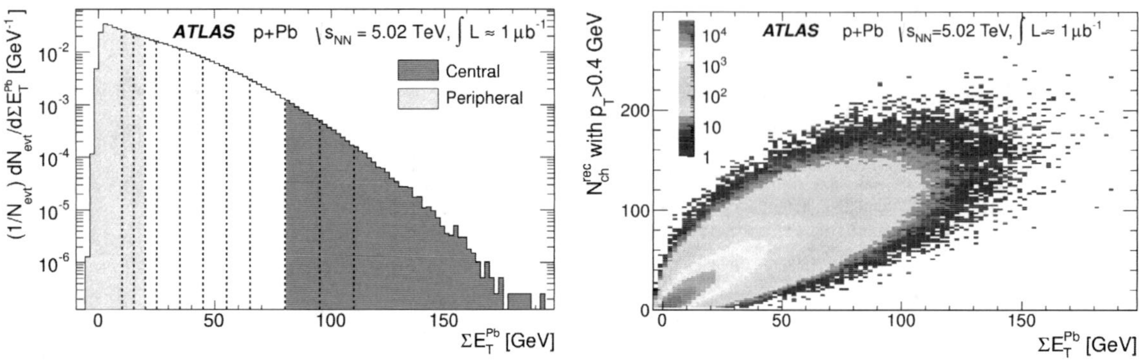

Figure 2. (left) Distribution of $\Sigma E_{\mathrm{T}}^{\mathrm{Pb}}$ for minimum-bias p+Pb events. Vertical lines indicate the boundaries of the event activity classes [10]. (right) Correlation between $\Sigma E_{\mathrm{T}}^{\mathrm{Pb}}$ and number of reconstructed charged particles with $p_{\mathrm{T}} > 0.4$ GeV and $|\eta| < 2.5$ in the same events [18].

value $\langle N_{\mathrm{ch}} \rangle$, calculated for each $\Sigma E_{\mathrm{T}}^{\mathrm{Pb}}$ event class, would have little auto-correlations with the 2PC measurements, and they are given in Table 1.

$\Sigma E_{\mathrm{T}}^{\mathrm{Pb}}$[GeV]	> 110	95-110	80-95	65-80	55-65	45-55	35-45	25-35
Fraction [%]	0.21	0.45	1.24	3.11	3.99	6.37	9.71	13.80
$\langle \Sigma E_{\mathrm{T}}^{\mathrm{Pb}} \rangle$[GeV]	122.4	101.2	86.4	71.4	59.6	49.7	39.7	29.7
$\langle N_{\mathrm{ch}} \rangle$	183±8	160±7	141±6	123±6	107±5	93.3±4.2	78.8±3.6	63.3±2.9
$\sigma_{N_{\mathrm{ch}}}$	37.0±2.1	33.1±1.9	31.5±1.8	29.6±1.7	27.6±1.6	25.9±1.5	24.1±1.4	21.8±1.2
$\Sigma E_{\mathrm{T}}^{\mathrm{Pb}}$[GeV]	20-25	15-20	10-15	< 10	> 80	55-80	25-55	< 20
Fraction [%]	8.67	10.11	11.98	30.36	1.90	13.47	29.88	52.45
$\langle \Sigma E_{\mathrm{T}}^{\mathrm{Pb}} \rangle$[GeV]	22.4	17.4	12.4	4.9	94.4	64.8	37.3	9.0
$\langle N_{\mathrm{ch}} \rangle$	51.0±2.3	41.8±1.9	31.7±1.5	15.9±0.7	150±7	114±5	74.7±3.4	24.5±1.1
$\sigma_{N_{\mathrm{ch}}}$	19.6±1.1	17.9±1.0	15.7±0.9	11.8±0.7	35.2±2.0	29.4±1.7	26.1±1.5	17.5±1.0

Table 1. A list of $\Sigma E_{\mathrm{T}}^{\mathrm{Pb}}$ classes, the associated percentage of events and average value $\langle \Sigma E_{\mathrm{T}}^{\mathrm{Pb}} \rangle$. For each class, the efficiency corrected average number of charged particles with $p_{\mathrm{T}} > 0.4$ GeV and $|\eta| < 2.5$ $\langle N_{\mathrm{ch}} \rangle$, standard deviation $\sigma_{N_{\mathrm{ch}}}$, and their systematic uncertainties are also listed [18]. Here $\sigma_{N_{\mathrm{ch}}}$ is the estimate of the physical fluctuation of N_{ch} for events selected in a given $\Sigma E_{\mathrm{T}}^{\mathrm{Pb}}$ range, the spread of N_{ch} due to detector inefficiency is much smaller and has been subtracted.

3. Correlation function and recoil subtraction

The two-particle correlation function is constructed as the ratio of the distribution for pairs in the same-event and mixed-event:

$$C(\Delta\phi, \Delta\eta) = \frac{S(\Delta\phi, \Delta\eta)}{B(\Delta\phi, \Delta\eta)}\,, \quad C(\Delta\phi) = \frac{S(\Delta\phi)}{B(\Delta\phi)}, \tag{1}$$

with each particle weighted by the inverse of the reconstruction efficiency. The mixed-pair distribution $B(\Delta\phi, \Delta\eta)$ is designed to remove residual structures due to detector acceptance and occupancy from the $S(\Delta\phi, \Delta\eta)$. The 1D distributions $S(\Delta\phi)$ and $B(\Delta\phi)$ are obtained by integrating $S(\Delta\phi, \Delta\eta)$ and $B(\Delta\phi, \Delta\eta)$, respectively, over $2 < |\Delta\eta| < 5$. The normalization of $C(\Delta\phi, \Delta\eta)$ is chosen such that the $\Delta\phi$-averaged value of $C(\Delta\phi)$ is unity.

Denoting N_a and N_b as the average number of trigger and associated particles per-event for given p_{T} selection, the meaning of the correlation function can be understood as follows. For a perfect detector, the mixed-event distribution describes the rate of combinatorial pairs $\pi B =$

N_aN_b, while the same event pairs contains both the correlated pairs and combinatorial pairs, i.e. $\pi S = J(\Delta\phi) + \xi_{\rm ZYAM} N_a N_b$. Factor $\xi_{\rm ZYAM}$ reflects the magnitude of a flat pedestal, estimated via the Zero-Yield at Minimum (ZYAM) procedure [3,20]. Hence $C(\Delta\phi) = J(\Delta\phi)/(N_aN_b) + \xi_{\rm ZYAM}$, and the correlated component is measured relative to the underlying event background.

Examples of the 2D correlation function are shown in Fig. 3. A clear long-range component in the right panel, on the order of a few percent of the background, is seen in the near-side. A similar long-range correlation, less flat in $\Delta\eta$, is also seen in the away-side. However, a significant fraction of this away-side correlation is due to recoil to the trigger particle, which complicates the extraction of the genuine long-range correlation.

The effects of recoil or momentum conservation, in a system with finite number of particles, have been studied previously [21, 22]. Under very general assumptions, these effects were shown to give a $\cos(\Delta\phi)$ correction to the correlation function:

$$\delta C(\Delta\phi) = -\frac{p_{\rm T}^{\rm a} p_{\rm T}^{\rm b}}{N\langle p_{\rm T}^2\rangle}\cos(\Delta\phi) \propto -\frac{p_{\rm T}^{\rm a} p_{\rm T}^{\rm b}}{N_b\langle p_{\rm T}^2\rangle}\cos(\Delta\phi) \tag{2}$$

where N is the average particle multiplicity, which is proportional to the average number of charged particles N_b in a given associated $p_{\rm T}$ range (typically 0.5-4 GeV in this analysis). In order to remove the recoil effects in p+Pb collisions, ATLAS uses the per-trigger yield defined as:

$$Y(\Delta\phi) = \left(\frac{\int B(\Delta\phi)d\Delta\phi}{\pi N_a}\right) C(\Delta\phi) - b_{\rm ZYAM} \tag{3}$$

For an ideal detector, $\int B(\Delta\phi)d\Delta\phi = N_aN_b$, and Eq. 3 reduces to

$$Y(\Delta\phi) = N_b\left(C(\Delta\phi) - \xi_{\rm ZYAM}\right) = N_bC(\Delta\phi) - b_{\rm ZYAM}, b_{\rm ZYAM} = N_b\xi_{\rm ZYAM}\,. \tag{4}$$

Hence the contribution of the recoil in Eq. 2 to the per-trigger yield is:

$$\delta Y(\Delta\phi) = N_b\delta C(\Delta\phi) \propto -\frac{p_{\rm T}^{\rm a} p_{\rm T}^{\rm b}}{\langle p_{\rm T}^2\rangle}\cos(\Delta\phi). \tag{5}$$

So as long as the values of $\langle p_{\rm T}^2\rangle$ are similar between different event classes, the recoil contribution can be estimated from peripheral events and subtracted from the central events.

Figure 4 overlays the per-trigger yield in various $\Sigma E_{\rm T}^{\rm Pb}$ classes with that for $\Sigma E_{\rm T}^{\rm Pb} < 20$ GeV. The latter has an approximately $-\cos\Delta\phi$ shape, suggesting that it reflects mainly the recoil effects. The subtracted distributions, $\Delta Y(\Delta\phi)$, obtained as:

$$\Delta Y(\Delta\phi) = Y(\Delta\phi)|_{\rm a\ given\ \Sigma E_{\rm T}^{\rm Pb}\ range} - Y(\Delta\phi)|_{\Sigma E_{\rm T}^{\rm Pb}<20{\rm GeV}}, \tag{6}$$

are nearly symmetric around $\Delta\phi = \pi/2$ and can be well described by a $a_0 + 2a_2\cos 2\Delta\phi$ function. Including a $2a_3\cos 3\Delta\phi$ term only slightly improves the agreement with the data. This observation indicates that the long-range component of the two-particle correlations can be approximately described by a recoil contribution plus a $\Delta\phi$-symmetric "double-ridge" component. The magnitude of the double-ridge is comparable to the recoil contribution for events with $\Sigma E_{\rm T}^{\rm Pb} > 55$ GeV (top 10% of events according to Table 1), but increases to about twice the recoil contribution for events with $\Sigma E_{\rm T}^{\rm Pb} > 110$ GeV (top 0.2% of events).

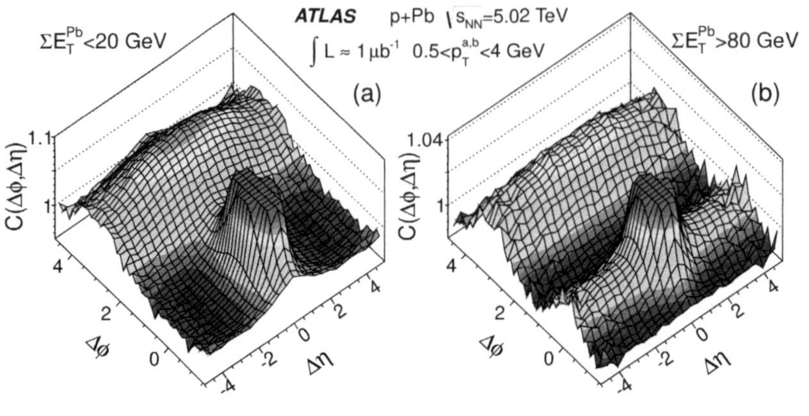

Figure 3. Two-dimensional correlation functions for peripheral events and central events, both with a truncated maximum to suppress jet fragmentation peak around $(\Delta\eta, \Delta\phi) = (0, 0)$ [10].

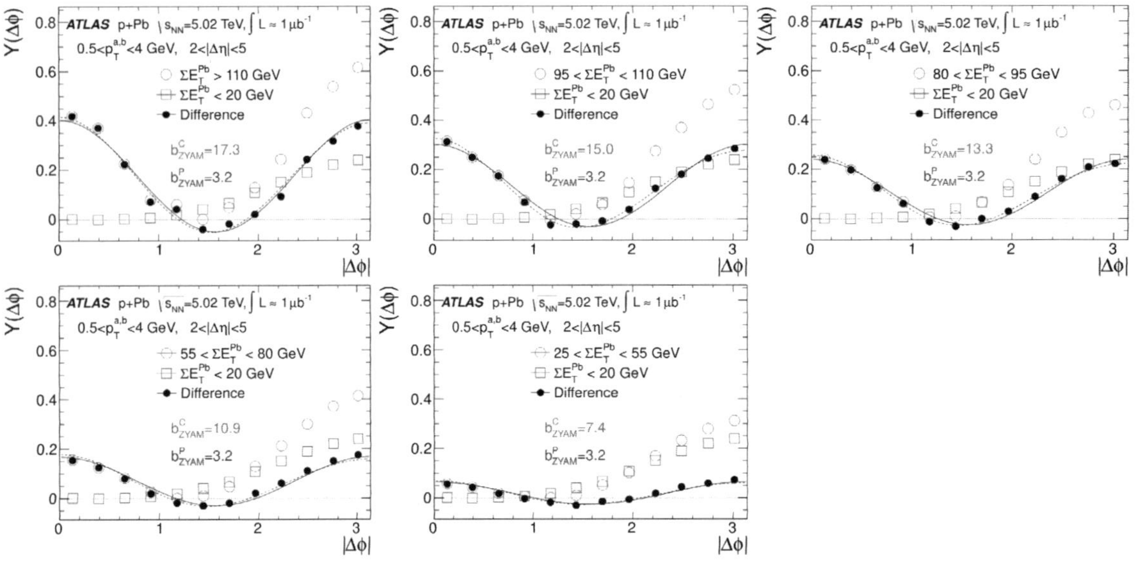

Figure 4. Distributions of per-trigger yield in a given ΣE_T^{Pb} class (open circles), the $\Sigma E_T^{Pb} < 20$ GeV class (open boxes), and their difference (solid circles), together with functions, $a_0 + 2a_2 \cos 2\Delta\phi$ (solid lines) and $a_0 + 2a_2 \cos 2\Delta\phi + 2a_3 \cos 3\Delta\phi$ (dashed lines), obtained via a Fourier decomposition [18]. The values for the ZYAM-determined pedestal levels are indicated on each panel for peripheral (b^{P}_{ZYAM}) and central (b^{C}_{ZYAM}) ΣE_T^{Pb} bins.

4. Properties of the "double-ridge"

The recoil subtraction procedure discussed above is repeated for each $\Delta\eta$ slice, and the resulting 2D or 1D distributions are then converted back into correlation functions, labelled as $C_\Delta(\Delta\phi, \Delta\eta)$ or $C_\Delta(\Delta\phi)$, respectively. The normalization of these distributions are fixed by requiring the average of $C_\Delta(\Delta\phi)$ defined for $2 < |\Delta\eta| < 5$ to be unity. Figure 5 compares the original 2D correlation functions $C(\Delta\phi, \Delta\eta)$ (left panels) with the recoil-subtracted correlation functions $C_\Delta(\Delta\phi, \Delta\eta)$ (right panels) in three ΣE_T^{Pb} classes. An away-side ridge is clearly present in all three recoil-subtracted distributions. This away-side ridge is flat to $|\Delta\eta| = 5$, with a magnitude similar to that for the near-side ridge. Interestingly, the recoil-subtraction procedure also suppress the near-side jet fragmentation peak around $(\Delta\phi, \Delta\eta) \sim (0, 0)$ to less than 10-15% of its original height. This implies that the properties of the double-ridge may be extracted for pairs with a modest $\Delta\eta$ gap, as is done by the ALICE collaboration [9] ($0.8 < |\Delta\eta| < 1.8$) and PHENIX collaboration [23] ($0.5 < |\Delta\eta| < 0.7$).

Figure 6 shows the recoil-subtracted 2D correlation functions separately for same-charge pairs and opposite-charge pairs. Despite the large differences in their residual short-range correlations, the extracted double-ridge are nearly identical.

The results discussed so far are obtained for a broad $p_{\rm T}$ range of 0.5–4 GeV. The double-ridge has also been extracted as a function of $p_{\rm T}$ as shown in Fig. 7. The distributions remain largely symmetric around $\Delta\phi = \pi/2$, however, a significant $\cos 3\Delta\phi$ component becomes apparent at $p_{\rm T}^{\rm a,b} > 2-3$ GeV.

To quantify the symmetry between the near- and away-side ridges, the per-trigger yields are integrated over $|\Delta\phi| < \pi/3$ and $|\Delta\phi| > 2\pi/3$, and plotted in Fig. 8 as a function of $p_{\rm T}^{\rm a}$ with $0.5 < p_{\rm T}^{\rm b} < 4$ GeV. According to Eq. 4, they directly reflect the $p_{\rm T}$ dependence of the fractional contribution of the ridge in the correlation function. The differences of the integrated yields between a given $\Sigma E_{\rm T}^{\rm Pb}$ class and the $\Sigma E_{\rm T}^{\rm Pb} < 20$ GeV class ($\Delta Y_{\rm int}$) are shown in the bottom panels. The values of $\Delta Y_{\rm int}$ show a similar magnitude and $p_{\rm T}^{\rm a}$ dependence between the near-side and away-side: they rise with $p_{\rm T}^{\rm a}$ and reach a maximum around 3–4 GeV. This pattern is visible for the near-side even before subtraction (top-left panel), but is less evident in the away-side before subtraction (top-right panel) due to the dominant contribution of the recoil component. The values of $\Delta Y_{\rm int}$ increase with $\Sigma E_{\rm T}^{\rm Pb}$, but remain symmetric between the near and away-side over the measured $p_{\rm T}^{\rm a}$ range.

5. Harmonic spectra and their factorization behavior

The amplitude of the $\cos n\Delta\phi$ modulation of $\Delta Y(\Delta\phi)$ relative to the underlying event, c_n, for $n = 2, 3$ is estimated using a_n, and the extracted value of $b_{\rm ZYAM}$ for central events (see Fig. 4):

$$c_n = a_n/(b_{\rm ZYAM}^{\rm C} + a_0). \tag{7}$$

Equivalantly, it can also be calculated as the Fourier coefficients of the recoil-subtracted correlation function $C_\Delta(\Delta\phi)$: $c_n = \langle C_\Delta(\Delta\phi) \cos n\Delta\phi \rangle$.

Figure 9 shows c_2 (top-left panel) and c_3 (bottom-left panel) as a function of $p_{\rm T}^{\rm a}$ for $0.5 < p_{\rm T}^{\rm b} < 4$ GeV in three $\Sigma E_{\rm T}^{\rm Pb}$ classes. The values of c_2 are much larger than those for c_3. The $p_{\rm T}^{\rm a}$ dependence of c_2 is similar to that for the $\Delta Y(\Delta\phi)$ in Fig. 8. The magnitude of the c_n is observed to increase for larger $\Sigma E_{\rm T}^{\rm Pb}$, opposite to the predictions from hydrodynamic calculations [12,24]. Using the techniques discussed in Ref. [4], c_n is converted into an estimate of s_n, the average $n^{\rm th}$-order Fourier coefficient of the event-by-event single-particle ϕ distribution, by assuming the factorization relation:

$$c_n(p_{\rm T}^{\rm a}, p_{\rm T}^{\rm b}) = s_n(p_{\rm T}^{\rm a}) s_n(p_{\rm T}^{\rm b}). \tag{8}$$

From this, $s_n(p_{\rm T}^{\rm a})$ is calculated as $s_n(p_{\rm T}^{\rm a}) = c_n(p_{\rm T}^{\rm a}, p_{\rm T}^{\rm b})/\sqrt{c_n(p_{\rm T}^{\rm b}, p_{\rm T}^{\rm b})}$, where $c_n(p_{\rm T}^{\rm b}, p_{\rm T}^{\rm b})$ is obtained from a_n calculated from Fig. 4. The $s_2(p_{\rm T}^{\rm a})$ values obtained this way exceed 0.1 at $p_{\rm T} \sim 2$–4 GeV, and the $s_3(p_{\rm T}^{\rm a})$ values remain less than 60% of the $s_2(p_{\rm T}^{\rm a})$ values over the measured $p_{\rm T}$ range.

The factorization relation used to compute $s_n(p_{\rm T}^{\rm a})$ is checked directly in Fig. 10 using three different sub-ranges of $p_{\rm T}^{\rm b}$ within 0.5–4 GeV. The factorization is found to be valid within 10%–20% for $s_2(p_{\rm T}^{\rm a})$, while the precision of $s_3(p_{\rm T}^{\rm a})$ data does not allow a quantitative test of the factorization. The analysis is also repeated for correlation functions separately constructed from like-sign pairs and unlike-sign pairs, and the resulting c_n and s_n coefficients are found to be consistent within their statistical and systematic uncertainties (see Fig. 6 and Fig. 9 at [18]).

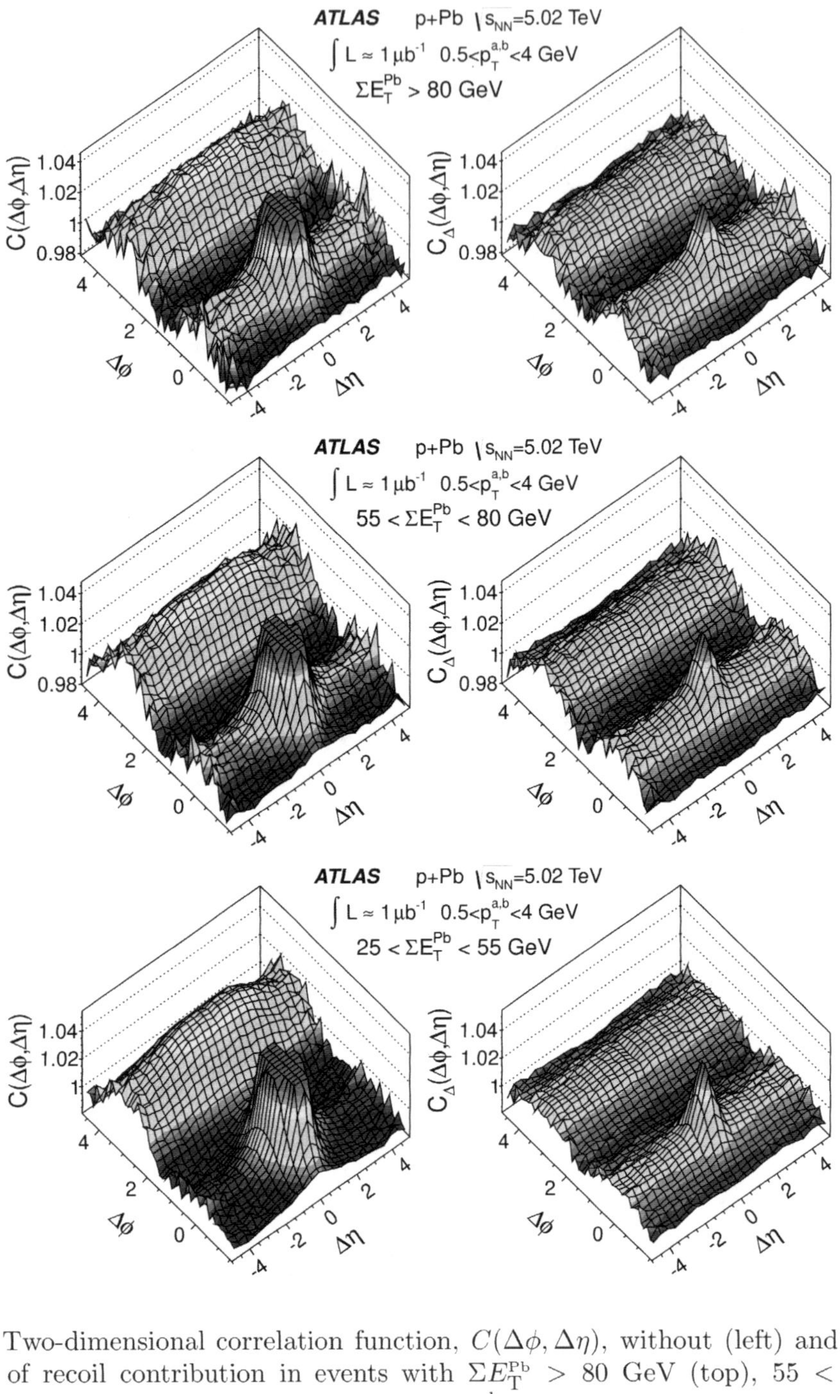

Figure 5. Two-dimensional correlation function, $C(\Delta\phi, \Delta\eta)$, without (left) and with (right) subtraction of recoil contribution in events with $\Sigma E_T^{Pb} > 80$ GeV (top), $55 < \Sigma E_T^{Pb} < 80$ (middle) and $25 < \Sigma E_T^{Pb} < 55$ (bottom) for $0.5 < p_T^{a,b} < 4$ GeV [18]. The recoil contribution is removed via a simple subtraction of the 2D per-trigger yield distribution in a given ΣE_T^{Pb} class by that in the peripheral class of $\Sigma E_T^{Pb} < 20$ GeV. The resulting per-trigger yield distribution plus original pedestal in this ΣE_T^{Pb} class is then re-normalized over $2 < |\Delta\eta| < 5$ to obtain the recoil-subtracted 2D correlation function $C_\Delta(\Delta\phi, \Delta\eta)$.

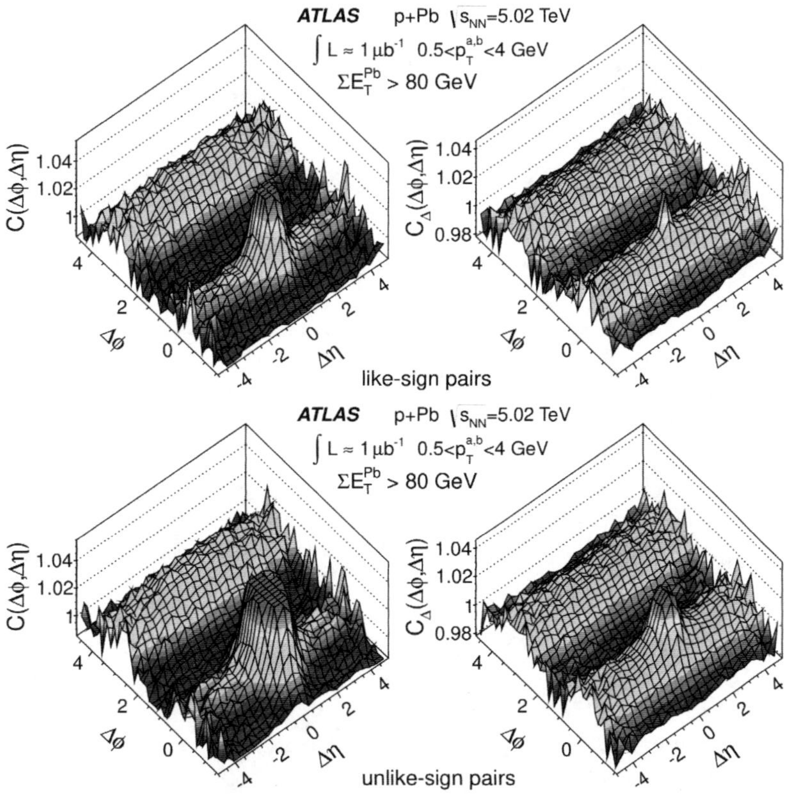

Figure 6. Two-dimensional correlation function, $C(\Delta\phi, \Delta\eta)$, without (left) and with (right) subtraction of recoil contribution in events with $\Sigma E_{\mathrm{T}}^{\mathrm{Pb}} > 80$ GeV and $0.5 < p_{\mathrm{T}}^{\mathrm{a,b}} < 4$ GeV for like-sign pairs (top) and unlike-sign pairs (bottom) [18].

6. Comment on event selection bias

The event class in the ATLAS analysis is defined by the $\Sigma E_{\mathrm{T}}^{\mathrm{Pb}}$ in $3.1 < \eta < 4.9$, well separated from the range $|\eta| < 2.5$ used to construct the correlation functions. The analysis is also repeated for event classes defined explicitly using the number of reconstructed charged particles $N_{\mathrm{ch}}^{\mathrm{rec}}$ in $p_{\mathrm{T}} > 0.4$ GeV and $|\eta| < 2.5$. As shown in Fig. 2, the correlation between $\Sigma E_{\mathrm{T}}^{\mathrm{Pb}}$ and $N_{\mathrm{ch}}^{\mathrm{rec}}$ is quite broad. The events selected in a $\Sigma E_{\mathrm{T}}^{\mathrm{Pb}}$ interval cover a broad range in $N_{\mathrm{ch}}^{\mathrm{rec}}$, and vice versa.

Figure 11 compares the integrated per-trigger yield as a function of $\Sigma E_{\mathrm{T}}^{\mathrm{Pb}}$ for events selected via $\Sigma E_{\mathrm{T}}^{\mathrm{Pb}}$ (left panel), as a function of $\langle N_{\mathrm{ch}} \rangle$ for events selected via $\Sigma E_{\mathrm{T}}^{\mathrm{Pb}}$ (middle panel) and as a function of N_{ch} for events selected via $N_{\mathrm{ch}}^{\mathrm{rec}}$ (right panel). The near-side ridge yields in the three cases show very similar trends and reach nearly the same value ($\sim$0.26) in the most central bin, suggesting that the three results are related to each other by an overall scale factor in the x-axis. The away-side yield, however, show quite different dependence between the right panel and the left two panels. In particular, the bend-over behavior at small N_{ch} and the increase at large N_{ch} can be attributed to auto-correlation bias associated with explicit selection of events based on N_{ch}: events required to have small number of N_{ch} preferably select events containing jets with smaller number of fragments, while events with large N_{ch} preferably select events containing jets with larger number of fragments. This bias can be significantly reduced if the yields are presented as a function of $\Sigma E_{\mathrm{T}}^{\mathrm{Pb}}$ or $\langle N_{\mathrm{ch}} \rangle$ for events classes based on the $\Sigma E_{\mathrm{T}}^{\mathrm{Pb}}$ (left and middle panels).

This bias is checked explicitly using HIJING simulation as shown in Fig. 12. HIJING has no

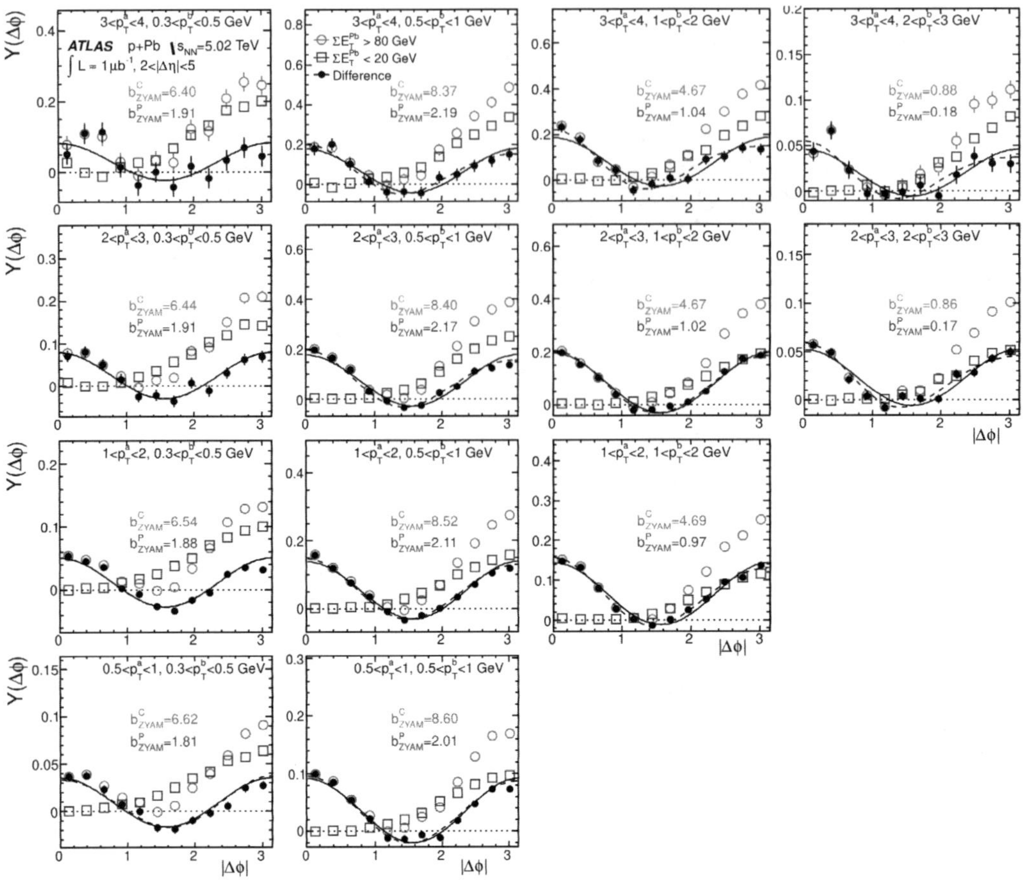

Figure 7. Distributions of per-trigger yield in the $\Sigma E_{\mathrm{T}}^{\mathrm{Pb}} > 80$ GeV event class (open circles), the $\Sigma E_{\mathrm{T}}^{\mathrm{Pb}} < 20$ GeV event class (open boxes), and their difference (solid circles) for various ranges of $p_{\mathrm{T}}^{\mathrm{a,b}}$, together with functions, $a_0 + 2a_2 \cos 2\Delta\phi$ (solid lines) and $a_0 + 2a_2 \cos 2\Delta\phi + 2a_3 \cos 3\Delta\phi$ (dashed lines), obtained via a Fourier decomposition [18].

flow physics, so all the structures in the 2D correlation function can be attributed to correlations involving a subset of the particles in the events. The per-trigger yields as a function of E_{T} over $3.1 < \eta < 4.9$ range show very weak variation, while the yields as a function of N_{ch} for events selected on N_{ch} show much stronger variation, very similar to the observation in the data.

7. Summary

ATLAS has measured two-particle correlations in $\sqrt{s_{\mathrm{NN}}} = 5.02$ TeV p+Pb collisions in broad ranges of event activity (quantified by $\Sigma E_{\mathrm{T}}^{\mathrm{Pb}}$, the E_{T} in $3.2 < \eta < 4.9$ in Pb-going side) and p_{T} (0.5-7 GeV). After subtracting the contributions from jet fragmentation around $(\Delta\phi, \Delta\eta) \sim (0, 0)$ and $\Delta\phi \sim \pi$, the resulting correlations show a significant long-range correlation that extends to $|\Delta\eta| = 5$ and are approximately symmetric around $\Delta\phi = \pi/2$. The $\Delta\phi$-shape of these correlations is well described by a $1 + 2c_2 \cos 2\Delta\phi + 2c_3 \cos 3\Delta\phi$ function over the measured ranges in p_{T}, $\Delta\eta$ and $\Sigma E_{\mathrm{T}}^{\mathrm{Pb}}$. The values of c_2 are found to increase with p_{T} to 3-4 GeV and fall at higher p_{T}, the values of the c_3 are found to be negligible at low p_{T} but increase to about 30% of the c_2 at high p_{T}. The overall magnitude of the c_n decreases by about 50% between events with $\Sigma E_{\mathrm{T}}^{\mathrm{Pb}} > 80$ and $25 < \Sigma E_{\mathrm{T}}^{\mathrm{Pb}} < 55$ GeV. The c_n parameters are converted to the Fourier

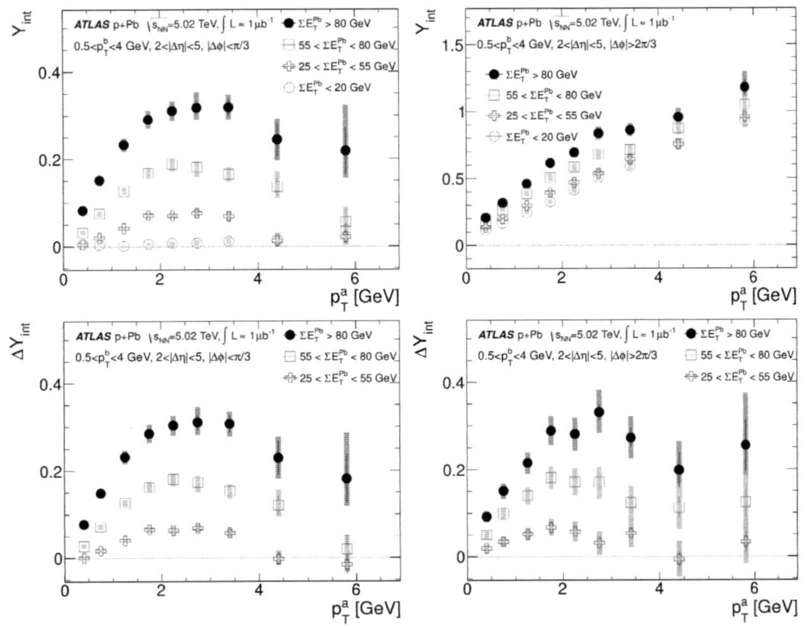

Figure 8. Per-trigger yields vs p_T^a for $0.5 < p_T^b < 4$ GeV in various ΣE_T^{Pb} event classes on the near-side (top-left panel) and away-side (top-right panel). The bottom panels show the difference of the yield from that in the $\Sigma E_T^{Pb} < 20$ GeV event class [18].

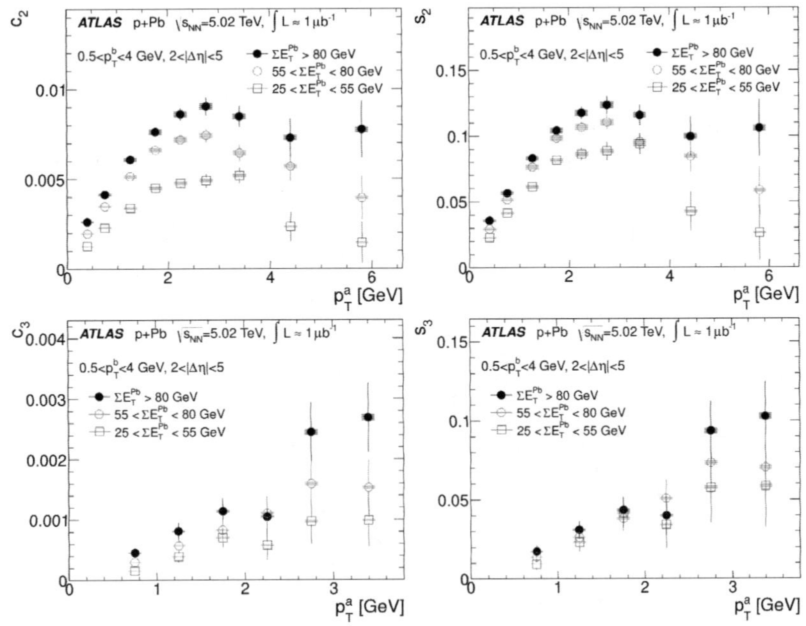

Figure 9. The c_n (left) and s_n (right) vs p_T^a for $0.5 < p_T^b < 4$ GeV in various ΣE_T^{Pb} event classes [18].

coefficients s_n for single-particle azimuthal distribution using a factorization ansatz Eq. 8. The magnitude of the extracted s_2 (s_3) reaches more than 0.1 (0.05) in 3-4 GeV for $\Sigma E_T^{Pb} > 80$ GeV,

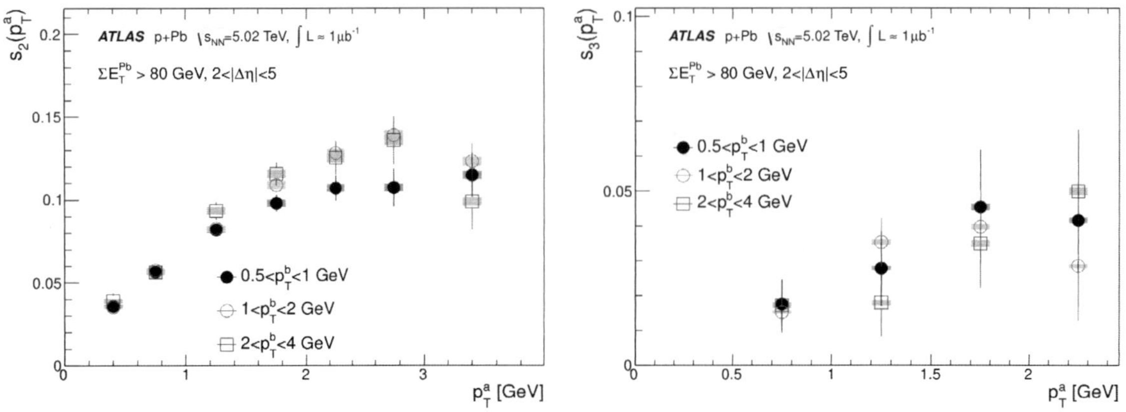

Figure 10. The $s_n(p_\mathrm{T}^\mathrm{a})$ in several p_T^b ranges. It quantifies the factorization relation $c_n(p_\mathrm{T}^\mathrm{a}, p_\mathrm{T}^\mathrm{b}) = s_n(p_\mathrm{T}^\mathrm{a})s_n(p_\mathrm{T}^\mathrm{b})$, i.e. checking how well $s_n(p_\mathrm{T}^\mathrm{a})$ extracted for different p_T^b ranges agree with each other [18].

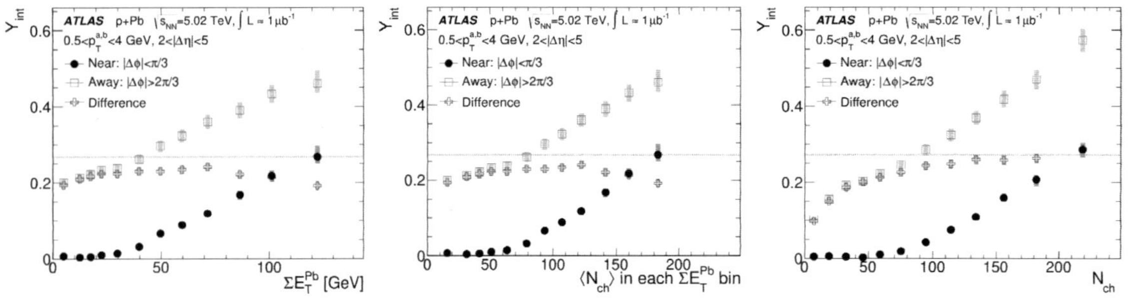

Figure 11. Integrated per-trigger yields for $0.5 < p_\mathrm{T}^{\mathrm{a,b}} < 4$ GeV as a function of $\Sigma E_\mathrm{T}^\mathrm{Pb}$ for events selected on $\Sigma E_\mathrm{T}^\mathrm{Pb}$ (left), as a function of $\langle N_\mathrm{ch}\rangle$ for events selected on $\Sigma E_\mathrm{T}^\mathrm{Pb}$ (middle) and as a function of N_ch for events selected on N_ch (right). The middle panel is same as left panel except for a change of x-axis using the $\langle N_\mathrm{ch}\rangle$ given in Table 1. The away-side yield in the right panel shows a clear drop for events explicitly required to have small N_ch, and a stronger increase for events explicitly required to have large N_ch, indicating the presence of auto-correlation bias [18].

but is about 20% lower for events with $25 < \Sigma E_\mathrm{T}^\mathrm{Pb} < 55$ GeV. These findings are consistent with final-state collective effects in high-multiplicity events [12, 14, 15], but they are also compatible with predictions in the Color Glass Condensate approach [16, 17].

This proceeding also addresses several technical aspects of the 2PC analysis, including the validity of the recoil-subtraction procedure and auto-correlation bias associated with the definition of event activity class. The recoil-subtraction procedure is demonstrated to work well (10-15%) in the relatively low p_T range of this analysis ($p_\mathrm{T} < 7$ GeV) where ridge structure is dominant. The definition of event activity class using $\Sigma E_\mathrm{T}^\mathrm{Pb}$ in rapidity range ($3.1 < \eta < 4.9$) not overlapping with particles used in the 2PC analysis ($|\eta| < 2.5$), is shown to reduce the auto-correlation bias. In contrast, this bias is clearly present when the particles involved in the correlation analysis are also used to define the event classes.

This work is in part supported by NSF under grant number PHY-1019387.

References

[1] STAR Collaboration 2009 *Phys. Rev. C* **80** 064912

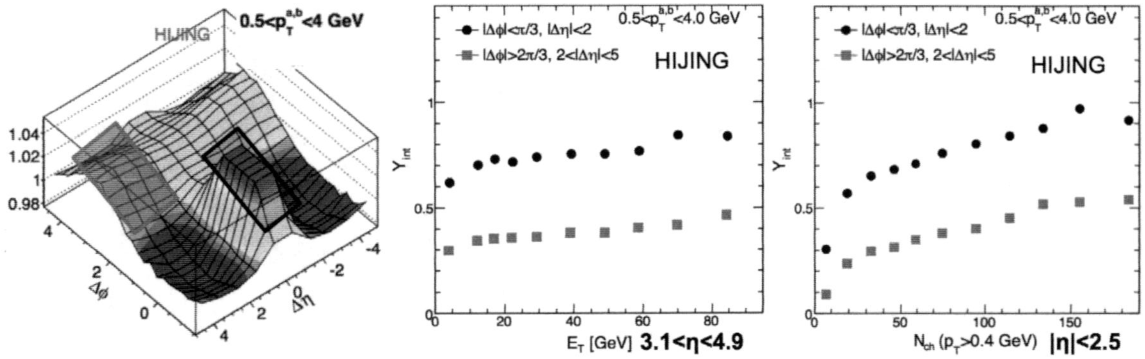

Figure 12. Illustration of event selection bias via HIJING simulation: (left) 2D correlation function and the two integration regions for per-trigger yield marked by the boxes, (middle) the yield vs forward E_T in $3.1 < \eta < 4.9$ for events selected with E_T, (right) the yield vs $N_{\rm ch}$ in $|\eta| < 2.5$ for events selected on $N_{\rm ch}$.

[2] PHOBOS Collaboration 2010 *Phys.Rev.Lett.* **104** 062301
[3] PHENIX Collaboration 2008 *Phys. Rev. C* **78** 014901
[4] ATLAS Collaboration 2012 *Phys. Rev. C* **86** 014907
[5] ALICE Collaboration 2012 *Phys. Lett.* **B708** 249
[6] CMS Collaboration 2012 *Eur. Phys. J.* **C72** 2012
[7] CMS Collaboration 2010 *JHEP* **09** 091
[8] CMS Collaboration 2013 *Phys.Lett.* **B718** 795
[9] ALICE Collaboration 2013 *Phys.Lett.* **B719** 29
[10] ATLAS Collaboration 2013 *Phys.Rev.Lett.* **110** 182302
[11] Voloshin S A, Poskanzer A M and Snellings R *Preprint* 0809.2949
[12] Bozek P 2012 *Phys.Rev.* **C85** 014911
[13] Bozek P and Broniowski W 2013 *Phys.Lett.* **B718** 1557
[14] Shuryak E and Zahed I *Preprint* 1301.4470
[15] Bozek P and Broniowski W *Preprint* 1304.3044
[16] Dusling K and Venugopalan R 2012 *Phys. Rev. D* **87** 054014
[17] Dusling K and Venugopalan R *Preprint* 1302.7018
[18] ATLAS Collaboration, https://atlas.web.cern.ch/Atlas/GROUPS/PHYSICS/PAPERS/HION-2012-13/
[19] ATLAS Collaboration 2008 *JINST* **3** S08003
[20] Ajitanand N *et al.* 2005 *Phys. Rev. C* **72** 011902
[21] Borghini N, Dinh P M and Ollitrault J Y 2000 *Phys.Rev.* **C62** 034902
[22] Borghini N, Dinh P, Ollitrault J Y, Poskanzer A M and Voloshin S 2002 *Phys.Rev.* **C66** 014901
[23] PHENIX Collaboration 2013 *Preprint* 1303.1794
[24] Bzdak A, Schenke B, Tribedy P and Venugopalan R *Preprint* 1304.3403

Journal of Physics: Conference Series **458** (2013) 012024 doi:10.1088/1742-6596/458/1/012024

The Color Glass Condensate, Glasma and the Quark Gluon Plasma in the Context of Recent pPb Results from LHC.

Larry McLerran
Physics Department, Brookhaven National Laboratory, Upton NY 11973, USA;
Riken Brookhaven Center, Brookhaven National Laboratory, Upton,, NY 11973, USA;
Physics Department, China Central Normal University, Wuhan, China.

Abstract. This paper describes some of the implications of the LHC pPb experimental results for understanding the Color Glass Condensate, the Glasma and the thermalized Quark Gluon Plasma. I attempt to outline the scientific issues, various possible explanations about what has been seen, and present ideas on how some of the alternative explanations might be resolved.

1. Introduction

The major goal of the heavy ion programs at RHIC and the LHC is to understand various forms of high energy density matter[1],[2],[3],[4],[5] . Matter in thermal equilibrium is the Quark Gluon Plasma[6]-[7]. High density, coherent gluonic matter in the colliding nuclei is the Color Glass Condensate[8],[9] [10],[11]. This matter controls the high energy limit of the interactions of hadrons, and provides initial conditions for the matter produced in such collisions. At early times after the collision, matter is in a Glasma[12],[13],[14][15]. This is highly coherent largely gluonic matter. It is strongly interacting matter even though the coupling is weak. The strength of the coupling is determined by a dimensional scale, the saturation momentum, which grows with increasing energy and increasing multiplicity of produced particles. The phase space occupation number of gluons is of order

$$\frac{dN}{d^2r_T dy d^2p_T} \sim \frac{1}{\alpha_s(Q_{sat})} \gg 1 \tag{1}$$

which enhances the effect of interactions due to coherence of the gluon field strength. (This is not unlike in gravity where an intrinsically weak gravitation field of nucleons is enhanced by the coherent interaction of many protons.)

The Glasma as it evolves ultimately will thermalize into a thermalized Quark Gluon Plasma[16]-[17]. Once thermalized, the QGP expands further producing particles, and evolving to a good approximation by the laws of perfect fluid hydrodynamics[18].

As the Glasma expands it interacts with itself, and will produce particles[19]-[20] . It will also generate flow patterns due to these interactions. In many ways, the transition between a Glasma and a thermalized Quark Gluon Plasma is more qualitative than abrupt. It should also be thought of as a Quark Gluon Plasma, but not a thermalized one. Electromagnetic plasmas need not be thermalized to be called plasmas and in this sense the Glasma is a Quark Gluon Plasma. Viewed from this perspective, the investigation of the Quark Gluon Plasma becomes an issue of understanding the properties of a Quark Gluon Plasma and focuses practical issues related to determining its properties of matter "as it is" rather than hard to define ideological issues about what matter "should be" .

In the remainder of this talk, I will discuss various aspects of the forms of matter described above in terms of what might be learned from experiment. I will comment on issues which are not yet so well resolved from either a theoretical or experimental view.

2. The Color Glass Condensate

The Color Glass Condensate is weakly coupled, but strongly interacting highly coherent gluonic matter in the wave function of a nucleus. It can be probed directly in ep and eA interactions at high energies. Assuming that final state interactions can either be computed or that they are in some circumstances small corrections, it can also be studied in high energy pp, pA and AA collisions.

The CGC determines the initial multiplicity of produced particles as a function of beam energy and centrality of collisions[21]. It is usually assumed that the multiplicity of produced gluons is equal to that of produced pions by using approximate entropy conservation during the evolution of the Glasma and the thermalized QGP. There are surely corrections associated with entropy production. Under this assumption, the saturation picture does a fair job of explaining pp and pA multiplicities at RHIC through LHC energies, Fig.1[23]. The CGC also predicts negative binomial distributions for produced particles, and provides a good description of the fluctuations in the pp multiplicities seen at CERN, Fig. 2[22].

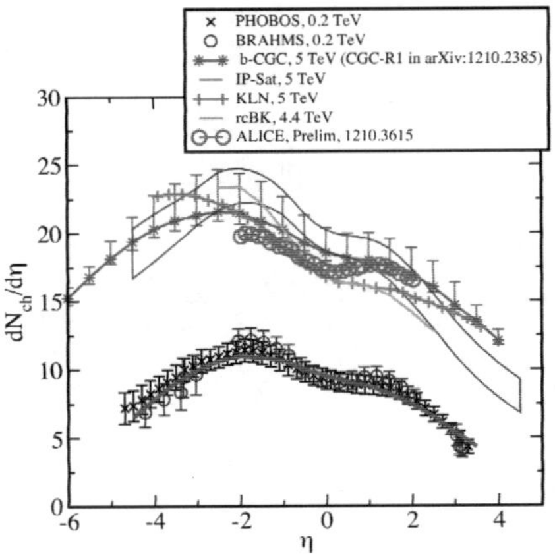

Figure 1. Multiplicity in minimum bias pPb collisions at LHC energy and various theoretical predictions.

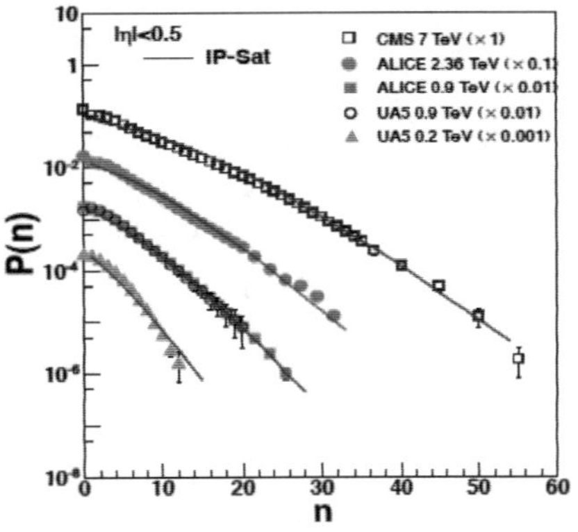

Figure 2. Distribution of multiplicity fluctuations in pp collisions at LHC energy.

The transverse momentum distributions for charged particle productions pp collisions at LHC energies are well described by the geometric scaling hypothesis motivated by the CGC description[24] . The transverse momentum broadening predicted by the CGC as a function of particle multiplicity is clearly seen in the data. There is also unpublished analysis of CMS data for identified particle productions which shows that such scaling works for the pp data on identified particles when the scaling variables is expressed in terms of M_T for massive particles[25],[26],[27]. Whether or not such scaling works for pp and pA collisions as a function of multiplicity for identified particles is not yet tested. It is more difficult to apply such reasoning to pA collisions because in general there are two saturation scales, that of the proton and that of the nucleus.

The behavior of single particle distributions at high multiplicity in pA and pp collisions might allow discrimination between the CGC hypothesis and hydrodynamic descriptions advocated by some. The CGC hypothesis predicts a growth of average p_T of produced particles with the saturation momentum, and the saturation momentum itself $Q_{sat}^2 \sim dN/dy/\pi R^2$ grows with multiplicity per unit area in the collision[24]. This is clearly seen in multiplicity fluctuations in the pp data, as shown in Fig. 3. There are also detailed predictions for the shape of the p_T distribution in pPb collisions as a function of centrality, energy and rapidity, an example of which is shown in Fig. 4 [29].

Another issue related to this is the measurement of radii of last interaction for high multiplicity pp and pA collisions[28]. Radii determined by HBT pion interferometry should grow as $R \sim (dN/dy)^{1/3}$. This is seen in nucleus-nucleus collisions. In pp collisions, there is only weak growth with multiplicity, and the radius is not much changed at high multiplicity compared to low multiplicity, which is a size typical of the proton, as shown in Fig. 5. Hydrodynamic simulations would suggest a trend similar to that of nuclear collisions, and a radius for high multiplicity events significantly larger than the proton size. It should be noted that if a hydrodynamic description works for the pA collisions at some multlplicity, it would be quite difficult to argue that it would not also work for pp collisions, since even if the production size region is initially smaller in a pp collision, eventually the matter will expand to a size typical of that of a larger system, and if it is already assumed that hydrodynamics describes the larger system, one would therefore expect it to describe it for a pp collisions at least from the point of time where the pp volume becomes equal to that of the pA.

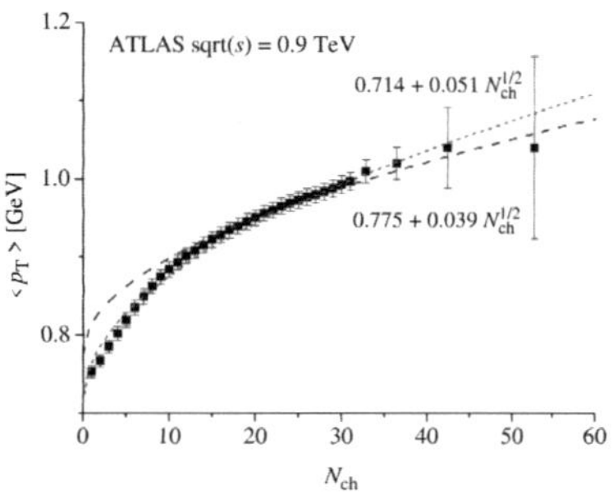

Figure 3. Tranvserse momenta of charged particles as a function of multiplicity compared to a simple model based on the CGC.

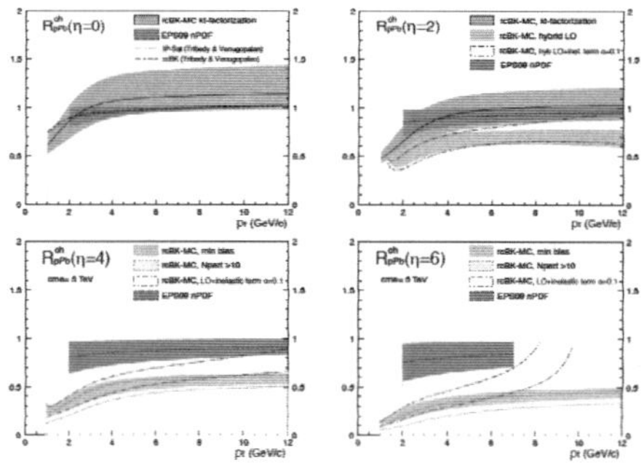

Figure 4. Predictions for the transverse momentum distributions in pPb collisions as a function of rapidity.

The CGC description predicts some broadening of transverse momenta distribution of produced particles as a function of centrality[30]. At RHIC energies, this was observed in the forward particle forward-backward angular correlation of charged particles[31][32], as shown in Fig. 6. In the backward direction there is a noticeable brooding and disappearance of the backward peak in dA collisions.

3. The Glasma

The Glasma is formed very quickly after the collision of two nuclei. It is initially an ensemble of lines of longitudinal color electric and color magnetic flux, as shown in Fig. 7. The typical

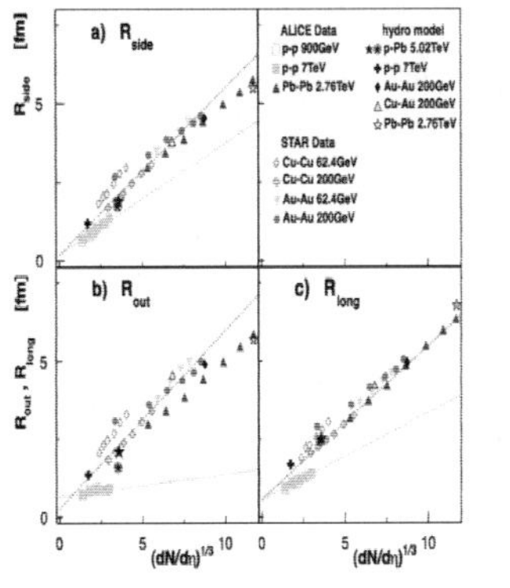

Figure 5. HBT radii in pp and AA collisions as a function of charged particle multiplicity.

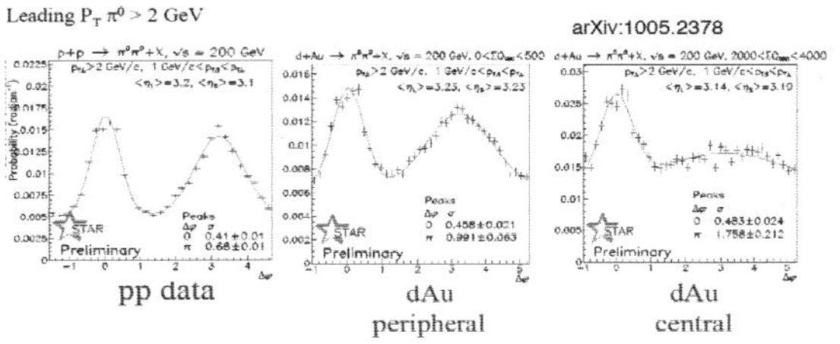

Figure 6. Two particle correlation at forward rapidity as measured in STAR..

transverse size scale of these flux lines is of the order of the inverse saturation scale (more properly the transverse correlation lengths for color and magnetic charge).

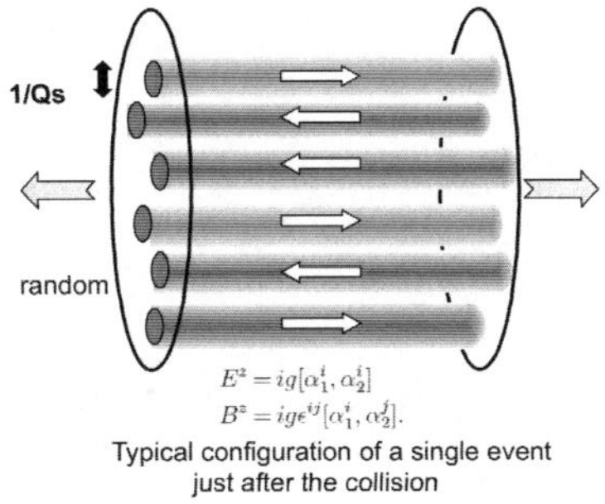

Figure 7. The colored fluc lines produced in hadron collisions immediately after a collision.

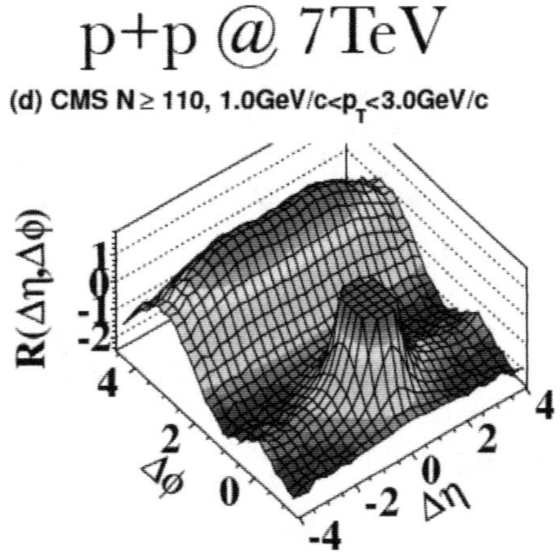

Figure 8. The ridge as seen in two particle correlations in pp collisions by the CMS experiment

These fields have a nonzero and maximal value of the topological charge density FF^d, and may be responsible for a number of interesting effects associated with charge correlations in the presence of a magnetic, field. The interested reader is referred to Refs. [33],[34],[35] for a discussion of the chiral magnetic effect and chiral magnetic wave.

One has observed two particle correlations over long distances in rapidity in pp, pA, dA and AA experiments[36],[37],[38],[39],[40],[41]. To have such an effect, one must both have fluctuations in the transverse positions of scattering centers, and a long rang structure such a as a flux line in the longitudinal space which is attached to the coordinate of the scattering center[42]. In AA collisions, such fluctuations in the transverse position might be generated by fluctuations in nucleon-nucleon scattering transverse positions[43]. In pp collisions, the fluctuations must be generated by quarks and gluons at the end of flux lines[44]-[45]. The CGC provides a theory of such fluctuations. Since this exists in pp collisions, there must at least be some component associated with quark and gluon fluctuations in AA collisions. Such fluctuations provide a good description of high order flow moments in AA collisions.

After such fluctuations have been formed in the initial state, they must somehow appear in the final state distribution of produced particles. This can happen either by initial state effects associated with the decay of a a flux tube, or by hydrodynamics effects which evolve initial energy density fluctuations into the momentum space distributions of produced particles, or perhaps by final state interactions which are not strong enough to generate fully developed hydrodynamic flow. In various regions of phase space, effects from of initial state or final state might dominate. For ridge measurements at low transverse momenta in AA collisions, almost certainly hydrodynamic effects play a dominant role. For pp collisions at high multiplicity, or high transverse momentum measurements, there is no consensus on this issue, and it is a subject of intense theoretical and experimental work, and needless to say, much controversy.

There are multiple issues to be settled in this controversy:

- Are the fluctuations that generate the ridge in pA collisions largely at the subnucleonic scale or at the lager scale sizes?
- How is the initial state structure which generates the ridge translated into particle distributions in the final state? Is this by initial state quantum mechanical correlations, by fully developed hydrodynamics, or by relatively weak final state interaction?
- What are the implications of understanding the ridge for our understanding of heavy ion collisions, and their interpretation as a nearly perfect fluid?

4. Summary and Conclusions

The pp and pA program at the LHC has generated much more scientific interest than anyone had originally envisaged. It may allow us to to resolve fundamental issues concerning the nature of interactions of high energy hadrons.

5. Acknowledgements

The research of L. McLerran is supported under DOE Contract No. DE-AC02-98CH10886. L. Mclerran gratefully acknowledges discussions with Adam Bzdak, Bjoern Schenke, Vladimir Skokov and Raju Venugopalan concerning the interoperation of LHC results about p-Pb collisions.

References

[1] M. Gyulassy and L. McLerran, Nucl. Phys. A **750**, 30 (2005) [nucl-th/0405013].
[2] I. Arsene *et al.* [BRAHMS Collaboration], Nucl. Phys. A **757**, 1 (2005) [nucl-ex/0410020].
[3] K. Adcox *et al.* [PHENIX Collaboration], Nucl. Phys. A **757**, 184 (2005) [nucl-ex/0410003].
[4] B. B. Back, M. D. Baker, M. Ballintijn, D. S. Barton, B. Becker, R. R. Betts, A. A. Bickley and R. Bindel *et al.*, Nucl. Phys. A **757**, 28 (2005) [nucl-ex/0410022].
[5] J. Adams *et al.* [STAR Collaboration], Nucl. Phys. A **757**, 102 (2005) [nucl-ex/0501009].
[6] N. Itoh, Prog. Theor. Phys. **44**, 291 (1970).
[7] N. Cabibbo and G. Parisi, Phys. Lett. B **59**, 67 (1975).
[8] L. V. Gribov, E. M. Levin and M. G. Ryskin, Phys. Rept. **100**, 1 (1983).
[9] A. H. Mueller and J. -w. Qiu, Nucl. Phys. B **268**, 427 (1986).
[10] L. D. McLerran and R. Venugopalan, Phys. Rev. D **49**, 2233 (1994) [hep-ph/9309289].
[11] L. D. McLerran and R. Venugopalan, Phys. Rev. D **49**, 3352 (1994) [hep-ph/9311205].
[12] A. Kovner, L. D. McLerran and H. Weigert, Phys. Rev. D **52**, 6231 (1995) [hep-ph/9502289].
[13] A. Kovner, L. D. McLerran and H. Weigert, Phys. Rev. D **52**, 3809 (1995) [hep-ph/9505320].
[14] A. Krasnitz and R. Venugopalan, Phys. Rev. Lett. **84**, 4309 (2000) [hep-ph/9909203].
[15] T. Lappi and L. McLerran, Nucl. Phys. A **772**, 200 (2006) [hep-ph/0602189].

[16] J. -P. Blaizot, F. Gelis, J. -F. Liao, L. McLerran and R. Venugopalan, Nucl. Phys. A **873**, 68 (2012) [arXiv:1107.5296 [hep-ph]].
[17] A. Kurkela and G. D. Moore, JHEP **1112**, 044 (2011) [arXiv:1107.5050 [hep-ph]].
[18] D. Teaney, J. Lauret and E. V. Shuryak, Phys. Rev. Lett. **86**, 4783 (2001) [nucl-th/0011058].
[19] F. Gelis, E. Iancu, J. Jalilian-Marian and R. Venugopalan, Ann. Rev. Nucl. Part. Sci. **60**, 463 (2010) [arXiv:1002.0333 [hep-ph]].
[20] J. Berges, K. Boguslavski, S. Schlichting and R. Venugopalan, arXiv:1303.5650 [hep-ph].
[21] D. Kharzeev and M. Nardi, Phys. Lett. B **507**, 121 (2001) [nucl-th/0012025].
[22] P. Tribedy and R. Venugopalan, Phys. Lett. B **710**, 125 (2012) [Erratum-ibid. B **718**, 1154 (2013)] [arXiv:1112.2445 [hep-ph]].
[23] M. Nicassio [ALICE Collaboration], AIP Conf. Proc. **1422**, 79 (2012).
[24] L. McLerran and M. Praszalowicz, Acta Phys. Polon. B **41**, 1917 (2010) [arXiv:1006.4293 [hep-ph]].
[25] A. Ortiz Velasquez [ALICE Collaboration], arXiv:1209.3553 [hep-ex].
[26] V. Khachatryan *et al.* [CMS Collaboration], JHEP **1002**, 041 (2010) [arXiv:1002.0621 [hep-ex]].
[27] Ference Sikler, private communication.
[28] K. Aamodt *et al.* [ALICE Collaboration], Phys. Rev. D **84**, 112004 (2011) [arXiv:1101.3665 [hep-ex]].
[29] J. L. Albacete, A. Dumitru, H. Fujii and Y. Nara, Nucl. Phys. A **897**, 1 (2013) [arXiv:1209.2001 [hep-ph]].
[30] C. Marquet, Nucl. Phys. A **796**, 41 (2007) [arXiv:0708.0231 [hep-ph]].
[31] E. Braidot [STAR Collaboration], arXiv:1005.2378 [hep-ph].
[32] A. Adare *et al.* [PHENIX Collaboration], Phys. Rev. Lett. **107**, 172301 (2011) [arXiv:1105.5112 [nucl-ex]].
[33] D. Kharzeev and A. Zhitnitsky, Nucl. Phys. A **797**, 67 (2007) [arXiv:0706.1026 [hep-ph]].
[34] D. E. Kharzeev, L. D. McLerran and H. J. Warringa, Nucl. Phys. A **803**, 227 (2008) [arXiv:0711.0950 [hep-ph]].
[35] D. E. Kharzeev and H. -U. Yee, Phys. Rev. D **83**, 085007 (2011) [arXiv:1012.6026 [hep-th]].
[36] L. Ray [STAR Collaboration], Nucl. Phys. A **854**, 89 (2011).
[37] D. Velicanu [CMS Collaboration], J. Phys. G **38**, 124051 (2011) [arXiv:1107.2196 [nucl-ex]].
[38] S. Chatrchyan *et al.* [CMS Collaboration], Phys. Lett. B **718**, 795 (2013) [arXiv:1210.5482 [nucl-ex]].
[39] S. Chatrchyan *et al.* [CMS Collaboration], arXiv:1305.0609 [nucl-ex].
[40] G. Aad *et al.* [ATLAS Collaboration], arXiv:1212.5198 [hep-ex].
[41] B. Abelev *et al.* [ALICE Collaboration], arXiv:1212.2001 [nucl-ex].
[42] A. Dumitru, F. Gelis, L. McLerran and R. Venugopalan, Nucl. Phys. A **810**, 91 (2008) [arXiv:0804.3858 [hep-ph]].
[43] B. Alver and G. Roland, Phys. Rev. C **81**, 054905 (2010) [Erratum-ibid. C **82**, 039903 (2010)] [arXiv:1003.0194 [nucl-th]].
[44] A. Dumitru, K. Dusling, F. Gelis, J. Jalilian-Marian, T. Lappi and R. Venugopalan, Phys. Lett. B **697**, 21 (2011) [arXiv:1009.5295 [hep-ph]].
[45] K. Dusling and R. Venugopalan, arXiv:1211.3701 [hep-ph].

Journal of Physics: Conference Series **458** (2013) 012025 doi:10.1088/1742-6596/458/1/012025

Recent results on jet physics from ALICE at the LHC

Alexandre SHABETAI (for the ALICE Collaboration)
SUBATECH - CNRS/IN2P3 - UMR 6457 - Laboratoire de physique subatomique et des technologies associées - 4 rue Alfred Kastler - BP 20722 44307 Nantes cedex 3, France
E-mail: `shabetai@in2p3.fr`

Abstract. An overview of recent results on jet physics from the ALICE collaboration is presented. We mainly discuss azimuthal correlations and nuclear modification factors for both hadrons and jets. The very low p_{T} tracking (> 150 MeV/c) capabilities of the ALICE detector associated to its electromagnetic calorimetry, allow to reconstruct both charged jets and full jets, with minimum bias on the jet fragmentation. A detailed measurement of the underlying event fluctuations will be discussed, as well as, jet spectra measurements, in pp and Pb-Pb collisions, using low constituent p_{T} cuts and hadron-recoil jet correlations, which give a controlled way to subtract the combinatorial background.

1. Physics motivations

Jets are produced in hard scatterings of high energy quarks and gluons in the early stage of heavy-ion collisions. This allows to probe the hot and dense matter formed in these collisions. Jets can be used to study in medium energy loss and its path length dependence thanks to their strong interaction with the medium, which leads to a modification of their structure and to a redistribution of their energy. Experimentally, this can be seen as a marked reduction of their energy measured in a given reconstruction cone. Their fragmentation pattern is also expected to be modified. This phenomenon is called "jet quenching" [1].

Many experimental observations such as the suppression of back-to-back azimuthal correlations, the suppression of inclusive hadron – or even jet – spectra can all be used in order to learn about in-medium energy loss. Comparing those to theoretical calculations provides insight into the density of the system.

Moreover, measurements of jets in proton-proton collisions provide the baseline needed for heavy-ion studies, and at the same time allow to test pQCD.

In these proceedings, an overview of recent results on jet physics from the ALICE collaboration is presented. We mainly discuss azimuthal correlations and nuclear modification factors, for both hadrons and jets.

2. Experimental setup

In ALICE, high p_T particles or jets are reconstructed at mid-rapidity using either two or three sub-systems:

Charged particles are measured using the central tracking system [2] which consists of the Time Projection Chamber (TPC) and of the Inner Tracking System (ITS). Photons (mainly from π^0 decay) and electrons are in turn measured by the ALICE Electromagnetic Calorimeter (EMCal) [3].

Journal of Physics: Conference Series **458** (2013) 012025 doi:10.1088/1742-6596/458/1/012025

The ALICE TPC measures the momentum of charged particle down to $p_{\rm T} = 150$ MeV/c over a pseudo-rapidity range of $|\eta| < 0.9$ and has a full azimuthal acceptance ($0 < \varphi < 2\pi$).

The electromagnetic component of the jet energy (which is on average about 1/3 of the total energy of a given jet) is measured by the ALICE EMCal. This is a SHASHLIK [2, 3] (Pb-scintillator) sampling electromagnetic calorimeter which is read-out by APDs. It covers $|\eta| < 0.7$ and has an energy resolution of $11\%/\sqrt{E} + 1.7\%$. It has been fully installed in 2011: its azimuthal acceptance was increased to $\Delta\phi = 107°$ allowing it to be used for jet physics.

3. Hadron $R_{\rm AA}$

The first observable that we will consider is the nuclear modification factor of charged hadrons.

This quantity is defined as the ratio of charged particle yields in Pb-Pb (to study final state effects) or in p-Pb (to study initial state effects) over the corresponding yield in pp collisions, normalize by $1/N_{\rm coll}$ calculated from the Glauber model [4]. In the absence of nuclear modifications, hard processes are expected to follow the $N_{\rm coll}$ scaling. The corresponding nuclear modification factor would then be unity ($R_{AA} = 1$).

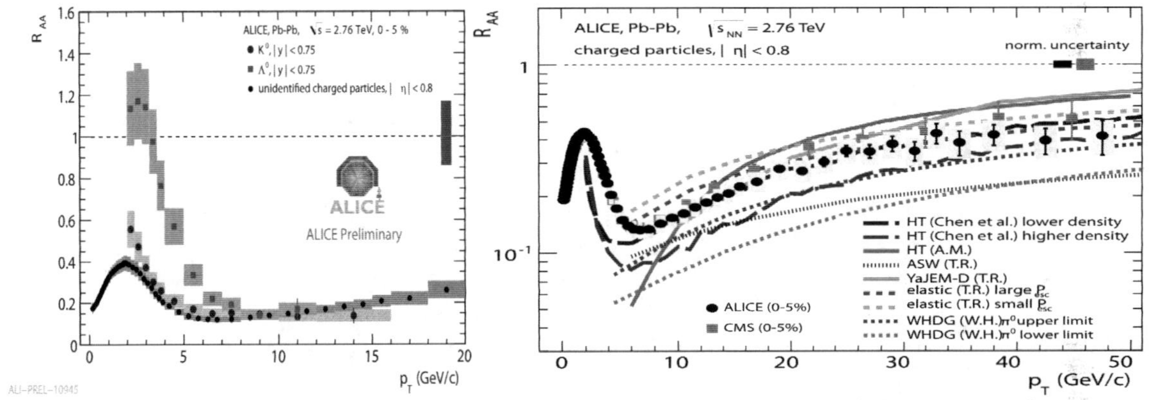

Figure 1. Left: Comparison of the nuclear modification factor for different particle species. Right: The hadron R_{AA} measured by ALICE and CMS.

Initial state effects have been quantified in p-A collisions: Using data at $\sqrt{s_{NN}} = 5$ TeV taken during the LHC p-A pilot run from September 2012, ALICE made a first measurement of R_{p-A} [5] (from 0.5 GeV/c < $p_{\rm T}$ <20 GeV/c) which is found to be consistent with binary scaling, indicating that the Cronin effect [6] at LHC is very small (if not null) compared to the one measured at RHIC [7, 8, 9, 10, 11].

In order to estimate final state effects in the medium, ALICE also measured the hadron R_{AA} [12] in Pb-Pb collisions at $\sqrt{s_{NN}} = 2.76$ TeV up to $p_{\rm T} = 50$ GeV/c. In central collisions ($0 - 5\%$ centrality) hadrons are losing a large part of their energy in the medium and are suppressed by a factor 6 at 7 GeV/c. However, the amount of suppression slowly decreases with increasing p_T. Above 40 GeV/c, hadrons are suppressed by a factor 2 and $R_{AA} \simeq 0.4 - 0.5$.

A comparison of the hadron nuclear modification factors (presented in the left panel of Fig. 1) for different particle species (unidentified particles, pions, Kaons and Lambdas) has also been made at the same center of mass energy of 2.76 TeV. Those measurements show a different behaviour at low p_T (corresponding to the soft regime): the Λ R_{AA} is close to unity at $p_T = 3$ GeV/c, whereas the Kaon and pion R_{AA} are smaller. The Λ/K ratio is found to be enhanced at intermediate p_T but starting at $p_T \sim 8$ GeV/c and above an agreement is found for all particles: the magnitude of the suppression is measured to be $R_{AA} = 0.2$ - 0.3 independent of the particle type.

The hadron R_{AA} measured by ALICE and CMS (see right panel of Fig. 1) [12, 13] are in good agreement and the trend of those measurements is reproduced by a large number of theoretical predictions. Many of the theory extrapolations from RHIC, however, predict too strong suppression at the LHC [14]. It has been suggested that taking into account the running of α_s reduces the suppression [15]. Overall, a full quantitative understanding of energy loss and the medium density requires further theoretical work [16].

4. Di-hadron azimuthal correlations

The second observable that we will consider are di-hadron azimuthal correlations. To build such correlations, a high p_T particle, called "trigger particle" is selected and associated with all other particle in the event.

One of the first experimental measurement of jet quenching was performed by the STAR collaboration at RHIC [9]. In Au-Au collisions, the disappearance of the away side peak of a di-hadron $\Delta\phi$ correlations was observed first by STAR. This indicated substantial interactions as the hard-scattered partons traverse the medium. The near side peak, on the other hand, was found to be very similar in pp and Au-Au collisions, one interpretation of these observations is that there is a strong "surface bias" effect, i.e. that the trigger particle selects jets with a short in-medium path length and/or little energy loss.

The ALICE collaboration measured long range di-hadron $\Delta\eta - \Delta\phi$ azimuthal correlations in Pb-Pb at $\sqrt{s_{NN}} = 2.76$ TeV (see Fig. 2) [17]. A different behaviour is found for low and high p_{T} regimes: At low p_{T} (3 GeV/$c < p_{\mathrm{T}}^t <$ 4 GeV/c and 2 GeV/$c < p_{\mathrm{T}}^a <$ 2.5 GeV/c), a clear near side ridge structure is seen and the away side is broad. A Fourier decomposition was made and the measured distribution is well reproduced by summing v_n harmonics with $n = 1$ to $n = 4$. This suggests that in this p_{T} range, the underlying physics is well described by hydrodynamics. At higher p_{T} (8 GeV/$c < p_T^t <$ 15 GeV/c and 6 GeV/$c < p_{\mathrm{T}}^a <$ 8 GeV/c) a different scenario is observed: the near side ridge is replaced by a "jet" peak and in the away side a small recoiling jet distribution can be seen. In addition, the previous global fit fails indicating that the physics is different in this higher p_{T} regime.

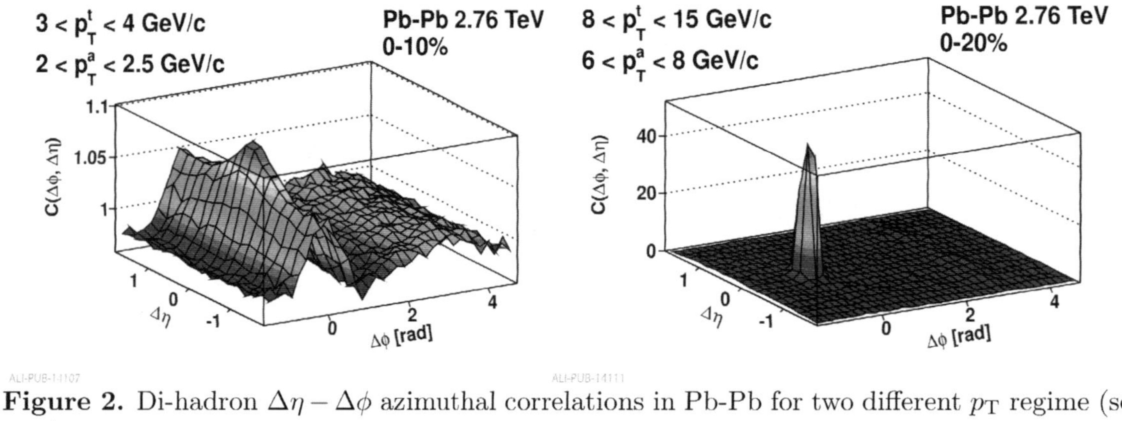

Figure 2. Di-hadron $\Delta\eta - \Delta\phi$ azimuthal correlations in Pb-Pb for two different p_{T} regime (see text).

However, those measurements do not give access to the full jet energy: they are limited by the fact that a high p_T leading particle at LHC energies only carries about 50% of the energy of the corresponding jet. To overcome this limitation, we have to measure full jets.

5. Full Jets measurements

A jet can be defined at several levels: A hard scattering process will lead into a partonic shower (which can be seen as a parton jet) which after hadronization will generate a hadronic shower (corresponding to a hadron jet). The charged tracks (particle jet) and the neutral energy of those hadrons will then be detected in the ALICE TPC and EMCal (see section 2). Experimentally, the jet energy has to be reconstructed. To do that, a simple algorithm consists in opening a cone, of a given radius R, around the jet axis and to sum the energy of all particles inside the cone. Sequential recombination algorithms, which have the advantage of collinear and infrared safety, can also be used. The k_T (for background clusters) and anti-k_T (for signal jets) algorithms [18, 19] from the FastJet package (with $0.2 \leq R \leq 0.4$) were used in the following analyses.

Analyses with charged jets using only tracking information (tracks with $p_T > 150$ MeV/c) and fully reconstructed jets including EMCal information (cluster energy with $E_T > 150\ MeV/c$) have been performed.

Jet-by-jet, to avoid double counting, we correct for the energy contribution from charged particles to the energy measured with EMCal (hadronic correction). The contribution from the underlying event is also subtracted (see section 5.2). Background fluctuations have a large impact on the measured jet spectrum in Pb-Pb. A response matrix $RM_{\delta p_T}$ containing the p_T smearing due to background fluctuations is constructed from the measured p_T distribution. The detector effects (efficiency, detector resolution) are corrected using a second response matrix RM_{det}. The two response matrices are combined to obtain the response matrix which will be needed for correcting the data using the unfolding procedure in order to obtain the results that will follow.

5.1. Inclusive differential jet cross-section in pp collisions

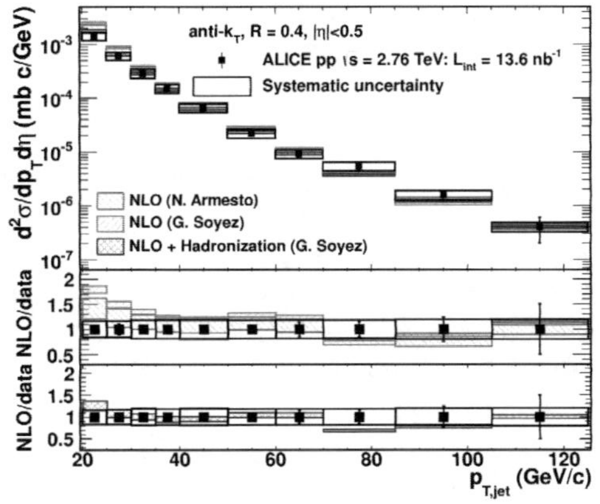

Figure 3. Upper panel: inclusive differential jet cross section for $R = 0.4$. The bands show the NLO pQCD calculations. Lower panels: ratio of NLO pQCD calculations to data.

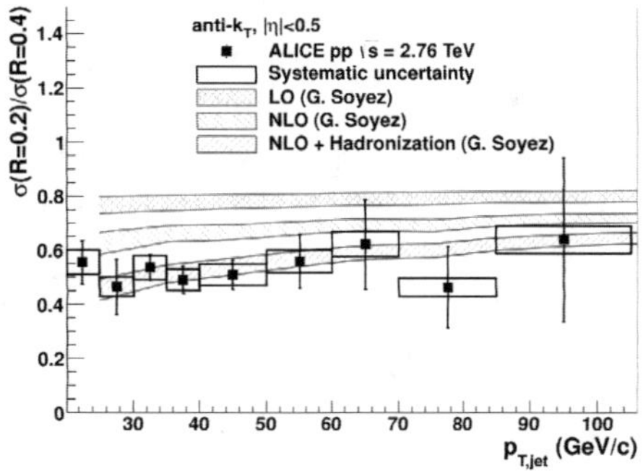

Figure 4. Ratio of inclusive differential jet cross sections for $R = 0.2$ and $R = 0.4$, with pQCD calculations.

As a baseline (and as a reference for Pb-Pb analyses), in pp collisions at $\sqrt{s_{NN}} = 2.76$ TeV, the inclusive differential jet cross-section was measured [20] at mid-rapidity (with an integrated luminosity of 13.6 nb^{-1}). Jets are measured over the transverse momentum range 20 to 125 GeV/c and are corrected using a bin-by-bin technique. Calculations based on Next-to-Leading Order perturbative QCD and PYTHIA 8 are in good agreement with the measurements (see Fig. 3). The ratio of inclusive jet cross-sections for jet radii $R = 0.2$ and $R = 0.4$ (see Fig. 4) is well reproduced by a Next-to-Leading Order perturbative QCD calculation when hadronization effects are included.

Journal of Physics: Conference Series **458** (2013) 012025 doi:10.1088/1742-6596/458/1/012025

5.2. Background estimation and fluctuations in Pb-Pb

One of the main experimental difficulties of measuring jets in heavy-ion collisions is the estimation and the subtraction of the fluctuating background (contribution from the underlying event).

In ALICE, the background energy density is estimated by clustering the whole event with the k_T-algorithm and calculating the density $\rho = \text{median}(\frac{p_T^{jet,i}}{A^{jet,i}})$ (where A is the area of a given jet i) for every jet except the two leading ones as proposed in [21]. The average backgroud density is then subtracted from the p_T of signal jets (found using anti-k_T) as $p_T^{sub} = p_T - \rho \times A$. ρ which was estimated for charged only (using 2010 data [22]) and full (charged + neutral) jets (using 2011 data [23]), scales with the event multiplicity [22]: $\rho \sim N \times < p_T >$.

In addition, point-to-point background fluctuations δp_T were estimated [22] by placing random cones in the measured events, or by embedding a known (high p_T) probe in the event and then looking at the collection of jets found by the anti-k_T algorithm and matched to the embedded probe, in this event, the corresponding $\delta p_T = p_{T_{jet}} - \rho \times A - p_{T_{probe}}$. The δp_T distribution is then fitted with a Gaussian and the width (σ) of the distribution is extracted. The corresponding σ_{ch} and σ_{ch+em} were estimated for a given jet radius from $R = 0.2$ to $R = 0.4$. For smaller reconstructed jets radii, the total energy inside the cone is smaller and consequently, the corresponding background fluctuations are also reduced. $\sigma_{ch} \simeq 10$ GeV/c was measured for $R = 0.4$ compared to $\sigma_{ch} \simeq 4.5$ GeV/c for $R = 0.2$. When neutral particles are included, the jet energy resolution increases, but the background fluctuations also become larger: $\sigma_{ch+em} > \sigma_{ch}$. For instance, for $R = 0.3$, $\sigma_{ch+em} \simeq 9$ GeV/c and $\sigma_{ch} \simeq 7$ GeV/c.

5.3. Jet spectrum in Pb-Pb at $\sqrt{s_{NN}}$ 2.76 TeV

Two analyses have been performed:

The first one [24] uses data from 2010 and is only using jets from charged particles. For this analysis jet radii of $R = 0.2$ and 0.3 were used. The minimum p_T of the jet constituents is $p_T^{const} = 0.15$ GeV/c. All jets with a jet axis within $|\eta| < 0.5$ were considered.

In order to reduce the number of combinatorial background jets, which helps to stabilize the unfolding procedure, the transverse momentum of the leading track of a given jet was required to be $p_T^{leading} > 5$ GeV/c. The effect of this selection on the corrected spectra was found to be small in the case of this analysis. To quantify it, the inclusive spectrum was compared to the ones obtained using $p_T^{leading} > 5$ GeV/c and $p_T^{leading} > 10$ GeV/c. The corrected spectra (using $R = 0.3$ jets) is shown in Fig. 5 (left panel). A centrality evolution (using 4 centrality classes from 0 - 10 to 50 - 80%) of the yield of jets, is observed. In addition, a ratio of the spectrum using $R = 0.2$ devided by the spectrum using $R = 0.3$ jets shows that there is no signifiquant jet broadening (from $R = 0.2$ to $R = 0.3$).

A second analysis was made [25], following the same strategy but using fully reconstructed (including both charged and neutral particles) $R = 0.2$ jets from 2011 Pb-Pb data. This full jet spectrum (shown Fig. 6 on the left panel) was obtained using the 10% most central events. As for the charged jet analysis (see above), the effect of the leading constituent track transverse momentum cut was studied using both $p_T^{leading} > 5$ GeV/c and $p_T^{leading} > 10$ GeV/c.

6. Jet nuclear modification factors

Using the charged jet spectrum measured in 4 centrality classes [24] from 2010 Pb-Pb data, a jet R_{CP}^{jet} was obtained in 3 centrality classes (from $0 - 10\%$ to $30 - 50\%$), using the spectrum measured in $50 - 80\%$ centrality as a reference (see Fig. 5 central panel).

A strong jet suppression, similar to the hadron R_{AA}, is observed for central events. For more peripheral events the suppression decreases. This implies that the full jet energy is not captured by jets with $R = 0.2$ and $R = 0.3$ in Pb-Pb collisions.

If we compare those measurements to other measurements from LHC (see Fig. 5 right panel), the jet R_{CP}^{jet} from ALICE, CMS and ATLAS [26] are consistent within systematic errors.

Using the full jet spectra [25] from 2011 data, combined with the differential cross-section measured in pp collisions (see section 5.1), a full jet nuclear modification factor R_{AA}^{jet} was built.
A strong suppression is observed (see Fig. 6 central panel) for $R = 0.2$ jets (in the $0-10\%$ centrality class). At low jet p_T this suppression decreases with p_T. At high p_T the jet $R_{AA} =$ 0.3-0.4. These results are consistent with the CMS $R = 0.2$ results from the 5% most central events (see Fig. 6 right panel) [27], despite the differences in the detector configurations as well as in jet finding and background subtraction procedures. As explained in section 2, in ALICE, jets can be found using tracking and electromagnetic calorimetry, whereas CMS uses particle-flow jets, which combine information from three detectors (tracker, electromagnetic and hadronic calorimeters).
In order to fully quantify the measured jet quenching, additional measurements are ongoing in order to study the radius and the centrality dependence of R_{AA}^{jet}.

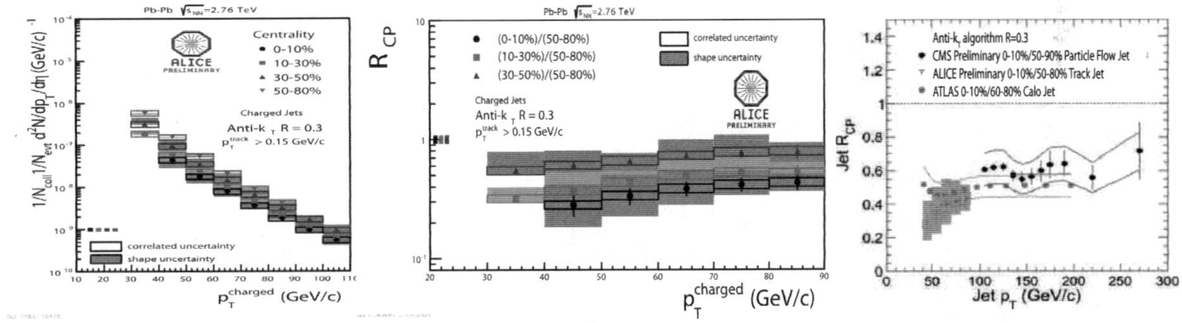

Figure 5. Left: Corrected jet spectrum and R_{cp}^{jet} (central panel) with charged tracks for jet radius $R = 0.3$. Right: Jet R_{CP}^{jet} from ALICE compared to CMS and ATLAS results.

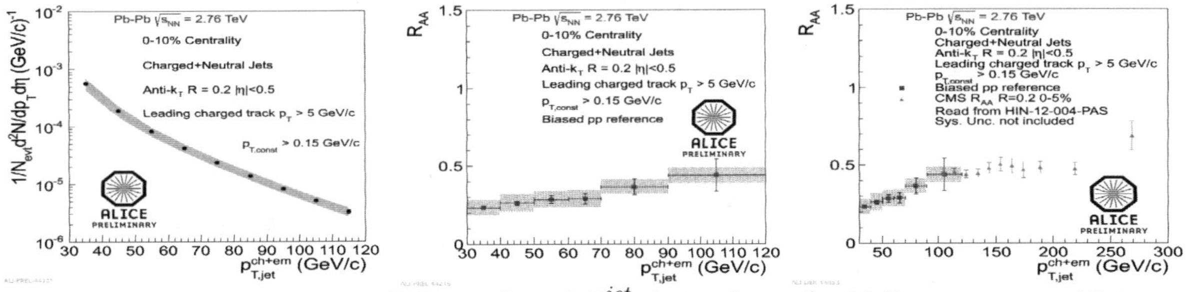

Figure 6. Corrected spectrum (left panel) and R_{AA}^{jet} (central panel) of fully reconstructed jets (reconstructed using $R = 0.2$ and requiring a high p_T leading track $p_T > 5$ GeV/c) in the $0-10\%$ centrality bin. Right: ALICE R_{AA}^{jet} compared to CMS $R = 0.2$ results from the 5% centrality bin.

7. Hadron-jet correlations

Both di-hadron and hadron-jet correlations are sensitive to the di-jet structure expected from hard scattering.

Model calculations show that a high p_T hadron trigger induces a geometrical bias [28], towards jets generated close to the surface of the fireball. The jet population recoiling from such a trigger is biased towards larger in-medium path length.

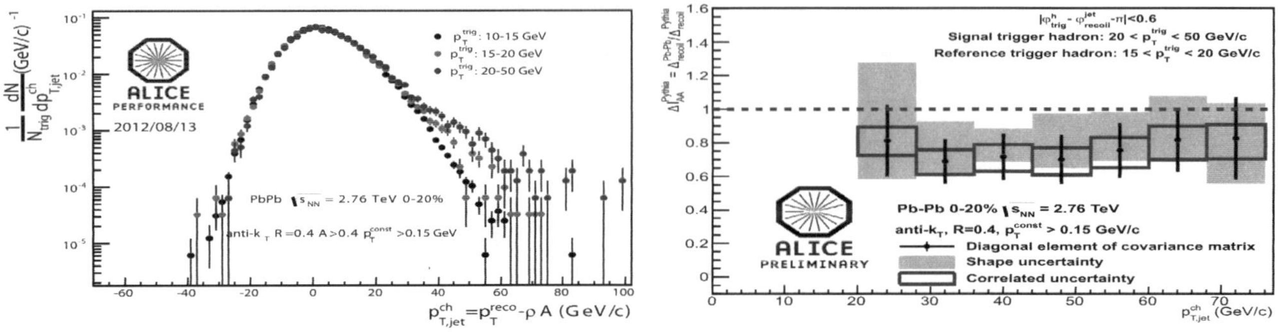

Figure 7. Left: Recoil jet distributions in $0-20\%$ central Pb-Pb collisions for $R=0.4$ and $p_T^{const}>0.15$ GeV$/c$. Right: ΔI_{AA}^{PYTHIA}.

To exploit that effect, a jet-hadron measurement was made [29], using 2010 Pb-Pb data, based on the semi-inclusive distribution of reconstructed charged particle jets (using anti-k_T with $R=0.2$ and 0.4) recoiling from a high p_T trigger hadron.

The distribution of recoil jets is measured by counting the number of jets in the event within $\phi(trig)-\phi(jet)<\pi-0.6$ and normalized to the corresponding number of triggers. Hadron triggers are selected with $p_T>10$ GeV$/c$ in order to select a single hard process in the collision.

The recoil jet distribution (measured in $0-20\%$ central Pb-Pb collisions), plotted as function of the jet $p_T^{sub}=p_T^{reco}-\rho\times A$, is shown Fig. 7 (left panel). For jet $p_T<20$ GeV$/c$ the shape of the distribution is identical for all choices of trigger p_T. This p_T region corresponds to combinatorial background jets. For jet $p_T>20$ GeV$/c$ a clear evolution with the trigger p_T can be seen. This region, dominated by high Q^2 events, corresponds to the "signal" part of the recoil jet spectrum which depends strongly on the trigger p_T (as a consequence of the trigger bias effect: the p_T of selected partons increases while increasing the trigger p_T).

To get rid of the combinatorial jets (in a purely data driven way [30]), the difference (called Δ_{recoil}) of two measured jet distributions (with hadron triggers in different p_T intervals) is used:

$$\Delta_{\rm recoil}(p^{\rm ch}_{T,\rm Jet})=Y(p^{\rm ch}_{T,\rm Jet};p_T^{\rm min},p_T^{\rm max})-Y(p^{\rm ch}_{T,\rm Jet};p^{\rm min}_{T,ref},p^{\rm max}_{T,ref}) \tag{1}$$

with

$$Y(p^{\rm ch}_{T,\rm Jet};p_T^{\rm min},p_T^{\rm max})=\frac{1}{N_{\rm tr}}\frac{{\rm d}N(p^{\rm ch}_{T,\rm Jet};p_T^{\rm min},p_T^{\rm max})}{{\rm d}p^{\rm ch}_{T,\rm Jet}} \tag{2}$$

In addition, this method does not impose any bias on the fragmentation of the recoil jets.

The resulting Δ_{recoil} distribution (which is free of combinatorial jets) is then unfolded using both χ^2 minimization and the Bayes theorem. The difference between the two methods contributes to the anti-correlated shape uncertainty shown in Fig. 7 (right panel).

The ratio $\Delta I_{AA}=\Delta_{recoil}^{Pb-Pb}/\Delta_{recoil}^{PYTHIA}$ of the measured recoil distribution over the same distribution from PYTHIA is shown Fig. 7 (right panel).

This measurement was made for $R=0.2$ and $R=0.4$ jets with $p_T^{const}>0.15$ GeV$/c$ and $p_T^{const}>2$ GeV$/c$. Overall, $\Delta I_{AA}\simeq 0.6-0.7$ with no visible broadening of recoil jets from $R=0.2$ to $R=0.4$. In addition, the comparison of these distributions (relative to PYTHIA) does not indicate a large energy redistribution towards lower p_T constituents.

ALICE has also measured [31] the hadron $I_{AA}=Y_{PbPb}/Y_{pp}$ (where Y is the yield measured as the integral of a given $\Delta\phi$ correlation peak) from di-hadrons azimuthal correlations. On the away side a suppression by a factor two is seen (redistribution of the energy of the high

Journal of Physics: Conference Series **458** (2013) 012025 doi:10.1088/1742-6596/458/1/012025

p_T particle in the medium) which can be compared to ΔI_{AA}^{jet} from hadron-jet measurements: $\Delta I_{AA}^{jet} \geq$ hadron I_{AA}. We can also note that on the near side a $\sim 20\%$ yield enhancement was measured (effect of fragmentation after energy loss).

8. Summary

The inclusive differential jet cross-section measured in pp collisions is in good agreement with NLO pQCD calculations and provides an important reference for Pb-Pb studies. In heavy-ion collisions, the background contribution and its fluctuations have been studied in details. Jets are strongly suppressed in Pb-Pb collisions. This suggests significant out-of-cone radiation. Overall, hadron and jet observables tend to be in agreement ($R_{AA}^{jet} \sim$ hadron R_{AA} and $\Delta I_{AA}^{jet} \geq$ hadron I_{AA}). Recoil jets are found to be significantly less suppressed than inclusive jets (a similar behavior exists for single particle i.e R_{AA} vs I_{AA}).

A qualitative agreement between experiments exists. However, the measurements are not directly comparable as they use different jet reconstruction techniques, background estimation, etc. An apple to apple comparison would help to build a coherent picture of jet quenching at LHC.

Comparing jet measurements to theory predictions is essential to learn about the medium properties but it is also a very challenging task, which requires close collaboration between theorists and experimentalists in order to find good observables or even to define a common framework. In the near future, more differential measurements using larger cone radii as well as the recent p-A run at LHC will provide some additional constraints.

References

[1] Wiedemann U A 2009 (*Preprint* 0908.2306)
[2] Aamodt K *et al.* (ALICE Collaboration) 2008 *JINST* **3** S08002
[3] Abeysekara U *et al.* (ALICE EMCal Collaboration) 2010 (*Preprint* 1008.0413)
[4] Alver B, Baker M, Loizides C and Steinberg P 2008 (*Preprint* 0805.4411)
[5] Abelev B *et al.* (ALICE Collaboration) 2012 (*Preprint* 1210.4520)
[6] Cronin J, Frisch H J, Shochet M, Boymond J, Mermod R *et al.* 1975 *Phys.Rev.* D **11** 3105
[7] Accardi A 2002 (*Preprint* hep-ph/0212148)
[8] Adler S *et al.* (PHENIX Collaboration) 2003 *Phys.Rev.Lett.* **91** 072303 (*Preprint* nucl-ex/0306021)
[9] Adams J *et al.* (STAR Collaboration) 2003 *Phys.Rev.Lett.* **91** 072304 (*Preprint* nucl-ex/0306024)
[10] Arsene I *et al.* (BRAHMS Collaboration) 2003 *Phys.Rev.Lett.* **91** 072305 (*Preprint* nucl-ex/0307003)
[11] Back B *et al.* (PHOBOS Collaboration) 2004 *Phys.Rev.* C **70** 061901 (*Preprint* nucl-ex/0406017)
[12] Abelev B *et al.* (ALICE Collaboration) 2013 *Phys.Lett.* B **720** 52–62 (*Preprint* 1208.2711)
[13] Chatrchyan S *et al.* (CMS Collaboration) 2012 *Eur.Phys.J.* C **72** 1945 (*Preprint* 1202.2554)
[14] Horowitz W and Gyulassy M 2011 *J.Phys.* G **38** 124114 (*Preprint* 1107.2136)
[15] Buzzatti A and Gyulassy M 2012 *Nucl.Phys.A* (*Preprint* 1210.6417)
[16] Armesto N, Cole B, Gale C, Horowitz W A, Jacobs P *et al.* 2012 *Phys.Rev.* C **86** 064904 (*Preprint* 1106.1106)
[17] Aamodt K *et al.* (ALICE Collaboration) 2012 *Phys.Lett.* B **708** 249–264 (*Preprint* 1109.2501)
[18] Cacciari M, Salam G P and Soyez G 2008 *JHEP* **0804** 063 (*Preprint* 0802.1189)
[19] Cacciari M and Salam G P 2006 *Phys.Lett.* B **641** 57–61 (*Preprint* hep-ph/0512210)
[20] Abelev B *et al.* (ALICE Collaboration) 2013 (*Preprint* 1301.3475)
[21] Cacciari M and Salam G P 2008 *Phys.Lett.* B **659** 119–126 (*Preprint* 0707.1378)
[22] Abelev B *et al.* (ALICE Collaboration) 2012 *JHEP* **1203** 053 (*Preprint* 1201.2423)
[23] Aiola S (Collaboration for the ALICE) 2013 (*Preprint* 1304.6668)
[24] Verweij M (ALICE Collaboration) 2012 (*Preprint* 1210.8264)
[25] Reed R (ALICE collaboration) 2013 (*Preprint* 1304.5945)
[26] ATLAS-CONF-2011-075 (ATLAS Collaboration) 2011
[27] CMS-PAS-HIN-12-00 (CMS Collaboration) 2012
[28] Renk T 2012 *Phys.Rev.* C **86** 061901 (*Preprint* 1204.5572)
[29] Cunqueiro L (ALICE Collaboration) 2012 (*Preprint* 1210.7610)
[30] de Barros G, Fenton-Olsen B, Jacobs P and Ploskon M 2012 (*Preprint* 1208.1518)
[31] Aamodt K *et al.* (ALICE Collaboration) 2012 *Phys.Rev.Lett.* **108** 092301 (*Preprint* 1110.0121)

Journal of Physics: Conference Series **458** (2013) 012026 doi:10.1088/1742-6596/458/1/012026

Quarkonium and B_c mesons from Pb + Pb at LHC energies

Edwin Norbeck, Jane Nachtman and Yasar Onel

Department of Physics and Astronomy, University of Iowa, Iowa City, IA 52242, USA

E-mail: edwin-norbeck@uiowa.edu

Abstract. The $b\bar{b}$ (ϒ) mesons appear to be produced in the initial PbPb collision at 2.76 TeV per nucleon pair followed by partial melting in the hot quark-gluon plasma. In sharp contrast, the $c\bar{c}$ (J/ψ) mesons seem more likely to be formed by recombination at the hadronization stage. The B_c mesons, with one quark of each kind are seldom seen in pp collisions because a particle-antiparticle pair requires the simultaneous production of four heavy quarks. Although a family of B_c mesons have been predicted, only the ground state has been seen. If the $c\bar{c}$ mesons are produced by recombination, it could be expected that B_c mesons would be abundant with PbPb. Because the quark and antiquark have different flavor, the B_c are relatively long lived, 0.45 ps (to be compared with about 1.5 ps for the lighter B mesons). They would be seen with PbPb reactions by $B_c^{\pm} \rightarrow J/\psi\,(\mu^+\mu^-)\pi^{\pm}$ looking at muons and pions from displaced vertices.

1. Introduction

Quarkonium is a meson made of a quark-antiquark pair, analogous to positronium. The term is usually applied to the heaviest quarks, c and b. (The top quark is too short-lived to combine with anything.) Quarkonia can self annihilate. Of particular interest here is the decay into a $\mu^+\mu^-$ pair. The mass and kinetic energy of a quarkonium can be determined from the direction and energy of the two muons. The $\mu^+\mu^-$ derived mass spectrum, measured with CMS [1], is shown in figure 1. This figure is for $\sqrt{s} =$ 7 TeV protons. The corresponding figure for PbPb at $\sqrt{s_{NN}}$ = 2.76 TeV/nucleon pair is similar but with poorer statistics. The behavior of $c\bar{c}$ (J/ψ) and $\mathbf{b\bar{b}}$ (Y) in the quark-gluon plasma of PbPb reactions are strikingly different. $\bar{b}c$ and $b\bar{c}$ mesons, $B_c^{\pm}$, are of mass between J/ψ and Y with a decay mode to J/ψ $\pi^{\pm}$, seen as $\mu^+\mu^-\ \pi^{\pm}$. If the $c\bar{c}$ mesons are formed by recombination in PbPb reactions, it could be expected that the $B_c^{\pm}$ are abundantly produced in PbPb collisions by recombination.

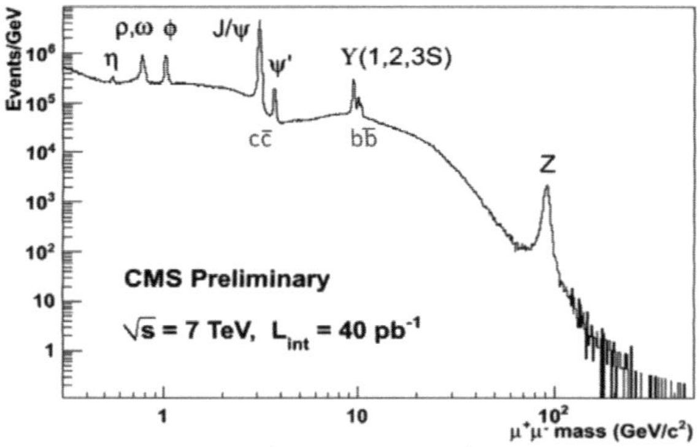

Figure 1. Invariant-mass spectrum from $\mu^+\mu^-$ measured with CMS.

2. Upsilon Υ(1S), Υ(2S), Υ(3S)

Melting of the excited states of quarkonia, $c\bar{c}$ (J/ψ) and $b\bar{b}$ (Υ), in the QGP is thought to provide a measure of the temperature in the region where they are formed [2]. They are formed early and provide a measurement of the maximum QGP temperature. A detailed discussion of suppression by color screening and enhancement by re-generation is given in [3]. The Υ(1S) is the most tightly bound quarkonium state, and so is the one with the highest dissociation temperature. As of this writing, the calculation of the temperature from the latest data has not been published. Figure 2 provides a visualization of these effects.

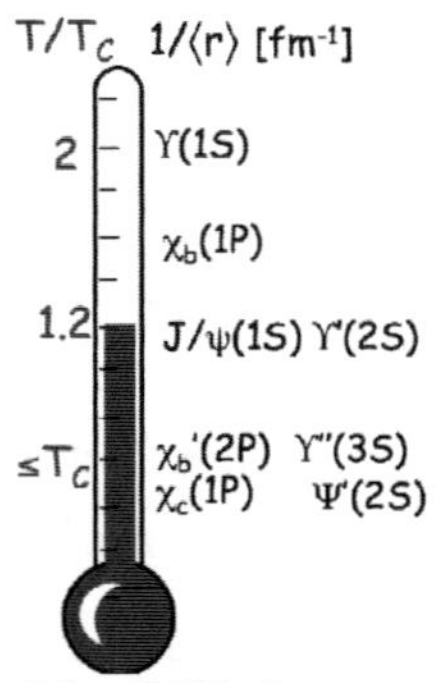

Figure 2. The QGP thermometer.

The prediction of the suppression pattern is complicated by various factors. These include feed-down contributions from higher-mass resonances into the observed quarkonium yields, as well as several competing nuclear and medium effects including recombination. These factors have played an important role in the interpretation of the charmonium measurements [4]. However, measurements of the excited state of the J/ψ, the ψ(2S), may require reconsideration of this analysis. The Υ states provide additional and theoretically cleaner probes of the QGP. The three Υ(nS) states are characterized by similar decay kinematics but distinct binding energies. They are useful for the measurement of relative state suppression because common experimental and theoretical factors cancel. The CMS detector clearly resolves the three Υ states below the open bottom threshold by their decay into muons making use of the excellent muon energy resolution of CMS [5].The ratios of the observed yields, figure 3, not corrected for differences in acceptance and efficiency, of the Υ(2S) and Υ(3S) states to the Υ(1S) state, in the PbPb and pp data, are [5]

$$\Upsilon(2S)/\Upsilon(1S)|_{pp} = 0.56 \pm 0.13 \text{ (stat.)} \pm 0.02 \text{ (syst.)},$$
$$\Upsilon(2S)/\Upsilon(1S)|_{PbPb} = 0.12 \pm 0.03 \text{ (stat.)} \pm 0.02 \text{ (syst.)},$$
$$\Upsilon(3S)/\Upsilon(1S)|_{pp} = 0.41 \pm 0.11 \text{ (stat.)} \pm 0.04 \text{ (syst.)},$$
$$\Upsilon(3S)/\Upsilon(1S)|_{PbPb} = 0.02 \pm 0.02 \text{ (stat.)} \pm 0.02 \text{ (syst.)} \quad (< 0.07 \text{ at } 95\% \text{ confidence level}),$$

where the systematic uncertainty arises from the fitting procedure. For the Υ(3S) to Υ(1S) ratio in PbPb, a 95% confidence level (CL) limit is set, based on the Feldman–Cousins statistical method [6]. There was a special pp run at $\sqrt{s} = 2.76$ TeV to allow a comparison of PbPb and pp cross sections. The shapes of the pp Υ spectra are essentially the same as at $\sqrt{s} = 7$ and 8 TeV.

The total Υ(nS) yields in PbPb and pp collisions can be compared after suitable corrections [5]. The corrected ratios, R_{AA}, for PbPb/pp are:

$$R_{AA}(\Upsilon(1S)) = 0.56 \pm 0.08 \text{ (stat.)} \pm 0.07 \text{ (syst.)},$$
$$R_{AA}(\Upsilon(2S)) = 0.12 \pm 0.04 \text{ (stat.)} \pm 0.02 \text{ (syst.)},$$
$$R_{AA}(\Upsilon(3S)) = 0.03 \pm 0.04 \text{ (stat.)} \pm 0.01 \text{ (syst.)} \quad (< 0.10 \text{ at } 95\% \text{ CL}).$$

As the Υ (3S) peak is not observed above the dimuon continuum (significance less than one standard deviation), only an upper limit is given. There is still no clear dependence on centrality for the Υ states in these data. The lifetimes of the three Υ states are long enough (of the order 5000 fm/c) that they get out of the collision region before they decay, but still so short that the decay products appear to come from the primary interaction region.

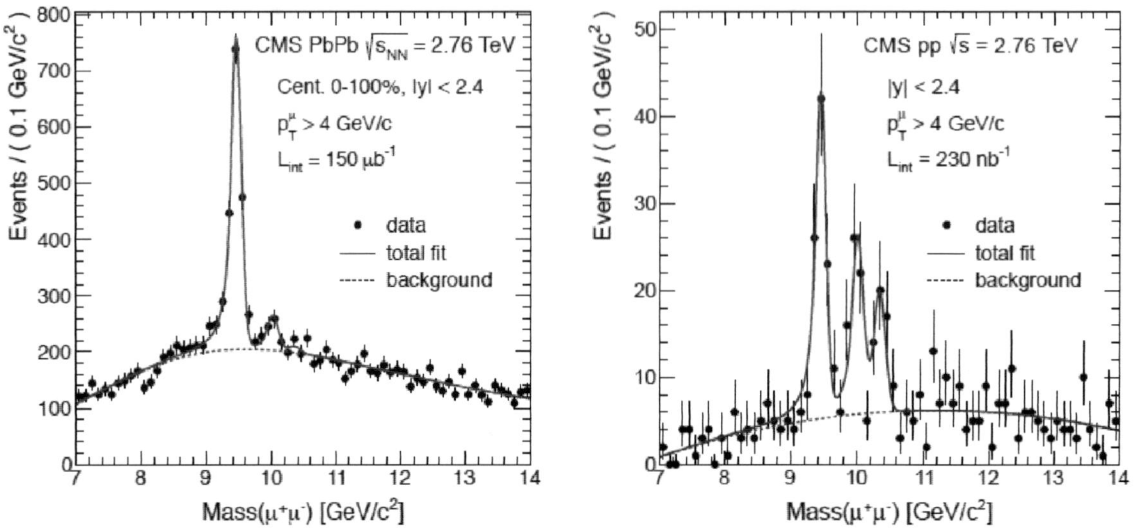

Figure 3. Dimuon invariant-mass distributions in PbPb (left) and pp (right) data at $\sqrt{s_{NN}}$ = 2.76 TeV. The same reconstruction algorithm and analysis selection are applied to both datasets, including a transverse momentum requirement on single muons of $p_T > 4$ GeV/c. The solid (signal + background) and dashed (background-only) curves show the results of the simultaneous fit to the two data sets [5].

3. J/ψ and ψ(2S)

Prompt and non-prompt J/ψ R_{AA} have been measured with CMS for both the 2010 and 2011 runs at $\sqrt{s_{NN}}$ = 2.76 TeV [7]. The non-prompt J/ψ are from various B hadrons with lifetimes sufficient for the vertex of the $\mu^+\mu^-$ from the J/ψ decay to be separated from the PbPb reaction vertex. About a quarter of the J/ψ were found to be non-prompt. The non-prompt fraction is found to vary slowly with p_T. Measurements have been made of the non-prompt fraction from pp at $\sqrt{s}$ = 7 TeV in the four LHC experiments with results varying from 10% to 60% depending on the J/ψ p_T [8-11]. Many more PbPb measurements, covering the complete range of p_T, will be needed before the results can be used as a measure of the relative attenuation of J/ψ and B mesons.

The ψ(2S), 3.683 GeV/c^2, is an excited state of the J/ψ, 3.097 GeV/c^2. It might be expected that the relative R_{AA} for the two states would provide a measure of the temperature, but there are complicating factors. The quantity of interest is the double ratio (N ψ(2S)/N J/ψ)PbPb/ (N ψ(2S)/N J/ψ)pp . The extent to which this ratio is less than one is a measure of the melting of the excited state, relative to the ground state, and is, therefore, a measure of the temperature. The instrumental effects cancel when all the measurements are made with the same detector. Measurements of the ratio were made with CMS using data from the 2011 PbPb run [12]. The results, figure 4, are considered preliminary because of uncertainties in the pp data.

One scenario is that the heavy-quark mesons are produced early by hard parton-parton collisions and that they then move through the nuclear medium. The attenuation in their number, given by R_{AA}, is caused by interactions in the medium. In this case R_{AA} would be less than one with the attenuation increasing as the size of the interaction region increases. The size of the interaction region is specified by N_{part}, the number of nucleons in the overlap region of the colliding nuclei. This can be related to the impact parameter, which is zero for the most central collisions. The less bound excited state would be attenuated more rapidly than the ground state, so that the ratio shown in figure 4 should decrease with N_{part}. Such predictions are in sharp disagreement with the data in figure 4. A different model assumes that the heavy quarks are produced early in primary hard collisions and do not combine into mesons until the system cools and hadronization occurs. With the large number of charm quarks and antiquarks produced in PbPb collisions (O(100)) at $\sqrt{s_{NN}}$ = 2.76 TeV, charmonia is produced at the hadronization stage of the PbPb collisions from initially uncorrelated c and $\bar{c}$ quarks produced in the initial inelastic collisions [13]. While no precise prediction of the relative probabilities to form a J/ψ or ψ(2S) state are provided, this mechanism could explain why ψ(2S) mesons were less suppressed than J/ψ in PbPb collisions. Such recombination effects would be contributing mostly at low charmonium p_T because the quarks are thermalized. The inclusive ψ(2S) yield is comprised of non-prompt ψ(2S) from b-hadron decays and directly produced ψ(2S). In contrast to prompt J/ψ, there is no significant feed down from heavier charmonium states. CMS has measured non-prompt contributions to the J/ψ and ψ(2S) inclusive yields in pp collisions at $\sqrt{s}$ = 7 TeV [14] that, in the forward region, are relatively similar and of the order of 20% for p_T around 6.5 GeV/c, increasing with p_T. For the data in figure 4, prompt and non-prompt J/ψ or ψ(2S) are not separated because of the limited ψ(2S) yield.

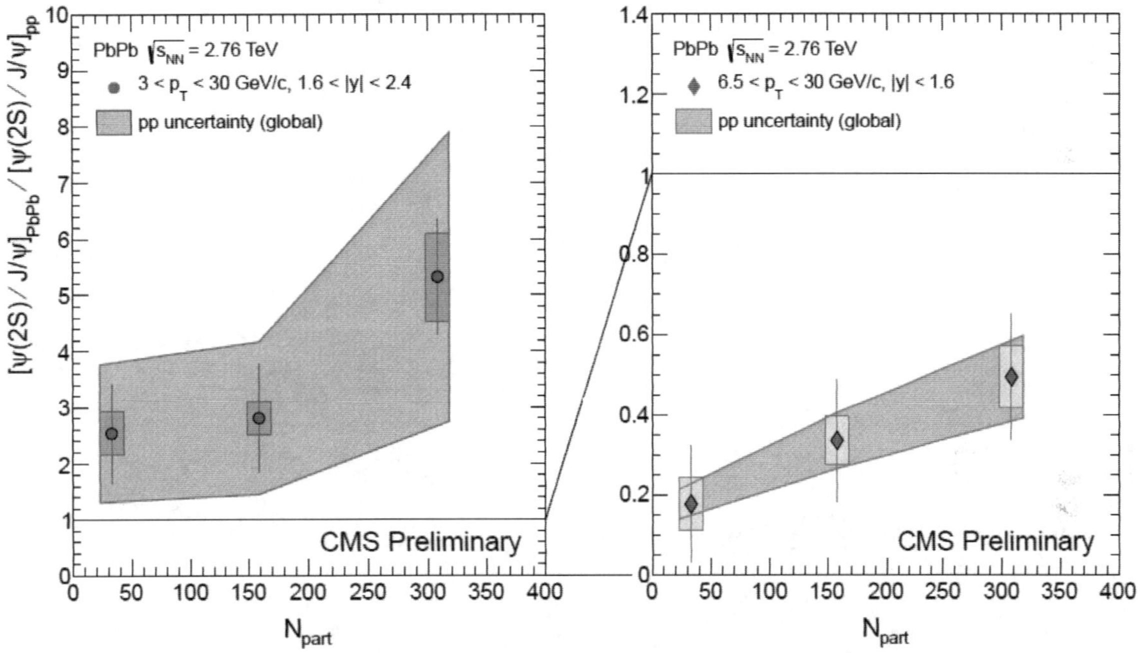

Figure 4. Measured yield double ratio $(N_{\psi(2S)}/N_{J/\psi})_{PbPb}/(N_{\psi(2S)}/N_{J/\psi})_{pp}$ as a function of the size of the interaction region. The p_T and rapidity bins are $3.0 < p_T < 30$ GeV/c and $1.6 < |y| < 2.4$ (left), and $6.5 < p_T < 30$ GeV/c and $|y| < 1.6$ (right). The error bars and boxes give the PbPb statistical and systematic uncertainties, respectively. The shaded band is the uncertainty on the pp measurement, common to all double-ratio points.

The increase in the (2S)/(1S) ratio with volume could be a reflection of the increase in the volume to surface ratio as the volume increases. If the hadronization occurs in a surface layer and the number of charmed quarks available for recombination is proportional to the volume, the concentration of charmed quarks at the surface would increase with volume thereby increasing the probability of recombination.

There is another possible explanation for the increase in the (2S)/(1S) ratio. A sufficiently energetic quark moving through the QGP is thought to produce a color charge analogue to Čerenkov radiation [15-16]. In this case the radiation is heavy quarkonium mesons rather than photons. This would provide a mechanism for the number of charmed quarks per unit volume to increase as the size of the reaction volume increases. The energetic light quarks that are producing the "radiation" have a larger distance available for travel.

4. B_c mesons

The $B_c^{\pm}$(1S)mesons have a mass of 6.273 GeV/c^2, intermediate between J/ψ at 3.097 GeV/c^2 and ϒ(1S) at 9.460 GeV/c^2 [17]. Because they cannot self annihilate, the life time is a relatively long 0.135 mm/c. The other B hadrons have even longer lifetimes of almost 500 mm/c. The B_c have a shorter lifetime because the weak decay can occur with either the c quark or the b quark. The B_c can be observed by the decay mode $B_c^{\pm} \rightarrow J/\psi + \pi^{\pm}$, with the J/ψ going to a muon pair that diverges from a point separated from the primary interaction (displaced vertex). They are difficult to produce with pp reactions because making a B_c pair requires the production of four heavy quarks. Of the positively charged B mesons from pp only 0.68% are B_c^+ [17]. The remaining 99.3% are $\bar{b}u$ with a mass of 5.279 GeV/c^2 and a lifetime of 0.492 mm/c.

B_c mesons have not been identified in PbPb studies, but if J/ψ is made primarily by recombination, it could be expected that recombination would also produce an abundance of B_c. The B_c mesons from recombination of thermalized quarks would have small p_T values. Prompt and non-prompt J/ψ decays to muon pairs have been measured for PbPb at = 2.76 TeV by CMS, figure 5. The fit assumes

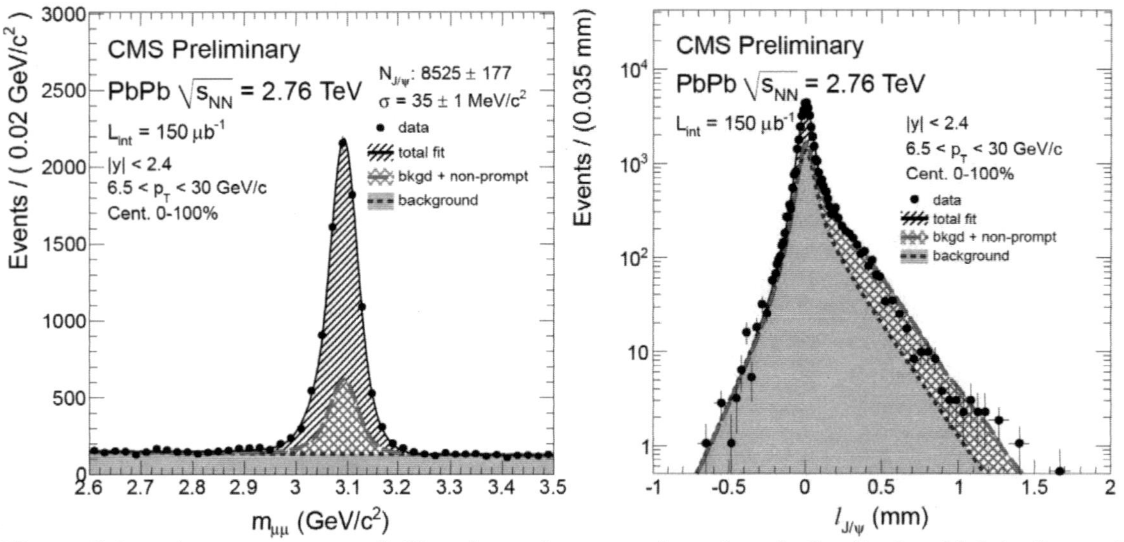

Figure 5. Invariant-mass spectra (left) and pseudo-proper decay length distribution (right) of opposite sign muon pairs. The background is shown in blue; just above that the non-prompt J/ψ are shown in red; and at the top are the prompt J/ψ decays [7].

that all of the non-prompt J/ψ result from decays of the various B mesons and baryons, all of which have about the same lifetime except for the B_c. The change in the slope at about 0.2 mm suggests the presence of a short-lived component, but it would be difficult to prove the existence of the B_c by decay rate vs. distance measurements alone, because of the large background that varies with distance.

If measurements of the invariant mass of μμπ events at small decay distances turned up even a few examples of masses close to that of the B_c, that would be sufficient to show that B_c are more abundantly produced with PbPb than with pp. There is a whole family of B_c, but only the ground state lives long enough to appear as a displaced vertex [18]. The pions and less than 500 MeV photons coming from the excited state B_c mesons, as they decay to the ground state, appear to come from the primary reaction vertex; for PbPb they would be buried in background.

The next PbPb run at the LHC will be at twice the energy. Because the production of heavy quarks increases faster than the background, the higher energy data should allow a more detailed study of the B_c mesons.

Journal of Physics: Conference Series **458** (2013) 012026 doi:10.1088/1742-6596/458/1/012026

5. Conclusions

It seems likely that PbPb reactions will make an abundance of B_c mesons, which are difficult to make with pp reactions. Heavy ion reactions should be considered as a possible source of new particles, both conventional and exotic particles.

References

[1] CMS Collaboration 2008 The CMS experiment at the CERN LHC *JINST* **3** S08004

[2] Matsui T, Satz H. 1986. J/ψ suppression by quark-gluon plasma formation. Phys. Lett. B178416

[3] Mócsy Á. 2009. Potential models for quarkonia. *Eur. Phys. J.* C **61**705–710

[4] Zhao X, Rapp R. 2011. Medium Modifications and Production of Charmonia at LHC. Nucl.Phys. A859 114-125

[5] Chatrchyan S, *et al.* (CMS Collab.) 2012. Observation of Sequential Υ Suppression in PbPb Collisions. *Phys. Rev. Lett.* **109** 222301

[6] Feldman GJ, Cousins RD. 1998. A unified approach to the classical statistical analysis of small signals. Phys. Rev. D 57 3873

[7] (CMS Collab.) 2012. Prompt and non-prompt J/ψ R_{AA} with 150 μb^{-1} integrated PbPb luminosity at $\sqrt{s_{NN}}$ = 2.76 TeV. CMS PAS HIN-12-014

[8] Book J (for ALICE Collab.) 2012. J/ψ Measurements with the ALICE Experiment at the LHC. *J. Phys.: Conf. Ser.* **389** 012028

[9] Aad G, *et al.* (ATLAS Collab.) 2011. Measurement of the differential cross-sections of inclusive, prompt and non-prompt J/ψ production in proton–proton collisions at $\sqrt{s}$ = 7 TeV. *Nucl.Phys.* B **850** 387-444

[10] Khachatryan V, *et al.* (CMS Collab.) 2011. Prompt and non-prompt J/ψ production in pp collisions at $\sqrt{s}$= 7 TeV. *Eur. Phys. J.* C **71** 1575

[11] Aaij R, *et al.* (LHCb Collab.) 2011. Measurement of J/ψ production in pp collisions at $\sqrt{s}$ = 7 TeV. *Eur. Phys. J.* C **71** 1645

[12] (CMS Collab.) 2012. Measurement of the ψ(2S) meson in PbPb collisions at $\sqrt{s_{NN}}$ = 2.76 TeV. CMS PAS HIN-12-007

[13] Andronic A, Braun-Munzinger P, Redlich K, Stachel J. 2007. Statistical hadronization of heavy quarks in ultra-relativistic nucleus–nucleus collisions. *Nucl. Phys.* A **789** 334-356

[14] Chatrchyan S, *et al.* (CMS Collab.) 2012. J/ψ and ψ(2S) production in pp collisions at $\sqrt{s}$ = 7 TeV. *JHEP* **02** 011

[15] Casalderrey-Solana J, Fernández D and Mateos D 2010 New Mechanism for Quark Energy Loss *Phys. Rev. Lett.* **104** 172301

[16] Mateos D 2011Guage/string duality applied to heavy-ion collisions: limitations, insights and prospects *J.Phys. G:Nucl.Part. Phys.* **38** 124030

[17] Aaij R *et al* (LHCb Collaboration) 2012 Measurements of B_c^+ Production and Mass with the $B_c^+ \rightarrow J/\psi + \pi^+$ Decay *Phys. Rev. Lett.* **109** 232001

[18] Godfrey S 2004 *Phys. Rev. D* **70** 054017

Journal of Physics: Conference Series **458** (2013) 012027 doi:10.1088/1742-6596/458/1/012027

Direct photon-hadron correlations with the PHENIX detector

J A Hanks for the PHENIX Collaboration
Department of Physics and Astronomy, Stony Brook University, Stony Brook, NY
E-mail: j.ali.hanks@gmail.com

Abstract. Direct photon tagged jets, in the form of photon-hadron correlations, are well suited to provide unique insight into how jets interact with the quark gluon plasma. Since photons do not interact strongly with medium produced at RHIC, at leading order the measured photon momentum approximately balances that of the away-side parton. The modification to the effective fragmentation function can be measured by comparing integrated away-side yields in direct photon-hadron correlations in Au+Au collisions to those in p+p. By varying the away-side integration range, the angular dependence of the observed modification can also be studied. Direct photon-hadron correlations have been measured with PHENIX in p+p and Au+Au using a statistical subtraction technique to remove the decay photon contribution from the inclusive photon-hadron correlations, with an additional isolation cut applied in p+p to reduce uncertainties. Recently published results showing a significant modification to the fragmentation function with an azimuthal angle dependence indicative of broadening are discussed in the context of related theoretical models and complementary results from other experiments.

1. Introduction

A variety of single-particle and two-particle correlation measurements illustrate well the suppression of high p_T hadrons, which is understood to be due to the energy loss of the parent parton as it traverses the medium produced in heavy-ion collisions. However, the details of parton-medium interactions going into such measurements are difficult to decouple due in part to lack of access to the original parton energy. The study of high p_T direct photons, which do not interact strongly with the medium and therefore escape unmodified and at leading order are produced back-to-back with a parton, predominately through the QCD Compton process, provides unique access to the energy of the parton (in this case a quark) prior to medium interaction. Through direct photon-hadron correlation measurements, the fragmentation function (FF) for the opposing jet, D(z) where $z = p_h/p_{jet}$, can be estimated by measuring $z_T = p_{h,T}/p_{\gamma,T} \approx z$. This is only an approximation due to initial state effects, such as the initial transverse momenta of the colliding partons in the nucleons, k_T, which can introduce a momentum imbalance between the photon and the opposing jet. Based on hadron-hadron correlation results, k_T has been measured to be ~ 3 GeV/c in both p+p and d+Au collisions, with no significant difference in the resulting smearing [1, 2]. This suggests that p+p should provide a good baseline estimate for the FF when studing medium modification to the effective fragmentation function in Au+Au.

Journal of Physics: Conference Series **458** (2013) 012027 doi:10.1088/1742-6596/458/1/012027

2. Method

Photon-hadron correlations are constructed by measuring the azimuthal angle, $\Delta\phi$, between trigger photons and associated charged hadrons in an event, with detector acceptance effects being corrected for by mixing trigger photons with associated hadrons not from the same event. The corrected correlations are then assumed to be a combination of jet constituents, the jet function, and a background of combinatorial pairs that are un-correlated, or correlated only due to event-level effects. In p+p collisions, where hard scattering processes dominate and event multiplicities are low, this background should result only in a small flat pedastal that can be estimated based on the zero-yield-at-minimum (ZYAM) assumption. In heavy-ion collisions, the high multiplicity of the event leaves a large background which includes an angular correlation between particles due to bulk anisotropies leading to flow. The total level of the background, b_0, is determined using the absolute normalization method [3]. The flow modulation can be decomposed into Fourier components, v_n, which can be measured independently [4] and included as additional terms in the background subtraction, as described in Eqn. 2. In the results discussed here only v_2 is included, with higher order effects estimated as an additional systematic uncertainty on the final results. The charged hadron efficiency, ϵ_a, is determined via a full GEANT simulation of the detector.

$$\frac{1}{N_t}\frac{dN^{pair}}{d\Delta\phi} = \frac{1}{N_t}\frac{N^{pair}_{real}}{\epsilon^a \int \Delta\phi}\left\{\frac{dN^{pair}_{real}/d\Delta\phi}{dN^{pair}_{mix}/d\Delta\phi} - b_0[1+2\langle v_2^t v_2^a\rangle cos(2\Delta\phi)]\right\}. \quad (1)$$

Ultimately, it is the yield of hadrons associated only with direct photons that will provide information about the fragmentations function. However, the trigger photons used to construct the jet functions, or per-trigger yields, are a combination of such direct photons and a large background of photons produced from meson decays, predominately from π^0 decays. To estimate this decay photon background, π^0-h correlations are measured and mapped to decay photon-hadron correlations according to the probability for a π^0 to decay into a photon within the relevant p_T range, as determined by a Monte Carlo study. The decay photon jet functions are then determined in the same way as the inclusive according to Eqn. 2. The γ_{direct}-h can then be extracted from the inclusive correlations using Eqn. 2

$$Y_{dir} = \frac{R_\gamma Y_{inc} - Y_{dec}}{R_\gamma - 1}, \quad (2)$$

where Y are the per trigger yields and R_γ is the ratio of the number of inclusive photons to the number of decay photons, which has been measured previously [5].

In p+p collisions, the significantly lower multiplicities make possible some additional event-by-event level cuts that can improve the signal to background for direct photons prior to estimating and subtracting the decay photon yield. First, photons are excluded which, when paired with another photon in the event, fall within the π^0 or η mass windows. Second, an isolation cut is applied requiring the total momentum and energy within a 0.3 rad cone around the trigger photon to be $< 10\% E_\gamma$. This restriction further reduces the contamination from decay photons, with the additional advantage that the contribution from fragmentation photons is greatly reduced [6]. These techniques greatly improve the measurement uncertainties while remaining consistent with the more straightforward statistical subtraction. Though an isolation cut is unfeasable in heavy-ion collisions, lack of a significant near-side yield, in conjunction with the consistency in the p+p results with and without this requirement, gives confidence that any observed modification is due to medium effects on the opposing jet.

3. Results

3.1. The p+p baseline

The yield on the awayside, $|\Delta\phi\pi| < \pi/2$, of direct photon-hadron correlations in p+p collisions is plotted as a function of $\xi = -ln(x_E)$ in Fig. 1, where $x_E = \frac{p_{h,T} cos\Delta\phi}{p_{g,T}} \approx z_T$. All the data points appear to lie on a universal curve, and agree well with the quark FF as measured by TASSO in e^+e^- collisions. Additionally, when plotted simply as a function of x_E and fit to an exponential, $dN/dz_T = Ne^{-bz_T}$, a slope of $b = 8.2 \pm 0.3$ is obtained, which is consistent with the expectation for quark jets of $b = 8$ [6]. These results support the claim that direct photon-hadron correlations indeed provide an effective measure of the quark FF. Furthermore, Fig. 2 shows the k_T vs p_T^{trig} values obtained from γ_{dir}-h correlations by fitting the $\sqrt{< p_{out}^2 >}$ results extracted from the $\Delta\phi$ distributions with a selection of LO+k_T smearing models. The measured k_T effect is consistent with that measured for π^0-h correlations and previous results, supporting the claim that any observed modification to the FF measured in p+p versus Au+Au will be dominated by medium effects.

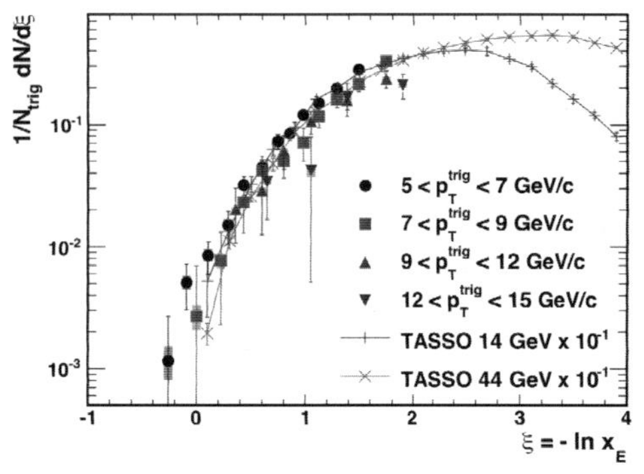

Figure 1. p+p integrated away-side yields as a function of $\xi(x_E)$ compared to e^+e^- TASSO data scaled to match the PHENIX acceptance.

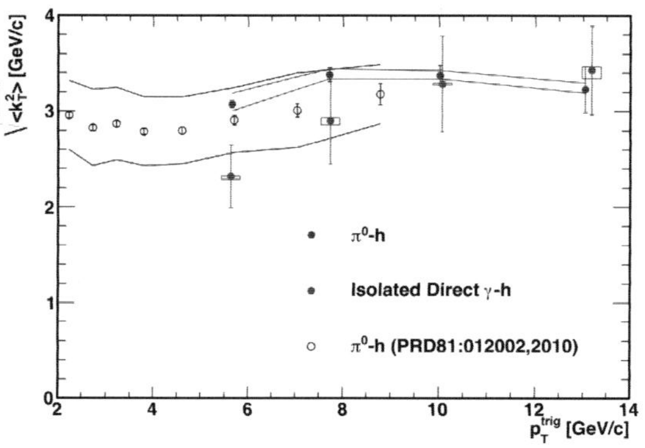

Figure 2. The k_T root-mean-square extracted from a LO+k_T smearing models for π^0-h (green) and γ_{dir}-h (red) correlations, compared to previous $\pi^0 - h$ results.

3.2. The modified FF

Based on the approximate x_E scaling in p+p, to improve statistics we use a wide trigger p_T bin of $5 - 9$ GeV/c for the purposes of comparing p+p and Au+Au results. The $\Delta\phi$ dependent per-trigger yields for p+p and Au+Au are shown in Fig. 3 as blue squares and open black circles respectively. Already a clear modification to the away-side distribution going from p+p to Au+Au is clear. On the other hand, the near-side distribution in Au+Au remains consistent with p+p, with an integrated yield within one sigma of zero, indicating that this measurement is not sensitive to the contribution from fragmentation photons. The ξ distribution (now defined based on z_T rather than x_E) obtained by integrating the full away-side ($|\Delta\phi - \pi| < \pi/2$ is shown in Fig. 4. At low ξ there is a clear suppression of the Au+Au relative to p+p, while at higher ξ there appears to be an enhancement. To further quantify this, the ratio of the yield in Au+Au to that in p+p, $I_{AA} = Y_{AA}/Y_{pp}$, is shown in the lower panel of Fig. 4. The suppression at low ξ is consistent with previous results [7], while the inclreased statistics and extended range acheived with these new results reveal a significant enhancement at high ξ. Also shown are two theoretical predictions based on models that track the redistribution of lost parton energy into

Journal of Physics: Conference Series **458** (2013) 012027 doi:10.1088/1742-6596/458/1/012027

enhanced soft particle production [8, 9]. The general trend anticipated by theory is consistent with what the data show, though quantitative comparison will require more direct calculations based on the relevant kinematic ranges acheivable with the available statistics.

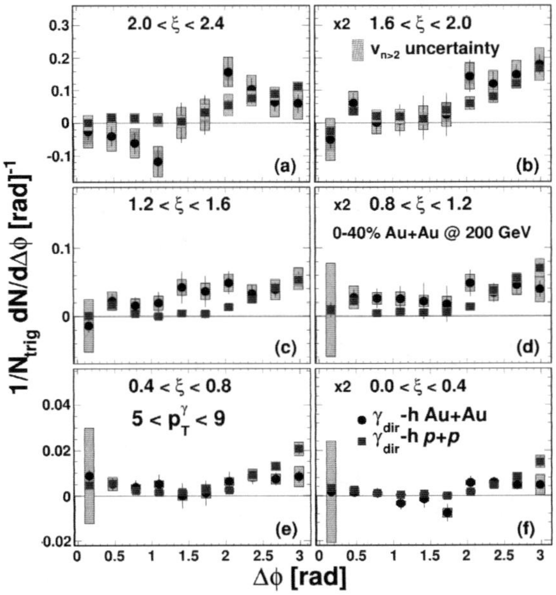

Figure 3. Direct photon-hadron $\Delta\phi$ dependent per-trigger yields in Au+Au (circles) and p+p (squares) in 0.4 ξ bins from 0 to 2.4. Systematic uncertainties due to potential unsubtracted flow effects shown separetly.

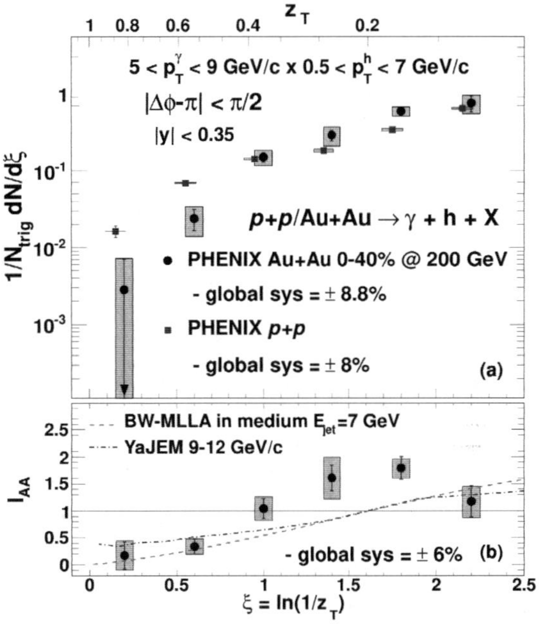

Figure 4. (a) The ξ dependent integrated away-side ($|\Delta\phi - \pi| < \pi/2$) per-trigger yields for Au+Au (circles) and p+p (squares). (b) The Au+Au to p+p ratio, I_{AA}, with comparison to BW-MLLA (dash) and YaJEM (dash-dot) calculations.

As Fig. 3 illustrates, the shape of the $\Delta\phi$ distribution in Au+Au appears modified relative to p+p. To quantify this effect beyond the examination of the integrated full away-side ($|\Delta\phi - \pi| < \pi/2$), two more restricted away-side integration ranges were selected, $|\Delta\phi - \pi| < \pi/3$ and $|\Delta\phi - \pi| < \pi/6$. The corresponding I_{AA} for these three integration ranges is shown in Fig. 5, as black circles blue squares and red triangles respectively. The suppression seen at low ξ is consistent for the three away-side definitions, while the relative enhancement seen at high ξ appears to be predominately at larger $\Delta\phi$, where the I_{AA} for the narrow ($|\Delta\phi - \pi| < \pi/6$) away-side remains consistent with one. The chosen integration ranges overlap, leading to correlation between the statistical and systematic uncertainties shown in the corresponding I_{AA} distributions. Therefore, to fully quantify the variation in I_{AA}, the ratio of the I_{AA} for the full away-side to the narrow away-side, in which these correlations will cancel, is shown in the bottom panel. A significant variation in the I_{AA} for $\xi > 1.0$ can be seen.

4. Conclusions

Evidence for modification to the jet fragmentation function due to medium interactions in heavy-ion collisions has been shown through comparison of direct photon-hadron correlations in $\sqrt{s_{NN}} = 200$ GeV Au+Au to p+p, using the photon as a proxy for the initial parton energy. Looking at the ratio of per-trigger yields we find a strong suppression at low ξ or high momentum fraction, z_T, consistent with previous results [7, 10]. The increased statistics and

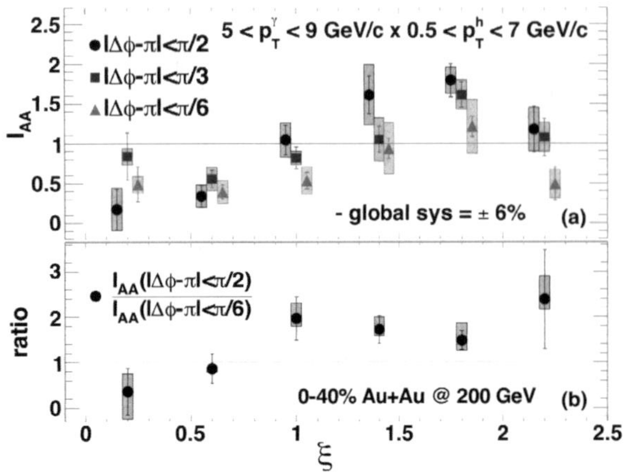

Figure 5. (a) The ξ dependent I_{AA} for different away-side integration ranges: $|\Delta\phi - \pi| < \pi/2$ (black circles), $|\Delta\phi - \pi| < \pi/3$ (blue squares), and $|\Delta\phi - \pi| < \pi/6$ (red triangles). (b) The ratio of the I_{AA} for the full ($|\Delta\phi - \pi| < \pi/2$) and narrow ($|\Delta\phi - \pi| < \pi/6$) away-side integration ranges.

larger z_T range accessed using a combination of data from 2007 and 2010 and going down to 0.5 in associated hadron p_T reveals an enhancement at high ξ indicative of increased soft particle production in response to the propagation of the initial parton through the medium. This enhancement is qualitatively consistent with the trend seen at the LHC through direct comparison of the fragmentation function for fully reconstructed jets in peripheral and central Pb+Pb [11, 12], and also with complementary measurements of jet-hadron correlations [13]. When the range of azimuthal angle integration of the away-side is restricted, this enhancement is reduced significantly, consistent with a broadening of the away-side distribution. Related measurements at LHC energies do not show a broadening of the distributions of fully reconstructed jets with respect to direct photon triggers [14]. However, evidence for the broadening of the jets themselves has been seen through studies of the jet profile, $\rho(r)$ [12], and the relative suppression of larger jet sizes (R=0.5,0.4,0.3) to a narrow baseline (R=0.2) [15]. Thus the angular dependence of the modification to I_{AA} shown here suggests that the medium enhances the production of soft jet fragments at large angles relative to the jet axis.

[1] Adler S S *et al.* (PHENIX Collaboration) 2006 *Phys. Rev. D* **74** 072002
[2] Adler S S *et al.* (PHENIX Collaboration) 2006 *Phys. Rev. C* **73** 054903
[3] Sickles A, McCumber M P and Adare A 2010 *Phys. Rev. C* **81** 014908
[4] Adare A *et al.* (PHENIX Collaboration) 2011 *Phys. Rev. Lett.* **107** 252301
[5] Afanasiev S *et al.* (PHENIX Collaboration) 2012 *Phys. Rev. Lett.* **109** 152302
[6] Adare A *et al.* (PHENIX Collaboration) 2010 *Phys. Rev. D* **82** 072001
[7] Adare A *et al.* (PHENIX Collaboration) 2009 *Phys. Rev. C* **80** 024908
[8] Borghini N and Wiedemann U Hep-ph/0506218 (2005)
[9] Renk T 2011 *Phys. Rev. C* **84** 067902
[10] Abelev B *et al.* (STAR Collaboration) 2010 *Phys. Rev. C* **82** 034909
[11] 2012 ATLAS-CONF-2012-115
[12] 2012 CMS-PAS-HIN-12-013
[13] Adamczyk L *et al.* (STAR Collaboration) ArXiv:1302.6184v1
[14] Chatrchyan S *et al.* (CMS Collaboration) 2013 *Phys. Lett. B* **718** 773
[15] Aad G *et al.* (ATLAS Collaboration) 2013 *Phys. Lett. B* **719** 220–241

Journal of Physics: Conference Series **458** (2013) 012028 doi:10.1088/1742-6596/458/1/012028

Prompt photon measurements with PHENIX's MPC-EX detector

Sarah Campbell for the PHENIX Collaboration
Building 510C, PHENIX, Brookhaven National Laboratoy, Upton, NY 11973
E-mail: campbels@iastate.edu

Abstract. The MPC-EX detector is a Si-W preshower extension to the existing Muon Piston Calorimeter (MPC). The MPC-EX consists of eight layers of alternating W absorber and Si mini-pad sensors. Located at forward rapidity, $3.1 < |\eta| < 3.8$, the MPC and MPC-EX will access low-x partons in the Au nucleus in p+Au collisions and high-x partons in the projectile in polarized p+p collisions. With the addition of the MPC-EX, the neutral pion reconstruction energy range extends to the luminosity limit, energies > 80 GeV, a factor of four improvement over current capabilities. Not only will the MPC-EX strengthen PHENIX's existing forward π^0 and jet measurements, it will provide sufficient prompt photon and π^0 separation to make a prompt photon measurement possible. Prompt photon yields at high p_T, $p_T > 3$ GeV/c, can be statistically extracted using the double ratio method. In transversely polarized p+p collisions, the measurement of the prompt photon single spin asymmetry, A_N, will resolve the sign discrepancy between the Sivers and twist-3 extractions of A_N. In p+Au collisions, the prompt photon R_{pAu} will quantify the level of gluon saturation in the Au nucleus at low-x, $x \sim 10^{-3}$, with a projected systematic error band a factor of four smaller than EPS09's current allowable range. The MPC-EX detector will expand our understanding of the gluon nuclear parton distribution functions, providing important information about the initial state of heavy ion collisions, and clarify how the valence parton's transverse momentum and spin correlates to the proton spin.

1. Introduction

The Muon Piston Calorimeter Extension (MPC-EX) is a new detector that will be located in front of PHENIX's existing Muon Piston Calorimeter (MPC), a lead-tungstate electromagnetic calorimeter. One of the primary goals of the MPC-EX upgrade is to measure forward prompt photons in transversely polarized p+p and p+A collisions. These measurements will provide information on the gluon distributions within nuclei and disclose how transverse momentum and spin within a valence parton relates to the spin of the proton or nucleus.

In these proceedings, potential measurements of forward prompt photons with the MPC-EX detector are presented. First, details of the detector design and functionality will be discussed. Then, the prompt photon measurements in p+Au collisions are described as they provide a mechanism to directly study the gluon parton distribution function inside the nucleus. Next, projections for prompt photon single spin asymmetry measurements, A_N, in transversely polarized p+p collisions are reported. These measurements will determine the sign of the Sivers effect. Finally, new opportunities to study the prompt photon A_N in polarized p+A collisions are detailed. This will quantify the changing levels of gluon saturation and the proportion of Sivers and Collins spin effects in various nuclei.

2. MPC-EX detector

The MPC-EX detector serves as both an electromagnetic preshower and charged particle tracker at forward rapidities, $3.1 < |\eta| < 3.8$. The duplicate functionality of the MPC-EX is possible because a dual gain SVX4 readout is used. The SVX4 readout provides the MPC-EX with the sensitivity to measure minimum ionizing peaks at 70 MeV and access full energy electromagnetic showers. The MPC-EX detector design is composed of eight layers of 2 mm tungsten absorber and $6.2 \times 6.2cm^2$ silicon mini-pad sensors with a total depth of around 50 mm. Each mini-pad sensor consists of a 4×32 array of mini-pads and each mini-pad has dimensions of $1.8 \times 15mm^2$. The mini-pad layers alternate in x- and y-orientation providing increased position resolution. The combined MPC and MPC-EX system will be able to provide charge track identification and reconstruct neutral pions out to the luminosity limit of energies greater than 80 GeV [1].

Existing forward neutral pion measurements in PHENIX are limited to energies below 20 GeV. Above this energy the pion's two decay photons merge into a single MPC cluster. With the MPC-EX, the pion reconstruction energy range will be extended by a factor of four to above 80 GeV. This is possible because of the improved position resolution of the early shower energy distribution as a result of the MPC-EX's mini-pad design. Given the detailed early shower information, the two centers of gravity for each photon shower in a π^0 decay can be identified and the mass is calculated for a single-track mass.

The increased pion reconstruction range provides the ability to separate high momentum neutral pions and prompt photons, making a forward prompt photon measurement possible at PHENIX. Forward prompt photons consist of three types of photons: direct photons produced primarily from quark-gluon Compton scattering, fragmentation photons emitted off a final state quark and QED radiation off of an initial state quark. Pythia simulations [2] are used to simulate these photon sources for the purpose of characterizing our detector response. In the case of direct and fragmentation photons, Pythia produces photons at a comparable level to next-to-leading order parton quantum chromo-dynamics (pQCD) calculations [3]. However, Pythia overestimates the production of QED radiation, which is included in the direct photon contribution in the next-to-leading order pQCD calculations. In the $3.1 < |\eta| < 3.8$ rapidity range, direct photons access high-x quarks in the polarized proton projectile with a range of $-0.4 < log_{10}(x_1) < 0$ and low-x gluons in the target proton or nucleus with a range of $-3.4 < log_{10}(x_2) < -1$.

3. Prompt photons in p+Au collisions

The gluon distribution in heavy nuclei, such as Au, show a strong suppression at low-x relative to the distributions in the proton called gluon shadowing. Gluon shadowing is thought to explain the suppressed production of forward mesons in d+Au collisions relative to p+p collisions [4]. Theoretical models calculate the gluon nuclear modification factor, R_G^A, to quantify this relative behavior according to Equation 1,

$$R_G^A\left(x, Q^2\right) = \frac{1}{A}\frac{xG_A\left(x, Q^2\right)}{xG_p\left(x, Q^2\right)} \tag{1}$$

where $G_A\left(x, Q^2\right)$ and $G_p\left(x, Q^2\right)$ are the gluon distribution function in the nucleus and the proton respectively and A is the number of partons in the nucleus.

Two competing pictures simultaneously describe these cold nuclear matter effects in p+Au collisions. These are non-perturbative extensions of pQCD and the color glass condensate (CGC) pictures. Non-perturbative extensions to pQCD use transverse momentum dependent parton distribution functions at low-x, a higher twist shadowing effect, initial state energy loss, absorption, modified structure functions and coherent effects to characterize the cold nuclear matter data [5]. The CGC picture uses a classical treatment of high gluon density characterized

Journal of Physics: Conference Series **458** (2013) 012028 doi:10.1088/1742-6596/458/1/012028

by a saturation scale, Q_{Sat}, with the assumption that α_S as a function of Q_{sat} is small. The CGC model predicts suppression at low-x, forward rapidity and central collision due to the resulting increased gluon density [6] [7]. Recent theoretical work has found an equivalence between the transverse momentum dependent parton distribution functions at low-x and the CGC model [8]. This provides hope that a similar equivalence may exist between CGC and higher twist shadowing effects. Reconciling these two pictures is important because they provide fundamental understanding of the partonic processes in nuclei and provide the initial conditions to heavy ion collisions at RHIC and the LHC.

We use EPS09 [9] as a baseline for comparison and to highlight the influence a MPC-EX measurement will contribute to our understanding of the gluon distribution at low-x. EPS09 is a next-to-leading order global analysis of nuclear parton distribution functions using cold nuclear matter data from SLAC, NMC, EMC, DIS, Drell Yan and PHENIX midrapidity π^0. Currently there are large uncertainties in the gluon nuclear parton distribution at low-x. This is due to a lack of data to characterize the gluon distribution at low-x. MPC-EX measurements at forward rapidities will clarify this low-x region in the EPS09 model and any other model of the behavior of low-x gluons.

By measuring prompt photons at forward rapidities in p+Au and p+p collisions, the nuclear modification factor, R_{pA}, can measure the level of gluon shadowing or saturation in nuclei. The prompt photon R_{pA}, Equation 2, is defined as the ratio of the prompt yield in p+Au collisions, γ^{pA}_{Prompt}, over the yield in p+p collisions, γ^{pp}_{Prompt}, scaled by the average number of binary collisions, N_{Coll}.

$$R_{pA} = \frac{1}{N_{Coll}} \frac{\gamma^{pA}_{Prompt}}{\gamma^{pp}_{Prompt}} \tag{2}$$

In particular, prompt photons provide a clear measurement of the gluon distribution as their production is dominated by quark-gluon processes as opposed to hadronic signals that are generated by a variety of processes. However, the prompt photon measurement in p+p and p+Au collisions must contend with large backgrounds from hadrons, neutral pions and other decay photons, primarily η decays. This analysis first removes as much of the hadron, π^0 and decay backgrounds as possible using the shower characteristics in the MPC and MPC-EX. Remaining backgrounds are removed through a double ratio calculation, Equation 3,

$$\gamma_{Prompt} = \gamma_{Incl} \times (1 - 1/R_\gamma) \tag{3}$$

where γ_{Incl} and π^0 are the inclusive photon and π^0 yields respectively. R_γ is the ratio of the measured inclusive photon-to-pion ratio over the known and simulated contributions from hadronic decays relative to the π^0, Equation 4. This method of background subtraction has previously been used in other analyses within the PHENIX collaboration [10].

$$R_\gamma = \frac{(\gamma_{Incl}/\pi^0)_{Meas}}{(\gamma_{Incl}/\pi^0)_{Sim}} \tag{4}$$

To simulate the MPC-EX and MPC system's ability to measure prompt photons, 868 million minimum bias p+p Pythia events are used with a 16.5 GeV MPC trigger and a 50 cm vertex cut. By limiting our prompt photon candidate sample to the highest energy cluster in the event and high transverse momentum clusters, $p_T > 3$ GeV/c, we reduce much of the neutral pion and hadronic backgrounds. Hadronic backgrounds are further removed by cutting on the energy deposited in the MPC-EX to reject minimum ionizing peaks at around 70 MeV. Clusters that have a single-track mass, described in Section 2, close to that of the pion are clearly single-track π^0 candidates and are rejected. Hadronic and single track π^0 showers in the MPC and MPC-EX tend to have wider shower shapes than prompt photon showers, by cutting on these distributions we can further discard these backgrounds. Additionally, isolation cuts are used to reduce the contribution of fragmentation photons in our prompt photon signal. A multivariate analysis was performed to optimize the MPC and MPC-EX shower shape and isolation cuts.

The resulting prompt photon candidates are shown in Figure 1. Prompt photons have a 30% efficiency, while π^0 have a 3% efficiency resulting in a 49% signal-to-π^0 ratio and an R_γ value of 1.34. Direct photons from quark gluon Compton scattering make up 57% of the prompt photon signal. Remaining fragmentation photons are produced by quark gluon processes 93% of the time and quark anti-quark processes 7% of the time. The direct photon concentration can be increased by tightening isolation and width cuts; this is discussed in further detail in Section 4.

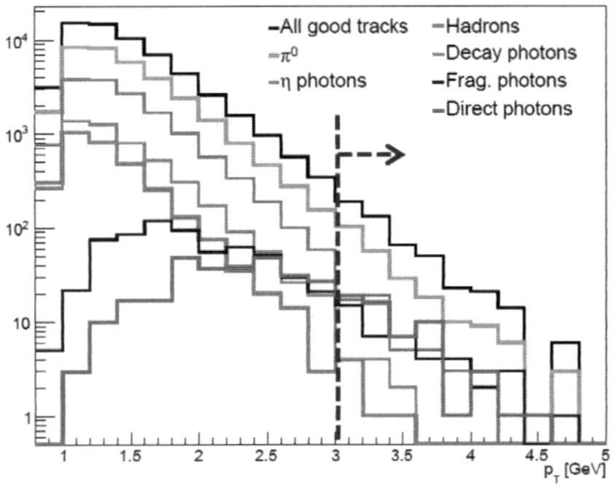

Figure 1. The p_T distribution of prompt photon candidates. The sum of all remaining candidates is shown in black. The remaining π^0 and η contributions are shown in bright green and olive green respectively. Remaining decay photons, primarily from ω and η' decays, are shown in light blue and hadrons are shown in pink. The prompt photons are shown separately for direct and fragmentation photons in red and blue respectively. The prompt photon measurement considers the projected yields at p_T above 3 GeV; this is noted on the plot with a blue dashed line.

3.1. Systematic errors

Our ability to measure gluon saturation or shadowing effects in forward prompt photon yields in p+Au and p+p collisions is limited not by statistics but by systematic errors. The double ratio method allows for the cancelation of many of these systematic errors. The systematic error in R_γ is calculated according to Equation 5,

$$\frac{\delta R_\gamma}{R_\gamma} = \frac{\delta \gamma_{Incl}}{\gamma_{Incl}} \oplus \frac{\delta \pi^0}{\pi^0} \oplus \frac{\delta MC_{Sim}}{MC_{Sim}} \tag{5}$$

where MC_{Sim} is the simulated inclusive photon-to-pion ratio, R_γ is defined in Equation 4, and γ_{Incl} and π^0 are the inclusive photon and neutral pion yields.

Given our experience with the MPC [11] and the MPC-EX's capabilities, the π^0 contribution can be measured to within an estimated systematic error of 6%, including an absolute energy scale uncertainty of 1 to 2%. A 4% relative systematic error in the Monte Carlo simulation of the inclusive photon-to-pion ratio is consistent with the systematic errors determined in PHENIX's direct photon measurement [10]. The error in γ_{Incl} is dominated by the subtraction of the remaining hadronic background. Because this subtraction is model dependent, a 20% error in the component is assumed, resulting in a 0.65% relative systematic error in γ_{Incl}. These errors combine to produce a relative error in R_γ of 7.22%. This is a correlated error between p+p and p+Au measurements. There are additional uncorrelated systematic errors in R_γ and γ_{Incl} of 1% and 2% respectively due to shifts in the energy scale in p+p and p+Au collisions. These systematic errors are summarized in Table 1.

These systematic errors characterize the MPC-EX's ability to constrain the allowable band of gluon modification from EPS09 [9]. Photon events are weighted by their x_2 and Q^2 according to EPS09 and R_{pAu} values are generated for each of the theoretical curves. The EPS09 central

Table 1. Correlated and uncorrelated systematic errors for the prompt photon measurement in p+p and p+Au collisions with the MPC-EX.

Quantity	Corr. errors	Uncorr. errors
R_γ	7.22%	1%
γ_{Inc}	0.65%	2%
N_{Coll}	–	3%

value of gluon suppression is assumed. The R_γ and γ_{Incl} in p+p and p+Au collisions are varied within their three sigma systematic errors in each direction. For each EPS09 curve the χ^2 is calculated with the simulated data. These χ^2 values are used to determine which curves are within one sigma and within the 90% confidence level of the simulated data. Figure 2 shows these one-sigma and 90% confidence level regions in for the current EPS09 results (hatched) and with the simulated MPC-EX measurement (dark and light blue). When the minimum or maximum EPS09 values are assumed smaller bands are found.

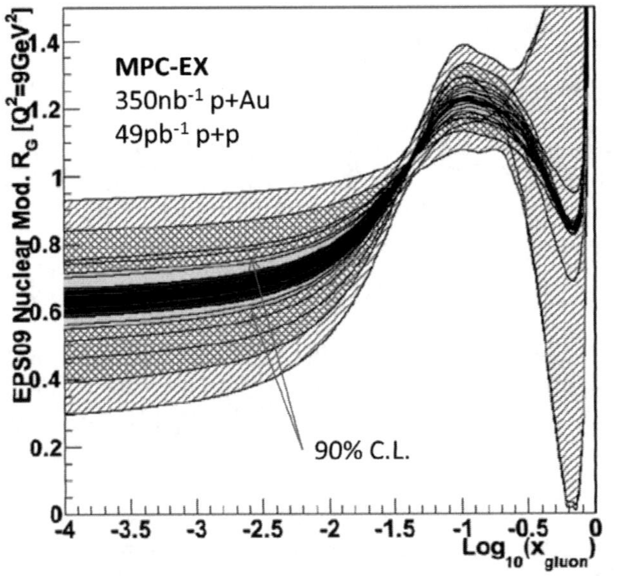

Figure 2. EPS09 exclusion plots in R_G vs $Log_{10}(x_{gluon})$ using the central EPS09 value. The outer hatched bands are the 90% confidence level region of curves. The light blue area represents the 90% confidence level region with the simulated MPC-EX measurement. The dark blue band is the one-sigma region with the simulated MPC-EX measurement.

4. Prompt photon A_N in transversely polarized p+p collisions

Large transverse single spin asymmetries have been measured at FermiLab [12] and RHIC [13] [14]. A particle's single spin asymmetry quantifies the left-right bias in particle production as defined in Equation 6,

$$A_N = \frac{1}{P}\frac{\sigma_L - \sigma_R}{\sigma_L + \sigma_R} \tag{6}$$

where σ_L and σ_R are the left and right particle cross-sections and P is the polarization. The mechanisms that generate these asymmetries are related to the initial state effects such as the Sivers Effect or final state effects such as the Collins Effect. The Sivers Effect is a left-right asymmetry associated with the correlation between the parton's transverse momentum and the nucleon's spin [15]. The Collins Effect is again a left-right asymmetry where the bias is generated by the quark's transverse spin-dependent fragmentation [16]. Semi-inclusive deep-inelastic scattering (SIDIS) results [17] [18] have constrained these two effects using transverse momentum dependent parton distribution extractions [19] [20]. An alternate extraction of these effects is determined using twist-3 transverse-spin-dependent correlations between quarks and gluons. An unanswered question in the field is the sign of the Sivers Effect. The transverse momentum dependent SIDIS extraction of the Sivers Effect produces negative values while with the twist-3 extraction positive values are produced [21] [22].

Preliminary PHENIX results using the MPC have measured large positive transverse single spin asymmetries in forward neutral clusters [23] and identified η mesons [24]. However, these measurements cannot isolate the sources of these single spin asymmetries because they sample multiple partonic processes. With the inclusion of the MPC-EX, the measurement of the prompt photon single spin asymmetry in 200 GeV transversely polarized p+p collisions is possible. The prompt photon single spin asymmetry is dominated by the scattering of a valence quark in the polarized projectile off of a gluon in the proton with only a small contribution for transversity and polarized fragmentation functions [22]. The prompt photon A_N measurement will provide a targeted measurement of the Sivers contribution to the single spin asymmetry. It will determine the sign of the Sivers Effect.

The prompt photon single spin asymmetry, $A_N^{\gamma Prompt}$, must be extracted from the inclusive photon asymmetry, $A_N^{\gamma Incl}$, by removing the non-zero asymmetry contributions from π^0, η and other decay photons. This is done according to Equation 7,

$$A_N^{\gamma Prompt} = (1 + 1/r)\, A_N^{\gamma Incl} - 1/r A_N^{BG} \tag{7}$$

where A_N^{BG} is the background asymmetry and r is the prompt photon-to-photonic background ratio. Equation 8 can determine our sensitivity to measure $A_N^{\gamma Prompt}$ with the MPC-EX by considering the statistical errors in the inclusive photon and background single spin asymmetries.

$$\left(\delta A_N^{\gamma Prompt}\right)^2 = (1 + 1/r)^2 \left(\delta A_N^{\gamma Incl}\right)^2 + \left(1/r \times \delta A_N^{BG}\right)^2 \tag{8}$$

For this calculation we use the projected statistical errors of the inclusive photon analysis detailed in Section 3 assuming 50 pb^{-1} of transversely polarized p+p collisions. The high p_T prompt photon measurement is divided into four p_T bins. Additionally, we assume that the MPC-EX can measure the background asymmetries with a minimum of twice the inclusive photon asymmetry precision in each bin, and we suppose a 60% polarization. The sensitivity of the MPC-EX high-x prompt photon A_N measurement as a function of x_F is shown in Figure 3 compared with the twist-3 extraction of the A_N and two semi-inclusive deep inelastic scattering extractions using alternate functional forms. With the MPC-EX's projected sensitivity, the sign of the Sivers Effect will be determined.

As the prompt photon A_N is measured, instead of the cleaner direct photon asymmetry, the contributions of fragmentation photons must be considered. By tightening isolation and shower width cuts in the simulated prompt photon analysis, we can increase the direct photon concentration in the prompt photon distribution from 57.4% to 78.6% while maintaining a relatively stable ratio of the prompt photon signal to hadronic decay photon background of between 0.34 and 0.31.

By looking at the prompt photon single spin asymmetry at low-x in 200 GeV transversely polarized p+p collisions, we can learn about the low-x gluon distribution in the proton. There are various models of the tri-gluon correlation functions, $O(x)$ and $N(x)$, that describe the gluon distribution's spin contribution [25]. These models will be primarily constrained by the open heavy flavor production at RHIC. However, the prompt photon A_N at low-x will serve as a complementary measurement that can further constrain these models.

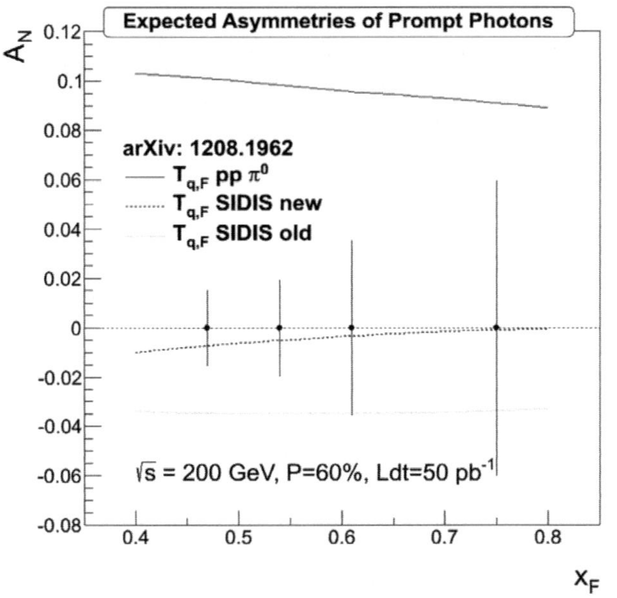

Figure 3. The sensitivity of high-x prompt photon A_N in $\sqrt{s_{NN}} = 200$ GeV transversely polarized p+p collisions as a function of x_F compared to the twist-3 (red solid line) and semi-inclusive deep inelastic scattering (blue dashed and black dotted lines) extractions of the Sivers functions using two different functional forms [21] [22]. The error bars contain the statistical errors and the uncertainties due to subtracting the decay photon background asymmetries.

5. Measurements in polarized p+A collisions

RHIC provides a unique opportunity to study transversely polarized p+A collisions [26], generating a new class of measurements that will provide information on the saturation scale, Q_{Sat}, and potential variations in the Collins and Sivers contributions in nuclei. The prompt photon A_N in polarized p+A collisions may have access to changes in the proportion of Sivers and Collins contributions to the single spin asymmetry. The single spin asymmetry of neutral pions in polarized p+A collisions, A_N^{pA}, relative to the p+p single spin asymmetry, A_N^{pp}, is expected to probe the saturation scales in p+A and p+p collisions in a color-glass framework. By varying the nuclei collided and the centrality of the collisions, the p_T and Q_{Sat}^p/Q_{Sat}^A dependence of the A_N^{pA}/A_N^{pp} ratio can be determined [27]. By measuring these single spin asymmetries at forward rapidities, the MPC-EX will access the kinematic region where p_T is close to Q_{Sat}.

6. Conclusion

The addition of the MPC-EX to the PHENIX experiment will provide a variety of new measurements in transversely polarized p+p and p+A collisions at RHIC. The prompt photon measurement in p+Au collisions will clarify the nuclear gluon parton distributions function at low-x. This will further our understanding of gluon saturation and shadowing in the nucleus and set the initial conditions for heavy ion collisions at RHIC and the LHC. In transversely polarized p+p collisions, the prompt photon single spin asymmetry will determine the sign of the Sivers Effect. This measurement not only will resolve a conflict in the extractions of the Sivers Effect but it also allows us to isolate the Sivers Effect from the other sources of single spin asymmetries. This measurement in polarized p+p collisions with the MPC-EX will deepen our understanding of the hadron structure and advance the application of fundamental QCD to p+p collisions. Finally, single spin asymmetry measurements in transversely polarized p+A collisions were briefly mentioned. While theoretical work in this area is ongoing, a systematic study of these single spin asymmetries in polarized p+A collisions will allow us to study the saturation scale and the spin contributions in nuclei. The MPC-EX provides access to an exciting new suite of measurements at RHIC.

References

[1] Campbell S *et. al.* 2012 A Proposal for the Muon Piston Calorimeter Extension (MPC-EX) to the PHENIX Experiment at RHIC arXiv:nucl-ex/1301.1096
[2] Sjostrand T *et. al.* 2001 High-energy-physics event generation with pythia 6.1 *Comput. Phys. Commun.*, 135:238-259. (arXiv:hep-ph/0010017)
[3] Werner Vogelsang. private communication, 2011.
[4] Arsene I *et. al.* 2004 *Nucl. Phys. A* **757** 1
[5] Nestor A 2006 *J. Phys. G* **32R** 367
[6] McLerran L D and Venugopalan R 1994 *Phys. Rev. D* **49** 2233
[7] Kharzeev D, Levin E and Nardi M 2004 *Nucl. Phys. A* **730** 448
[8] Dominguez F, Marquet C, Xiao B-W and Yuan F 2011 *Phys. Rev. D* **83** 105005
[9] Eskola K J, Paukkunen H and Salgado C A 2009 *JHEP* **0904** 065
[10] Adare A *et. al.* 2012 *Phys. Rev. Lett.* **109** 122302
[11] Meredith B 2011 A Study of Nuclear Effects Using Forward-Rapidity Hadron Production and Di-Hadron Angular Correlations in d+Au and p+p Collisions with the PHENIX detector at RHIC *Ph.D. Thesis, University of Illinois at Urbana-Champaign*
[12] Adams D L *et. al.* 1991 *Phys. Lett. B* **264** 462
[13] Abelev B I *et. al.* 2008 *Phys. Rev. Lett.* **101** 222001
[14] Arsene I *et. al.* 2008 *Phys. Rev. Lett.* **101** 042001
[15] Sivers D 1990 *Phys. Rev. D* **41** 83
[16] Collins J C 1993 *Nucl. Phys. B* **396** 161
[17] Airapetian A *et. al.* 2005 *Phys. Rev. Lett.* **94** 012002
[18] Airapetian A *et. al.* 2010 *Phys. Lett. B* **693** 11
[19] Anselmino M 2005 *Phys. Rev. D* **72** 094007
[20] Anselmino M 2009 *Nucl. Phys. Proc. Suppl.* **191** 98
[21] Kang Z-B, Qui J-W, Vogelsang W and Yuan F 2011 *Phys. Rev. D* **83** 094001
[22] Gamberg L and Kang Z-B 2012 Single transverse spin asymmetry of prompt photon production arXiv:hep-ph/1208.1962
[23] Koster J 2010 Measurment of Transverse Spin Asymmetries in Polarized p+p Collisons, and the Realization of New Electromagnetic Calorimeters for Forward Physics *Ph.D. Thesis, University of Illinois at Urbana-Champaign*
[24] Kleinjan D 2011 *AIP Conf. Proc.* **1441** 241
[25] Koike Y and Yoshida S 2012 *Phys. Rev. D* **85** 034030
[26] Tepikian S and Trbojevic D CAD Study CA/AP/#447
[27] Kang Z-B and Yuan F 2011 *Phys. Rev. D* **84** 034019

29th Winter Workshop on Nuclear Dynamics (WWND2013) IOP Publishing
Journal of Physics: Conference Series **458** (2013) 012029 doi:10.1088/1742-6596/458/1/012029

Dihadron Correlations Relative to the Event Plane in 200 GeV Au+Au Collisions from STAR

Fuqiang Wang (for the STAR Collaboration)
Department of Physics, Purdue University, West Lafayette, Indiana 47907, USA
E-mail: fqwang@purdue.edu

Abstract. Dihadron correlations with a high-p_T trigger particle are analyzed by STAR relative to the event plane in Au+Au collisions at 200 GeV (arXiv:1010.0690v1). The elliptic and quadrangular flow anisotropies are subtracted. The remaining dihadron correlation signals are found to be composed of a near-side peak, which is separated into jet-like and "ridge"-like components, and an away-side correlation structure. The ridge-like structure is found to decrease with the trigger particle azimuthal angle relative to the event plane from in-plane to out-of-plane. The away-side structure is found to evolve from single-peak for in-plane triggers to double-peak for out-of-plane triggers. Is the dihadron correlation signal just a manifestation of the triangular and higher-order harmonic flows? This talk addresses this question, exploring the answers using limiting-case assumptions, and discusses the implications of the data regarding the ridge and the double-peak structure.

1. Introduction

Collisions at the Relativistic Heavy Ion Collider (RHIC) of Brookhaven National Laboratory have created a hot and dense medium with properties that resemble a nearly perfect fluid of strongly interacting quarks and gluons (sQGP) [1, 2, 3, 4]. Energetic partons produced by rare hard-scattering processes traverse and interact with the sQGP, losing energies resulting in suppression of high transverse momentum (p_T) particle yields and modifications to jet-like correlations [5, 6]. Several novel phenomena have been observed in jet-like correlations with high-p_T trigger particles. On the near-side of the high-p_T trigger particle, correlated particles in $\Delta\phi$ are found to extend to large pseudo-rapidities (η) [6, 7, 8, 9]. On the away-side of the trigger particle, the correlation function is found to be double-peaked away from the back-to-back direction of $\Delta\phi = \pi$ [6, 10, 11, 12].

The effect of jet-medium interactions is expected to depend on the path-length of the sQGP medium the away-side jet traverses. STAR has analyzed dihadron correlations as a function of the trigger particle relative to the event plane in arXiv:1010.0690v1 [13]. Indeed the correlations are found to be dependent on the trigger particle direction. In particular, the ridge correlation magnitude is strongest for in-plane trigger particles and decreases with increasing ϕ_s from in-plane to out-of-plane. The away-side correlation is found to be single peaked at $\Delta\phi = \pi$ for in-plane triggers, but become double-peaked for out-of-plane triggers. The results indicate that the ridge is dominated by in-plane triggers and the away-side double-peak is dominated by out-of-plane triggers.

The major backgrounds to jet-like correlations are two-particle correlations caused by single-particle anisotropic flows. The elliptic and quadrangular flow backgrounds have been subtracted

Journal of Physics: Conference Series **458** (2013) 012029 doi:10.1088/1742-6596/458/1/012029

in dihadron correlations as a function of ϕ_s in [13]. It has been recently suggested, however, that there should be triangular anisotropic flow [14, 15]. The triangular anisotropy is a result of hydrodynamics evolution from initial-state geometry triangularity due to event-by-event fluctuations [16, 17, 18]. It is demonstrated illustratively by the NeXSPheRIO model [19, 20, 21] that initial energy density fluctuations (hot spots) with subsequent hydro evolution may generate a near-side ridge and a double-peak correlation on the away side. Triangular anisotropic flow would result in three peaks at $\Delta\phi = 0$, $2\pi/3$, and $4\pi/3$ in two-particle correlations. These peaks would qualitatively explain the observed ridge and double-peak structure in dihadron correlations.

The natural question is then, whether hydrodynamic triangular flow can explain entirely the observed ridge and double-peak correlations? This talk attempts to answer this question.

The current understanding of anisotropic flow backgrounds, up to the fourth-order harmonic, is given by:

$$\frac{dN}{d\Delta\phi} = B\Big[1 + 2v_2^{(a)}v_2^{(t,R)}\cos(2\Delta\phi) + 2v_4^{(a)}\{\psi_2\}v_4^{(t,R)}\{\psi_2\}\cos(4\Delta\phi) + 2v_3^{(a)}v_3^{(t)}\cos(3\Delta\phi) + 2V_4\{\text{uc}\}\cos(4\Delta\phi)\Big]. \quad (1)$$

Here $v_n^{(t)}$ and $v_n^{(a)}$ are the trigger and associated particle n^{th} harmonic flow anisotropy, respectively; $v_2^{(t,R)} = \langle\cos 2(\phi_t - \psi_2)\rangle^{(R)}$ and $v_4^{(t,R)} = \langle\cos 4(\phi_t - \psi_2)\rangle^{(R)}$ are, respectively, the trigger particle second and fourth harmonic anisotropies with respect to the second harmonic plane, ψ_2 and averaged with a given ϕ_s range (see [13] for details); and $V_4\{\text{uc}\}$ is the component of the fourth harmonic anisotropy uncorrelated to ψ_2. It is assumed in Eq. (1), and verified by experiments [22, 23], that v_3 fluctuation effects are uncorrelated to ψ_2.

In the following sections we discuss the flow backgrounds of Eq. (1) in detail. We focus on the dihadron correlations at large $|\Delta\eta| > 0.7$ in the interest to study the near-side ridge correlations as well as the away-side correlations.

2. Systematic uncertainty of v_2

For the dihadron correlation data presented in [13], the maximum systematic uncertainty bound of v_2 is obtained by $\langle\cos 2\Delta\phi\rangle$ averaged over the away-side of two-particle correlations between the particle of interest and a reference particle in the flow measurements in order to avoid near-side small angle nonflow contributions. The results [13] are shown in the first row of Fig. 1. This would be the correct maximum systematic bound of v_2 if only even harmonic flows are present. With non-vanishing odd harmonic flows, this estimate of the maximum systematic bound would not necessarily be the maximum v_2. Therefore, we now use the $\langle\cos 2\Delta\phi\rangle$ averaged over the entire 2π range as the upper bound of the v_2 systematic uncertainty. This is larger than the previous estimate because of the inclusion of large nonflow contributions from near-side correlations. The two-particle correlations to obtain the maximum v_2 are restricted to $|\Delta\eta| > 0.7$ because the dihadron correlations we study are from the same $|\Delta\eta| > 0.7$ region. The new estimated systematic upper bound is the maximum possible v_2 that should be subtracted from raw dihadron correlations. The low systematic bound of v_2 is still the four-particle cumulant measurement as in [13]. The default v_2 is the average of $v_2\{4\}$ and $v_2\{2\}$. The dihadron correlation results with the new estimate of v_2 systematic uncertainties are shown in the second row of Fig. 1.

3. Subtraction of v_3

The v_3 of trigger and associated particles are measured by the two-particle cumulant method with a reference particle of $0.15 < p_T < 2$ GeV/c [23]. A η-gap (η_{gap}) of 0.7 is applied between the particle of interest and the reference particle.

Journal of Physics: Conference Series **458** (2013) 012029 doi:10.1088/1742-6596/458/1/012029

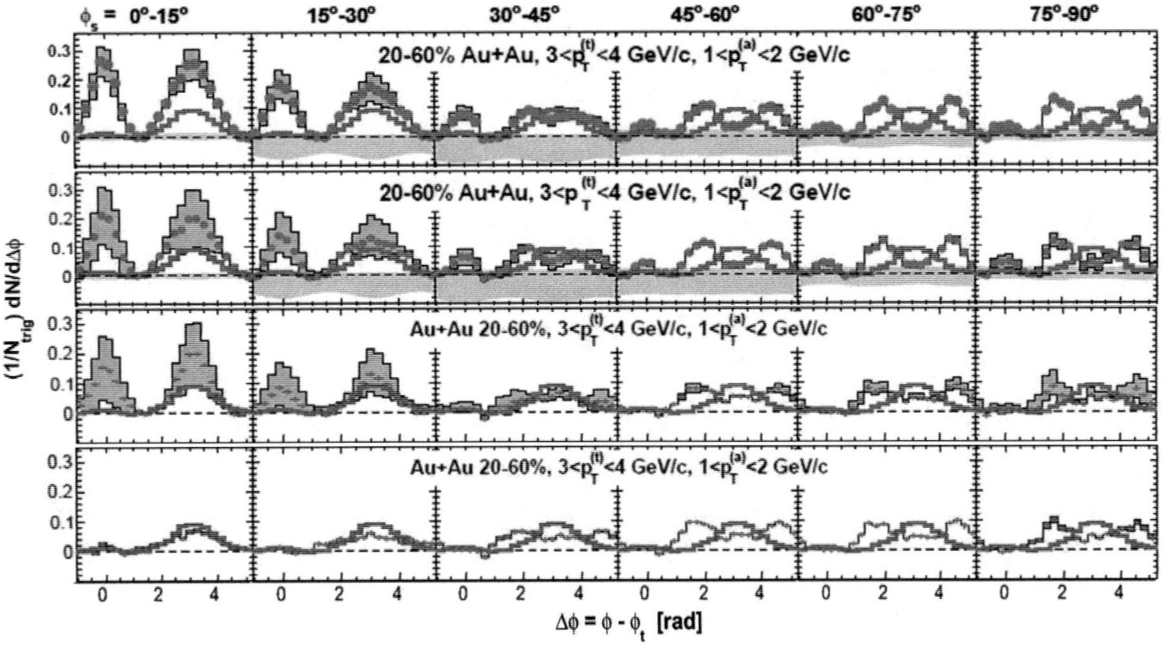

Figure 1. (Color online) Anisotropic background subtracted dihadron correlations with trigger particles in six slices of azimuthal angle relative to the event plane, $\phi_s = |\phi_t - \psi_{\rm EP}|$, with a cut on the trigger-associated pseudo-rapidity difference of $|\Delta\eta| > 0.7$. The triangle two-particle $\Delta\eta$ acceptance is not corrected. The trigger and associated particle p_T ranges are $3 < p_T^{(t)} < 4$ GeV/c and $1 < p_T^{(a)} < 2$ GeV/c, respectively. The data points are from minimum-bias 20-60% Au+Au collisions. The backgrounds due to elliptic anisotropy v_2 and quadrangular anisotropy relative to the second harmonic plane, $v_4\{\psi_2\} = 1.15v_2^2$ are subtracted. Shown in the thin histograms embracing the shaded area are systematic uncertainties due to flow subtraction, the two- and four-particle cumulant v_2 (with the average used for the data points). The systematic uncertainty due to ZYAM normalization are shown as the horizontal shaded band around zero in the top two rows. For comparison, the inclusive dihadron correlations from d+Au collisions are superimposed as the thick (green) histograms. First row: the upper systematic uncertainty of v_2 is $\langle\cos 2\Delta\phi\rangle$ average over the away-side of two-particle cumulant. Second row: the upper systematic uncertainty of v_2 is $\langle\cos 2\Delta\phi\rangle$ average over the entire 2π range. Third row: an additional v_3 background, obtained from two-particle cumulant with η-gap of $\eta_{\rm gap} = 0.7$, is subtracted. Fourth row: The ϕ_s-dependent $v_2\{p_T\text{-}p_T\}$ measured by two-particle cumulants with $\eta_{\rm gap} = 0.7$ and 1.2 are used (the thin histograms embracing the shaded area), with their average shown in the data points. An additional, but small ψ_2-uncorrelated $V_4\{\rm uc\}$ is subtracted. The subtracted v_3 is as same as the third row. The data should be compared to the lower systematic bounds in the third row where the subtracted v_2 is the ϕ_s-independent two-particle cumulant measurement.

The third row of Fig. 1 shows the dihadron correlation functions for $|\Delta\eta| > 0.7$ obtained with $v_3\{2\}$ included in the background subtraction. The change from the second row is the additional subtraction of the $v_3\{2\}$ contribution. The qualitative features of the correlation functions are unchanged. The away-side double-peak structure out-of-plane remains prominent. The decreasing trend of the ridge magnitude from in-plane to out-of-plane is unaffected because a constant v_3 contribution over ϕ_s is subtracted. This indicates that the main features of the

Journal of Physics: Conference Series **458** (2013) 012029 doi:10.1088/1742-6596/458/1/012029

measured near-side ridge and away-side double-peak in the dihadron correlations with high p_T trigger particles, whether or not integrated over ϕ_s, are not entirely due to the triangular flow contributions, but other physics mechanisms.

4. Subtraction of uncorrelated v_4

We have so far subtracted the $v_4\{\psi_2\}$ background correlated with the second harmonic plane, ψ_2. We have used the parameterization to the previous $v_4\{\psi_2\}$ measurement [24],

$$v_4\{\psi_2\} = 1.15v_2^2\,. \tag{2}$$

There is an additional contribution to v_4 that is uncorrelated to ψ_2 and arises from fluctuations. The uncorrelated component can be obtained by

$$V_4\{\mathrm{uc}\} = v_4^{(t)}\{2\}v_4^{(a)}\{2\} - v_4^{(t)}\{\psi_2\}v_4^{(a)}\{\psi_2\}\,. \tag{3}$$

where $v_4\{2\}$ is the v_4 measured by the two-particle cumulant method with $\eta_{\mathrm{gap}} = 0.7$. Because $V_4\{\mathrm{uc}\}$ is small, its effect on dihadron correlation is negligible. The dihadron correlation results with $V_4\{\mathrm{uc}\}$ subtraction is effectively as same as those shown in the third row of Fig. 1.

5. Effect of possible ϕ_s-dependent v_2

One can always attribute all azimuthal dependence to Fourier harmonics. In fact, Luzum [25] argued that our $|\Delta\eta| > 0.7$ correlation data can be fitted by Fourier harmonics up to the 4th order and the fitted coefficients are consistent with features expected from anisotropic flows. This is not surprising because nonflow effects, which must be contained in the fitted Fourier coefficients, are relatively small compared to the flow contributions in our kinematic regions. If the observed ϕ_s-dependent ridge is due to anisotropic flow, then the harmonic flows must be ϕ_s-dependent. This may not be impossible, however, as the requirement of trigger particles in a particular ϕ_s bin from the event plane reconstructed from particles in $0.15 < p_T < 2$ GeV/c could preferentially select events with associated particle v_2 displaced from the average. In the following, we analyze the two-particle cumulant v_n in events of different ϕ_s values separately, and subtract them from the dihadron correlations.

Since reference particles are used to reconstruct the event plane to determine the ϕ_s, one cannot calculate v_n from cumulant of the associated particle and a reference particle in event sample selected according to ϕ_s. Instead, we form two-particle cumulant from particles in a given associated $p_T^{(a)}$ bin, applying η-gap of 0.7. The v_n of the associated particles are simply

$$v_n\{p_T\text{-}p_T\}(\phi_s) = \sqrt{V_n\{p_T\text{-}p_T, \eta_{\mathrm{gap}}{=}0.7\}(\phi_s)}\,. \tag{4}$$

Here $V_n\{p_T\text{-}p_T, \eta_{\mathrm{gap}}{=}0.7\}$ indicates the two-particle cumulant with particle pairs from the same p_T bin. We use $v_n\{p_T\text{-}p_T, \eta_{\mathrm{gap}}{=}0.7\}$ or simply $v_n\{p_T\text{-}p_T\}$ to stand for the resultant anisotropy measurement. Figure 2 shows the obtained $v_n\{p_T\text{-}p_T\}$ of $1.5 < p_T < 2$ GeV/c as a function of ϕ_s of trigger particles of $3 < p_T^{(t)} < 4$ GeV/c. The $v_2\{p_T\text{-}p_T\}$ decreases with ϕ_s. The decrease is a consequence of the decreasing ridge with increasing ϕ_s. The $v_4\{p_T\text{-}p_T\}$ is found to be smallest with $\phi_s = 45°$ and largest with $\phi_s = 0°$ and $90°$. On the other hand, the $v_3\{p_T\text{-}p_T\}$ is independent of ϕ_s, consistent with the expectation that the third and second harmonic planes are uncorrelated. The $v_3\{p_T\text{-}p_T\}$ from the cumulant of same-p_T bin pairs is consistent with that obtained from the cumulant with a reference particle, $v_3\{2\}$.

Although the measured $v_4\{p_T\text{-}p_T\}$ is ϕ_s-dependent, the contribution of the ψ_2-uncorrelated v_4 to flow background is negligibly small, as discussed in Sec. 4. We therefore use the ϕ_s-independent $v_4\{2\}$ measured by two-particle cumulant with a reference particle, as in Sec. 4. We have checked our results using the ϕ_s-dependent $v_4\{p_T\text{-}p_T\}$ and found no observable difference.

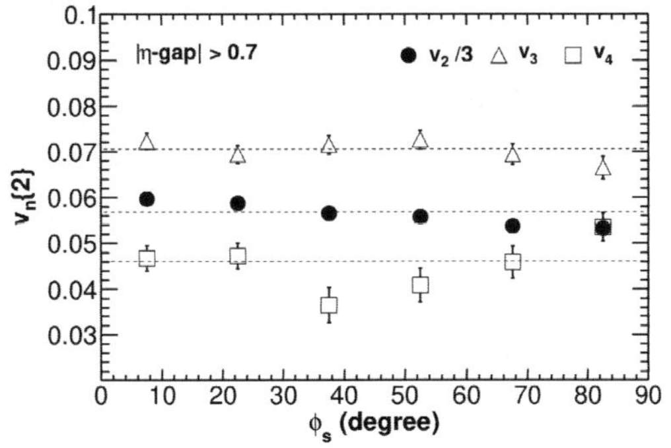

Figure 2. Harmonic $v_n\{p_T\text{-}p_T\}$ of associated particles of $1.5 < p_T^{(a)} < 2$ GeV/c as a function of ϕ_s of trigger particles of $3 < p_T^{(t)} < 4$ GeV/c. Note the $v_2\{p_T\text{-}p_T\}$ is scaled down by a factor of 3 to fit into the plot coordinate range. The $v_n\{p_T\text{-}p_T\}$ is measured by the two-particle cumulant method with particle pairs from the same associated $p_T^{(a)}$ bin and with $\eta_{\rm gap} = 0.7$. The data are from minimum-bias 20-60% Au+Au collisions. Error bars are statistical. The horizontal lines are to guide the eye.

We subtract flow background using the ϕ_s-dependent $v_2\{p_T\text{-}p_T\}(\phi_s)$. The trigger particle v_2 is still given by the two-particle cumulant flow obtained with a reference particle. This is because the trigger v_2 is the second harmonic modulation of trigger particles which determines the ϕ_s. The bottom row of Fig. 1 shows the dihadron correlation results with subtraction of ϕ_s-dependent $v_2\{p_T\text{-}p_T\}$. The change from the lower systematic bound in the third row is the subtraction of the ϕ_s-dependent v_2 in place of the ϕ_s-independent one.

As seen from figure, the near-side ridge is diminished, maybe as expected, because the large $\Delta\eta$ ridge is presumably included in the subtracted v_n. However, it is important to point out that it is not automatically guaranteed that the ridge will be gone just because the v_n's are measured by two-particle cumulant either with a reference particle or with a particle from the same p_T region. This is because they are not simply measured by the trigger-associated particle pair at $|\Delta\eta| > 0.7$. If they were, then the correlation will be strictly zero everywhere, both on the near side and on the away side.

It is interesting to note, despite of the diminished near-side ridge, that the away-side correlation is not diminished. It still evolves from a single peak with in-plane trigger particles to a double peak with out-of-plane trigger particles. The observation of the away-side double-peak structure for out-of-plane triggers seems robust against the wide range of flow background subtraction.

Since the ridge is diminished after subtraction of ϕ_s-dependent v_2 from two-particle cumulant at large $\Delta\eta$, can we conclude that the physics origin of the ridge is hydrodynamic v_n flow? The answer is no because any non-hydrodynamic origin of v_n is also included in the two-particle v_n measurements. In other words, any ridge signal (whatever its physics origin might be) is included in v_n, and the ridge would be subtracted after subtraction of v_n. However, one also cannot rule out the ridge being part of hydrodynamic flow. This is because it is still possible that hydrodynamic flow of the underlying event is biased by the selection of the trigger particle orientation, and all the long-range $\Delta\eta$ correlation may be indeed due to flow.

5.1. Implications on inclusive dihadron correlations

If v_2 depends on ϕ_s, then there is an important implication to the inclusive dihadron correlation (i.e., without cutting on ϕ_s). For inclusive dihadron correlation, a flow background

Journal of Physics: Conference Series **458** (2013) 012029 doi:10.1088/1742-6596/458/1/012029

$\langle v_2^{(t)}\{2\}\rangle \cdot \langle v_2^{(a)}\{2\}\rangle$ has been used in place of $\langle v_2^{(t)}\{2\} \cdot v_2^{(a)}\{2\}\rangle$. (Note, for clarity, we have omitted the $\langle ... \rangle$ notation throughout the article except here.) This is correct because fluctuations are already included in the two-particle cumulant flow measurement of $\langle v_2\{2\}\rangle$. However, if v_2 depends on trigger particle orientation ϕ_s, then the equality $\langle v_2^{(t)}\{2\}(\phi_s) \cdot v_2^{(a)}\{2\}(\phi_s)\rangle = \langle v_2^{(t)}\{2\}(\phi_s)\rangle \cdot \langle v_2^{(a)}\{2\}(\phi_s)\rangle$ is no longer valid. The left side will be always larger than the right side. This means that the inclusive dihadron flow background is underestimated by $\langle v_2^{(t)}\{2\}\rangle \cdot \langle v_2^{(a)}\{2\}\rangle$. In fact, because $v_2^{(t)}\{2\}(\phi_s)$ is positive for $\phi_s \sim 0$ and negative for $\phi_s \sim \pi/2$, the true background magnitude for inclusive dihadron correlation is even larger than that for the $\phi_s = 0$ dihadron correlation, which has the largest background magnitude of all ϕ_s bins. Namely $\langle v_2^{(t)}\{2\}(\phi_s) \cdot v_2^{(a)}\{2\}(\phi_s)\rangle > \langle v_2^{(t)}\{2\}\rangle \cdot v_2^{(a)}\{2\}(\phi_s = 0)$.

Figure 3 illustrates the effect. The left panel shows the raw dihadron correlation for $3 < p_T^{(t)} < 4$ GeV/c and $1 < p_T^{(a)} < 1.5$ GeV/c together with two flow background curves both ZYAM-normalized. The blue histogram is from a traditional inclusive dihadron correlation analysis with the v_2 modulation calculated from $\langle v_2^{(t)}\{2\}\rangle \cdot \langle v_2^{(a)}\{2\}\rangle$. The red histogram is that calculated from the ϕ_s-dependent $v_2\{2\}(\phi_s)$ by $\langle v_2^{(t)}\{2\}(\phi_s) \cdot v_2^{(a)}\{2\}(\phi_s)\rangle$ which is the correct flow background provided $v_2(\phi_s)$ is the real flow. (The v_3 and v_4 contributions are included in both flow background histograms). As seen from Fig. 3, the traditional flow background is underestimated. The right panel of Fig. 3 shows the dihadron correlation signals after subtraction of the traditional background in the histogram and by the correct flow background in the data points, respectively. The signal from the traditional average flow background subtraction is less double-peaked. This means, if the ridge is entirely due to flow that must be ϕ_s-dependent, then all the inclusive dihadron correlation analyses have under-subtracted the flow background, resulting in a more peaked away-side correlation signal.

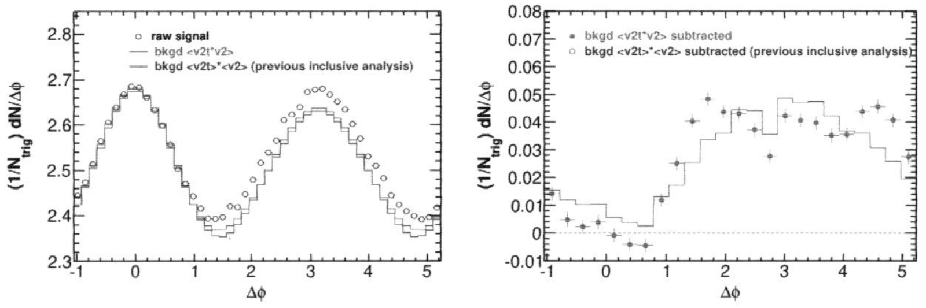

Figure 3. (Color online) Effect of possible ϕ_s-dependent elliptic flow anisotropy on inclusive dihadron correlations. Left panel: raw $\Delta\phi$ correlation together with flow background obtained by two different ways, one by the average of the product of the trigger $v_2^{(t,R)}$ and the associated particle $v_2^{(a)}(\phi_s)$ as from this analysis (red histogram), and the other by the product of the average trigger and associated v_2 as from the standard inclusive dihadron correlation analysis (blue histogram). Right panel: the correlation signals subtracted by the background from this analysis (red points) and by the standard background from inclusive dihadron correlation analysis (blue histogram). The data are from minimum-bias 20-60% Au+Au collisions. The trigger and associated particle p_T ranges are $3 < p_T^{(t)} < 4$ GeV/c and $1 < p_T^{(a)} < 1.5$ GeV/c, respectively. A $|\Delta\eta| > 0.7$ cut is applied to the trigger-associated pairs. Error bars are statistical.

6. Effect of dipole fluctuations

We have neglected the effect of dipole fluctuations (rapidity-even v_1) in flow background subtraction of Eq. (1). Preliminary data [26] indicate that the dipole fluctuation effect changes sign at $p_T \approx 1$ GeV/c, negative at lower p_T and positive at higher p_T. For $p_T^{(a)}$ = 1-2 GeV/c shown in Fig. 1, the dipole fluctuation effect is approximately zero and may be neglected. The qualitative conclusions on the near-side and away-side correlations are unaffected by the potential dipole fluctuations.

7. Effect of possible biases in event-plane reconstruction

In our analysis, the event plane is reconstructed by particles excluding those within $|\Delta\eta| < 0.5$ of the trigger particle. Question remains how much is the effect of possible biases to the reconstructed event plane by particles correlated to the trigger, especially on the away side. One way to estimate this possible effect is to analyze dihadron correlations relative to the event plane reconstructed from particles without excluding those within $|\Delta\eta| < 0.5$ of the trigger, thereby maximizing the biases from jet-correlations. Our study shows that introducing a stronger bias in event-plane reconstruction causes a relatively small change in the correlation signals. This suggests that possible event-plane biases in our default results in Fig. 1 may be also relatively small.

8. Conclusions

In this talk, we revisit our dihadron correlations relative to the event plane in arXiv:1010.0690v1. We focused on the $\Delta\phi$ correlations at $|\Delta\eta| > 0.7$ to study the ridge and away-side correlations. We subtract background from triangular anisotropy measured by the two-particle cumulant method with a η-gap (η_{gap}) of 0.7. The qualitative feature of the correlation data is unchanged. The ridge, with a reduced magnitude, is still present for in-plane trigger particles. It decreases from in-plane to out-of-plane.

We consider the effect of a v_n that is dependent on the trigger-particle ϕ_s. We analyze the two-particle cumulants $v_n\{p_T\text{-}p_T\}$ in events with different trigger particle ϕ_s separately. The second harmonic $v_2\{p_T\text{-}p_T\}$ is found to decrease with increasing ϕ_s. This is synonymous to the decreasing ridge magnitude with ϕ_s. The fourth harmonic $v_4\{p_T\text{-}p_T\}$ is found to also depend on ϕ_s, but its effect on dihadron correlation is negligible. The third harmonic $v_3\{p_T\text{-}p_T\}$ is found to be independent of ϕ_s. The dihadron correlations are studied relative to the event plane with the subtraction of the two-particle cumulant anisotropies, $v_3\{2, \eta_{\text{gap}}{=}0.7\}$, $v_4\{2, \eta_{\text{gap}}{=}0.7\}$ and the ϕ_s-dependent $v_2\{p_T\text{-}p_T, \eta_{\text{gap}}{=}0.7\}(\phi_s)$. With this exploratory subtraction of the v_n values, the ridge is found to be eliminated. However, this result does not enlighten the origin of the ridge because the measured v_n have likely already included the ridge; whether the ridge is due to flow or nonflow is undetermined.

With the wide range of flow subtraction, the away-side double-peak structure for out-of-plane triggers remains robust. This is true even with the subtraction of ϕ_s-dependent $v_n\{p_T\text{-}p_T, \eta_{\text{gap}}{=}0.7\}$. This may indicate path-length dependent modifications to the away-side jet traversing the sQGP medium.

References

[1] I. Arsene *et al.* (BRAHMS Collaboration), Nucl. Phys. **A757**, 1 (2005).
[2] B. B. Back *et al.* (PHOBOS Collaboration), Nucl. Phys. **A757**, 28 (2005).
[3] J. Adams *et al.* (STAR Collaboration), Nucl. Phys. **A757**, 102 (2005).
[4] K. Adcox *et al.* (PHENIX Collaboration), Nucl. Phys. **A757**, 184 (2005).
[5] C. Adler *et al.* (STAR Collaboration), Phys. Rev. Lett. **90**, 082302 (2003).
[6] J. Adams *et al.* (STAR Collaboration), Phys. Rev. Lett. **95**, 152301 (2005).
[7] B. I. Abelev *et al.* (STAR Collaboration), Phys. Rev. C **80**, 064912 (2009).
[8] B. I. Abelev *et al.* (STAR Collaboration), Phys. Rev. Lett. **105**,022301 (2010).

[9] B. Alver *et al.* (PHOBOS Collaboration), Phys. Rev. Lett. **104**, 062301 (2010).
[10] S. S. Adler *et al.* (PHENIX Collaboration), Phys. Rev. Lett. **97**, 052301 (2006).
[11] M. M. Aggarwal *et al.* (STAR Collaboration), Phys. Rev. C **82**, 024912 (2010).
[12] A. Adare *et al.* (PHENIX Collaboration), Phys. Rev. C **78**, 014901 (2008).
[13] H. Agakishiev *et al.* (STAR Collaboration), arXiv:1010.0690v1.
[14] Mishra *et al* , Phys. Rev. C **77** (2008).
[15] B. Alver and G. Roland, Phys. Rev. C **81**, 054905 (2010).
[16] H. Petersen *et al.*, Phys. Rev. C **82**, 041901 (2010).
[17] B. Schenke, S. Jeon, and C. Gale, Phys. Rev. Lett. **106**, 042301, (2011).
[18] Z. Qiu and U. W. Heinz, Phys. Rev. C **84**, 024911, (2011).
[19] J. Takahashi *et al.*, Phys. Rev. Lett. **103**, 242301 (2009).
[20] Y. Hama *et al.*, Nonlin. Phenom. Complex Syst. **12**, 466 (2009), arXiv:0911.0811 [hep-ph].
[21] R. Andrade *et al.*, J. Phys. G **37**, 094043 (2010), arXiv:0912.0703 [nucl-th].
[22] A. Adare *et al.* (PHENIX Collaboration), Phys. Rev. Lett. **107**, 252301 (2011).
[23] L. Adamczyk *et al.* (STAR Collaboration), arXiv:1301.2187.
[24] J. Adams *et al.* (STAR Collaboration), Phys. Rev. C **72**, 014904 (2005).
[25] M. Luzum, Phys. Lett. B **696**, 499 (2011).
[26] Y. Pandit (STAR Collaboration), arXiv:1211.7162 [nucl-ex].

Journal of Physics: Conference Series **458** (2013) 012030 doi:10.1088/1742-6596/458/1/012030

Kadanoff-Baym Approach to Thermalization of Quantum Fields

Akihiro Nishiyama[1] **and Yoshitaka Hatta**[2]

[1]Maskawa Institute for Science and Culture, Kyoto Sangyo University, Kyoto 603-8555, Japan,
[2] Graduate School of Pure and Applied Sciences, University of Tsukuba, Ibaraki 305-8571, Japan

E-mail: nishiaki@het.ph.tsukuba.ac.jp

Abstract. Numerical results of nonequilibrium field theory are shown for $O(N)$ scaler model in 2+1 dimensions with longitudinal expansion. We adopt Next-to-Leading Order of $1/N$ expansion of self energy with background classical fields to cover all of time evolution of instability. We show differences between quantum dynamics and classical statistical approximation in strongly coupled regimes. As a result we show that we can achieve Boltzmann tail of the final distribution function only in quantum dynamics.

1. Introduction

Recent experiments are devoted to create Quark-Gluon Plasma (QGP) by colliding two nuclei at center-of-energy 200GeV at RHIC and 2.76TeV at LHC. With respect to QGP hydrodynamics with small viscosity succeeds in tracing dynamics after equilibration of Glasma. The Glasma is given as classical longitudinal color electric and magnetic fields with vacuum quantum fluctuations. The success of hydrodynamics is based on rapid equilibration of Glasma $t_{\rm eq} = 0.6$-$1.0{\rm fm}/c$ [1]. This equilibration time seems to be comparable with the formation time of gluon partons [2]. Then normal parton picture fails to describe equilibration of Glasma, as is shown in Ref. [3]. Hence we need to trace nonequilibrium processes beyond simple normal parton picture. In addition other approaches based on classical statistical lattice theories do not describe late-time Boltzmann tail in the distribution function. Thus we adopt approaches which describe late time true equilibration. As a candidate of approaches, we adopt nonequilibrium quantum field theoretical approach, which is based on Kadanoff-Baym(KB) equation [4] with equation of motion of classical fields. It can go beyond normal parton picture and describe final Bose-Einstein distribution.

The first merit of solving these equations is that we can trace field-particle conversion. Once particle production occurs from classical fields, they collide each other, so that Bose-Einstein distribution is realized at late time. The second merit is that presence of finite decay width in spectral function induces rapider change of distribution function due to binary 2-to-2 collisions than binary collisions in Boltzmann equation. They might contribute to rapid equilibration of gluons [5].

In section 2 we introduce time evolution equation for classical fields and quantum fluctuations. Section 3 is devoted to give numerical results. We summarize our work in Sec. 4.

2. Time evolution equation

In this section we show KB equation and time evolution equation of classical fields in $O(N)$ scalar model. First the action of this model is given by,

$$S = \int d^{d+1}x \left[\frac{1}{2}\partial\phi_a\partial\phi_a - \frac{1}{2}m^2\phi_a\phi_a - \frac{\lambda}{24N}(\phi_a\phi_a)^2 \right], \tag{1}$$

where particle components a runs over $1, \cdots, N$ and d represents the spatial dimension. The merit of adopting this model is that we can cover all time evolution of instability by use of $1/N$ expansion. Next we show the equations of motion of classical fields $\bar{\phi}_a \equiv \langle\phi_a\rangle = \bar{\phi}\delta_{a1}$ and quantum fluctuations $F_{ab}(x,y) \equiv \frac{1}{2}\langle\{\tilde{\phi}_a(x), \tilde{\phi}_b(y)\}\rangle$ $\rho_{ab}(x,y) \equiv \langle[\tilde{\phi}_a(x), \tilde{\phi}_b(y)]\rangle$ (that are Fourier transformed), where $\tilde{\phi}_a = \phi_a - \bar{\phi}_a$. They are

$$\left[\partial_\tau^2 + \frac{1}{\tau}\partial_\tau + m^2 + \frac{\lambda}{6N}\left(\bar{\phi}(\tau)^2 + F_{11}(\tau,\tau) + \sum_{b\neq 1} F_{bb}(\tau,\tau)\right)\right]\bar{\phi}(\tau)$$
$$= -\int_{\tau_0}^{\tau} \tau' d\tau' \Sigma_{11}^{\rho}(\tau,\tau')\bar{\phi}(\tau'), \tag{2}$$

$$G_0^{-1}F(\tau,\tau',p) = -\int_{\tau_0}^{\tau} \tau'' d\tau'' \Sigma^{\rho}(\tau,\tau'',p)F(\tau,\tau',p) + \int_{\tau_0}^{\tau'} \tau'' d\tau'' \Sigma^{F}(\tau,\tau'',p)\rho(\tau'',\tau',p), \tag{3}$$

$$G_0^{-1}\rho(\tau,\tau',p) = -\int_{\tau'}^{\tau} \tau'' d\tau'' \Sigma^{\rho}(\tau,\tau'')\rho(\tau,\tau',p), \tag{4}$$

$$G_0^{-1} \equiv \left[\frac{\partial^2}{\partial\tau^2} + \frac{1}{\tau}\frac{\partial}{\partial\tau} + \frac{p_\eta^2}{\tau^2} + p_T^2\right]\delta_{ab} + M_{ab}^2(\bar{\phi}).$$

Here in order to trace time evolution in longitudinally expanding system, we use proper time $\tau = \sqrt{t^2 - z^2}$, rapidity $\eta = \tanh^{-1}\frac{z}{t}$ and its Fourier transformed p_η (we adopt spatial homogeneity) and set initial time τ_0, contribution of local tadpole mass and classical fields $M_{ab}(\bar{\phi})$ and Next-to-Leading Order self-energy $\Sigma^{F,\rho}$ of $1/N$ expansion which includes binary (2-to-2) collisions non-perturbatively [8].

We set $\bar{\phi}_a(\tau_0) = \sqrt{\frac{6N}{\lambda}}\sigma\delta_{a1}$ with vacuum quantum fluctuations for F and ρ as an analogy of initial longitudinal classical fields in the Glasma. We have done numerical simulations in 2+1 dimensions for weakly coupled $\lambda = 10^{-4}$ and strongly coupled $\lambda = 10$ regimes in an expanding system. Then we make a comparison between quantum dynamics and classical statistical approximation. In the classical approximation, $\rho\rho$ terms in self-energy are omitted [5, 6, 7].

3. Numerical analyses in 2+1 dimension

This section is devoted to show time evolution of classical field $\bar{\phi}(\tau)$ (Fig. 1 and 3) and statistical functions $F_{11}(\tau,\tau,p)$ (Fig. 2 and 4). In Fig.1 we show time evolution of classical field in weakly coupled regimes. Classical fields damps with oscillation due to $\frac{\partial_\tau\bar{\phi}}{\tau}$ term in Eq. (2) in expanding metric. Then parametric resonance instability (primary instability) occurs as shown in $\tau/\tau_0 < 20$. At late times particle production is accelerated by secondary instability at $20 < \tau/\tau_0 < 40$ which is due to field induced 1-to-2 processes in NLO self-energy of $1/N$ expansion. (Particle production stops at $F = O(1/\lambda) = 10^4$, then coupling expansion of self-energy with λ breaks down since nth-loop order self-energy with coupling expansion is $\lambda^n F^{2n-1} = O(\lambda^{1-n})$. Thus we need to adopt $1/N$ expansion to trace all non-perturbative regions

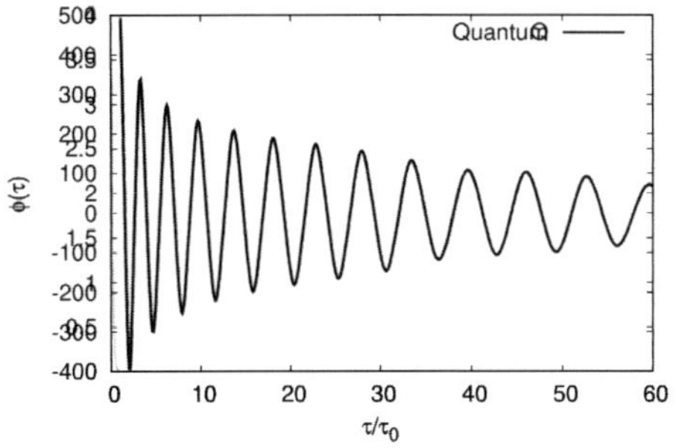

Figure 1. Time evolution of the classical field $\bar{\phi}$ for quantum (black solid line)
and classical statistical approximation
in weakly coupled regimes.

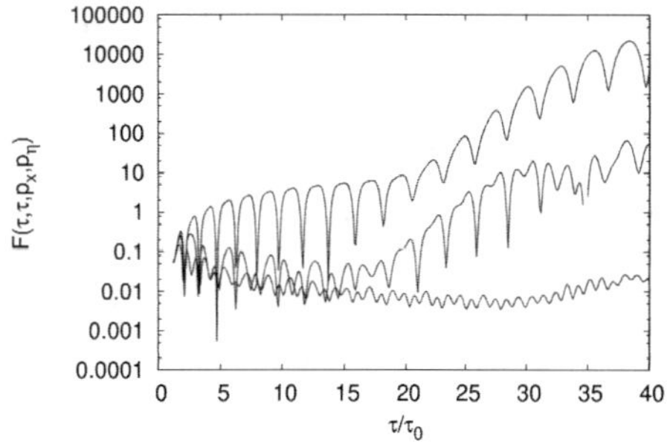

Figure 2. Time evolution of
$F_{11}(\tau, \tau, p_\eta = 0, p_T = 0, \pi\sigma/8, \pi\sigma/4)$
for for quantum (black solid line)
and classical statistical approximation
(red dashed line) in weakly coupled regimes.

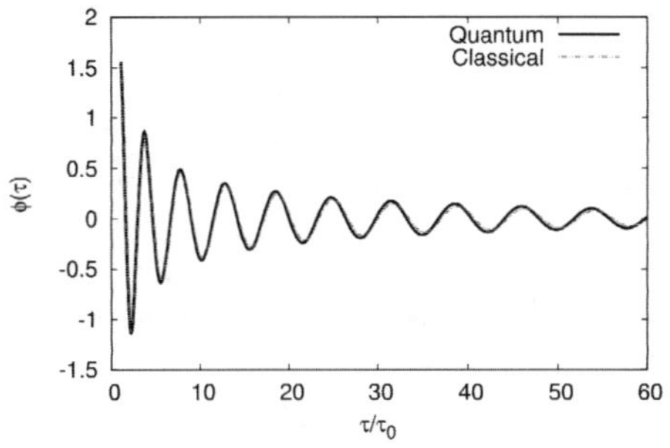

Figure 3. Time evolution of the classical field $\bar{\phi}$ for quantum (black solid line)
and classical statistical approximation
(red dashed line) in strongly coupled regimes.

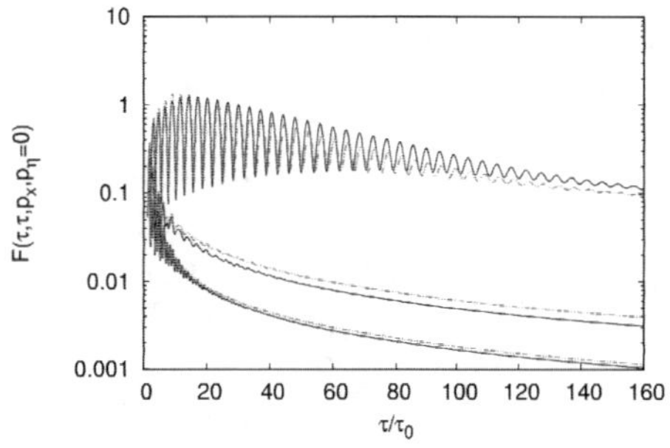

Figure 4. Time evolution of
$F_{11}(\tau, \tau, p_\eta = 0, p_T = 0, \pi\sigma/8, \pi\sigma/4)$
for for quantum (black solid line)
and classical statistical approximation
(red dashed line) in strongly coupled regimes.

in the evolution of the system.) Then no difference appears between quantum dynamics and classical approximation in weakly coupled regimes. This means that there is no hint of quantum equilibration. However in strongly coupled regimes differences appear in quantum dynamics and classical approximation. In Fig.3 classical fields damps with oscillation due to $\frac{\partial_\tau \bar{\phi}}{\tau}$ term in Eq. (2) in expanding metric. Here field particle conversion starts to occur. Time evolution of F in Fig.4 shows this particle production, where F changes from $O(1)$ to $O(1/\lambda)$. Finally we show the distribution function n_p (Fig. 5) by use of functional fit $F = \frac{C}{\tau\sqrt{m_{\rm eff}^2 + p_T^2}}\left(n_p + \frac{1}{2}\right)$ where C and $m_{\rm eff}$ are fitting parameter. In quantum dynamics the distribution function shows Boltzmann tail ($n_p = \frac{1}{e^{p_T/T} - 1}$ with temperature T) and late time Bose-Einstein distribution , while it gives power law $n_p = \frac{T}{p_T} - \frac{1}{2}$ in classical statistical approximation. True equilibration is not achieved in classical statistical approximation.

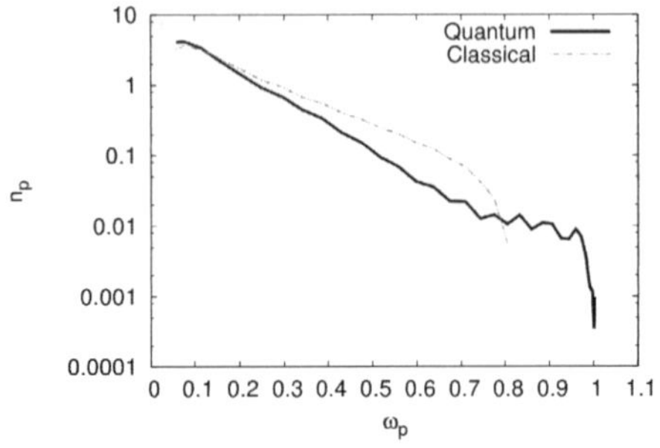

Figure 5. Number distribution function n_p at late time $\tau/\tau_0 = 150$.

4. Summary

In this work we have shown the equilibration process by solving the Kadanoff-Baym equation and time evolution equation of classical fields for $O(N)$ scalar model in a spatially homogeneous expanding system in 2+1 dimensions. We have adopted NLO nonlocal self-energy contributing to binary (2-to-2) collisions by use of $1/N$ expansion which covers all time evolution of instability of F from $O(1)$ to $O(1/\lambda)$. In weakly coupled regimes $\lambda = 10^{-4}$, no difference appears between quantum dynamics and classical statistical approximation in time evolution with instability , which means no true quantum equilibration occurs. In strongly coupled regimes $\lambda = 10$, quantum dynamics shows Boltzmann tail in the high momentum regimes. On the other hand, the result with classical statistical approximation shows power law behavior in distribution function, which is not true equilibration. Thus we need to adopt quantum dynamics to achieve the final Bose-Einstein distribution.

[1] U. W. Heinz and P. F. Kolb, Nucl. Phys. **A702** (2002), 269; U. W. Heinz, AIP Conf. Proc. **739** (2005), 163.
[2] B. Muller and A. Schafer, Phys. Rev. C **73** (2006) 054905 [arXiv:hep-ph/0512100].
[3] R. Baier, A. H. Mueller, D. Schiff and D. T. Son, Phys. Lett. B **502** (2001) 51 [arXiv:hep-ph/0009237].
[4] L.P. Kadanoff and G. Baym, *Quantum Statistical Mechanics* (Benjamin, New York, 1962); G. Baym, Phys. Rev. **127** (1962) 1391.
[5] Y. Hatta and A. Nishiyama, Phys. Rev. D in press.
[6] J. Berges, S. Scheffler and D. Sexty, Phys. Rev. D **77**, 034504 (2008) [arXiv:0712.3514 [hep-ph]].
[7] T. Epelbaum and F. Gelis, Nucl. Phys. A **872**, 210 (2011) [arXiv:1107.0668 [hep-ph]].
[8] G. Aarts, D. Ahrensmeier, R. Baier, J. Berges and J. Serreau, Phys. Rev. D **66**, 045008 (2002)

Journal of Physics: Conference Series **458** (2013) 012031 doi:10.1088/1742-6596/458/1/012031

Building a Hydrodynamics Code with Kinetic Theory

Irina Sagert[1], Wolfgang Bauer[1,2], Dirk Colbry[2], Rodney Pickett[2], Terrance Strother[3]

[1]Department of Physics and Astronomy, Michigan State University, East Lansing, Michigan, 48824, USA

[2]Institute for Cyber-Enabled Research, Michigan State University East Lansing, Michigan 48824, USA

[3]XTD-6, Los Alamos National Laboratory, Los Alamos, New Mexico 87545, USA

E-mail: sagert@nscl.msu.edu

Abstract. We report on the development of a test-particle based kinetic Monte Carlo code for large systems and its application to simulate matter in the continuum regime. Our code combines advantages of the Direct Simulation Monte Carlo and the Point-of-Closest-Approach methods to solve the collision integral of the Boltzmann equation. With that, we achieve a high spatial accuracy in simulations while maintaining computational feasibility when applying a large number of test-particles. The hybrid setup of our approach allows us to study systems which move in and out of the hydrodynamic regime, with low and high particle densities. To demonstrate our code's ability to reproduce hydrodynamic behavior we perform shock wave simulations and focus here on the Sedov blast wave test. The blast wave problem describes the evolution of a spherical expanding shock front and is an important verification problem for codes which are applied in astrophysical simulation, especially for approaches which aim to study core-collapse supernovae.

1. Introduction

Hydrodynamics is a frequent approach to describe the evolution of matter via its thermodynamic properties. Hereby, the physical particles the medium is comprised of have mean-free-paths λ which are much smaller than a characteristic length scale of the studied system L. If the so-called Knudsen number K_n is small with $K_n = \lambda/L \ll 1$, the continuum approximation can be applied and the evolution of matter is described by the Navier-Stokes equations. The latter are a set of coupled differential equations and are derived from the conservation of mass, momentum, and energy density as well as the assumption of local thermal equilibrium [1]. The Navier-Stokes equations can be solved numerically either by the Eulerian, i.e. grid-based, method or the Lagrangian approach, also known as Smooth Particle Hydrodynamics [2]. Numerical hydrodynamics is a widely applied tool in science and engineering. However, if the mean-free-path of physical particles becomes large, resulting in $K_n \gg 1$, the continuum approximation breaks down. As a consequence, the Navier-Stokes equations cannot be applied anymore to evolve the system and transport equations have to be solved instead.

2. Kinetic Theory

In transport theory the many-body problem is solved via modeling the evolution of the particle phase-space density function $f(\vec{x}, \vec{p}, t)$ with position $\vec{x}$, momentum $\vec{p}$, and time t. Hereby, the equation of motion of the n-body density matrix is rewritten into a one-particle evolution, coupled to an equation of motion for the n-body correlation function. Truncation of the latter to the three-body level leaves a system of coupled non-linear equations of motion for the one-particle density function and the two-body correlation function. To derive the transport equations, the Wigner transform is applied to the particle density function and is then expanded to sums over all possible single particle states. Rearranging the equation of motion to a collisionless one-body propagation and collision terms results in the so-called Boltzmann transport equation [3]:

$$\frac{\partial f}{\partial t} + \frac{\vec{p}}{m} \cdot \vec{\nabla}_r f + \vec{F} \frac{\partial f}{\partial \vec{p}} = \left. \frac{\partial f}{\partial t} \right|_{\mathrm{Coll}} . \tag{1}$$

The left-hand side of eq.(1) represent the collisionless motion of an ensemble of particles in an external force field $\vec{F}$ while the change in the particles' phase space distribution function $f = f(\vec{x}, \vec{p}, t)$ due to binary collisions is given by the collision integral on the right hand side. The probabilities of particle interaction depend hereby on their interaction cross-sections σ which are connected to the particle mean-free-paths via $\sigma \propto 1/\lambda$. Numerically, the Boltzmann transport equation can be solved by the test-particle method. Hereby, the phase space distribution function is represented by a sum of delta functions [4]:

$$f(\vec{x}, \vec{p}, t) = \sum_{i=0}^{N} \delta^3 \left(\vec{x} - \vec{x}_i(t)\right) \delta^3 \left(\vec{p} - \vec{p}_i(t)\right), \tag{2}$$

whereas N is the number of test-particles in the simulation. Inserting $f(\vec{x}, \vec{p}, t)$ into eq.(1) results in $2N$ coupled first-order differential equations of motion:

$$\frac{d}{dt}\vec{p}_i = \vec{F}(\vec{r}_i) + \vec{C}(\vec{p}_i), \ \frac{d}{dt}\vec{r}_i = \frac{\vec{p}_i}{m_i}, \ i = 1, \ldots, N. \tag{3}$$

Hereby, $\vec{C}(\vec{p}_i)$ represents the effects of two-body collisions on the ith test-particle's momentum. Kinetic methods are typically applied for rarefied gases with large Knudsen numbers. Examples for transport models applications are studies of hypersonic flow [5], nano-scale devices [6], particle production in heavy ion collisions [7, 8, 9, 10], the dynamics of inertial confinement fusion (ICF) capsules [11], and astrophysics [12]. However, application areas can be extended to describe systems that can move in and out of the continuum limit. Especially heavy-ion collisions and core-collapse supernovae contain baryons which are in hydrodynamic equilibrium, as well as particles with mean-free-paths which can be very large. In heavy-ion collisions the latter are e.g. π mesons while for core-collapse supernovae, neutrinos can be trapped, free-streaming, or diffusing in the baryonic matter. In core-collapse supernovae, neutrinos are typically evolved by coupling solvers for the neutrino transport equations to the evolution of the hydrodynamics equations. However, while for one-dimensional core-collapse supernova simulations, the Boltzmann transport equation can be solved exactly [13], multi-dimensional calculations are currently computationally not feasible and can only be performed with approximations to the neutrino transport [14]. On the other hand, Monte Carlo neutrino transport has been suggested as an alternative approach since it is expected to scale better for multi-dimensional simulations [15].

Transport codes are able to handle matter in different states, from low density rarefied gases to high density matter in the continuum regime [16]. However, since the simulation of matter in the continuum regime requires the frequent interaction of particles with each other, large values of N are computationally challenging. One of the most efficient approaches to solve the collision integral is the Direct Simulation Monte Carlo (DSMC) method [17]. Hereby, the simulation space is partitioned into a scattering grid and interaction partners are chosen randomly amongst particles within a cell. The resulting simulation times scale as $O(N \log N)$ and can be reduced even further by performing the calculations in parallel on multiple processors. However, a consequence of the probabilistic choice of interaction partners is that spatial details cannot be resolved to smaller scales than the size of a grid cell. Furthermore, the finite separation between collision partners and the instantaneous character of interactions can lead to causality violations in a relativistic regime [10]. An approach which minimized the distance between interaction partners is the Point-of-Closest-Approach (PoCA) method. Hereby, scattering partners are determined as test-particles whose path crosses within one timestep. The PoCA method has been successfully applied in modeling physical systems with limited number of constituents, for example the simulation of heavy-ion collisions [18, 19, 20]. While such simulations have very good spatial resolution, the PoCA method typically requires the comparison of each particle with every other particle of the simulation and has a computational scaling of $O(N^2)$. The time to search for interaction partners can be reduced by a spatial sort of particles. Nevertheless, for large values of N, PoCA methods are computationally very expensive.

3. Combination of the DSMC and PoCA methods

In our approach we combine the advantages of the DSMC and the PoCA techniques to solve the collision integral of the transport equations. Our final goal is to study astrophysical systems such as core-collapse supernovae and matter in ICF capsules via the kinetic scheme. Similar to the DSMC method we divide the simulation space into a grid. However, interaction partners are not chosen randomly but are determined from test-particles in a grid cell and the neighboring cells via the PoCA method. With that, we ensure a spatial accuracy which is higher than in usual DSMC simulations. To avoid a computationally expensive spatial sorting we connect and propagate particles in their corresponding cell via a linked lists. Furthermore, the setup of the scattering grid allows us to use multiple processors and determine interaction partners for different cells in parallel. With that, though computationally more expensive than DSMC methods, our algorithm has a much smaller computational time than traditional PoCA methods and can thereby describe systems with a large number of interacting particles. Collision partners are determined via three steps [21]. First, possible interaction partners are selected as particles reaching their point of closest approach during the timestep Δt. For that we determine the sign of the crossing number χ:

$$\chi = (\vec{r}_{\rm rel}(t) \cdot \vec{v}_{\rm rel}(t))(\vec{r}_{\rm rel}(t+\Delta t) \cdot \vec{v}_{\rm rel}(t+\Delta t)). \tag{4}$$

Hereby, $\vec{r}_{\rm rel}$ and $\vec{v}_{\rm rel}$ are the relative position and velocity vectors for particles A and B at the current as well as the next timestep:

$$\vec{r}_{\rm rel}(t) = \vec{r}_A(t) - \vec{r}_B(t), \quad \vec{v}_{\rm rel}(t) = \vec{v}_A(t) - \vec{v}_B(t) \tag{5}$$

A negative value of χ indicates that within Δt, particles A and B reach their distance of closes approach and their paths cross. As a next step we determine the collision time via [21]:

$$t_{c\,1,2} = \frac{1}{|\vec{v}_{\rm rel}|^2}\left[-(\vec{v}_{\rm rel} \cdot \vec{r}_{\rm rel})^2 \pm \sqrt{(\vec{v}_{\rm rel} \cdot \vec{r}_{\rm rel})^2 - |\vec{v}_{\rm rel}|^2\left(|\vec{r}_{\rm rel}|^2 - (r_{{\rm eff},A} + r_{{\rm eff},B})^2\right)}\right]. \tag{6}$$

If the distance of closest approach between particles A and B is larger than the sum of their effective radii $r_{\rm eff}$ the collision time t_c has imaginary values and the interaction does not take place. These first two steps typically leave several potential interaction partners for one particle of interest. The final collision partner is then determined from this sample as the particle with the shortest distance to the particle of interest. After collision partners have been assigned to each other, we perform the scattering in the center of mass frame of each collision pair. Previous approaches showed that a limitation of interactions to unambiguously determined scattering partners can lead to a significant underestimate of the collision rate [10]. In our work, we allow different particles to have the same collision partner and perform the corresponding interactions sequentially during one time-step Δt [21]. We find that typically all test-particles interact with each other when matter is in the hydrodynamic regime. Once all collisions have been performed each particle's position is updated according to :

$$\vec{x}(t+\Delta t) = \vec{x}(t) + \vec{v}_{\rm old} t_{\rm min} + \vec{v}_{\rm new}(\Delta t - t_{\rm min}), \quad \text{with } t_{\rm min} = -\left(\vec{r_{\rm rel}} \cdot \vec{v_{\rm rel}}\right) / \left|\vec{v_{\rm rel}}\right|^2 . \qquad (7)$$

Hereby, t_{min} is the time when particles reach their distance of closest approach. This is an improvement of the approach which was used in [21]. Large particle densities and small mean-free-paths can lead to negative collision times t_c (see eq.(6)). In [21] this case was treated as an instantaneous collision. However, by allowing particles to reach their point of closest approach during Δt, we achieve a better localization of shock fronts which is important in e.g. astrophysical simulations of core-collapse supernovae.

4. Sedov shock test

Shock wave studies are very well suited to demonstrate a code's ability to model hydrodynamic behavior. To reach the continuum regime, i.e. $K_n \ll 1$, we set the particle mean-free-path to be very small with respect to the width of a scattering grid cell. Since analytic shock wave solutions are generally obtained for an ideal gas equation-of-state, they require a simple description of particle interactions in from of elastic scattering. Hereby, to facilitate a simple treatment of particle interactions we assign particle degrees of freedom f via the number of dimensions in the simulation. This affects the equation-of-state of the studies matter via its heat capacity ratio $\gamma = 1 + 1/f$ which becomes $\gamma = 2$ for a two dimensional simulation and $\gamma \sim 1.67$ in the case of three dimensions. It has already been demonstrated that the kinetic approach can reproduce the Sod and Noh shock problems [21]. The current work is focused on the Sedov blast wave test. While the Sedov test is a standard test for hydrodynamic codes it has not been discussed for kinetic schemes. The test describes an expanding spherical shock front and is therefore of special interest for codes which aim to simulate astrophysical problems, such as core-collapse supernovae. The initial setup of the Sedov test is a simulation space filled with matter at uniform density $n_{\rm in}$ and vanishing pressure. An explosion is set up via a point-like deposition of energy $E_{\rm blast}$ [22, 23]. The result is a spherical shock wave that moves outwards leaving behind matter with decreasing densities towards the center. While the bulk velocity of matter has a radial dependence $v_b \sim r/t$, with t being the time and r the radial distance from the center of the simulation space, the position of the blast wave can be written as:

$$r_{\rm shock} = \left(E_{\rm blast}\, t^2/\alpha\, n_{\rm in}\right)^{1/(2+d)}, \qquad (8)$$

with a peak density of $n(r) = n_{\rm in}\,(\gamma+1)/(\gamma-1)$. Here, d is the number of dimensions in the simulation and α is a constant of the order one. As the shock wave moves away from the center it leaves behind matter at vanishingly low densities. This can be a challenging problem for simulations operating with a finite number of particles since densities cannot become arbitrarily small [24]. Furthermore, as in all numerical studies, a general problem is the finite spatial resolution of a simulation which typically leads to a broadening of shock fronts. For the Sedov test, the latter impacts the density profile reducing its peak values. We perform the Sedov test in two dimensions with $N = 3.5 \times 10^7$ test-particles distributed randomly in 250×250 bins over $-0.04375 \leq x, y \leq 0.04375$. The blast wave is initialized around $x = y = 0$ with a radius of $r_{\rm blast} = \Delta x$. The mean-free-path is chosen to $\lambda = 0.001\,\Delta x$. While the particle velocities outside the blast region are initialized with absolute values of $v_{\rm in} = 3$, we deposit $E_{\rm blast}$ by

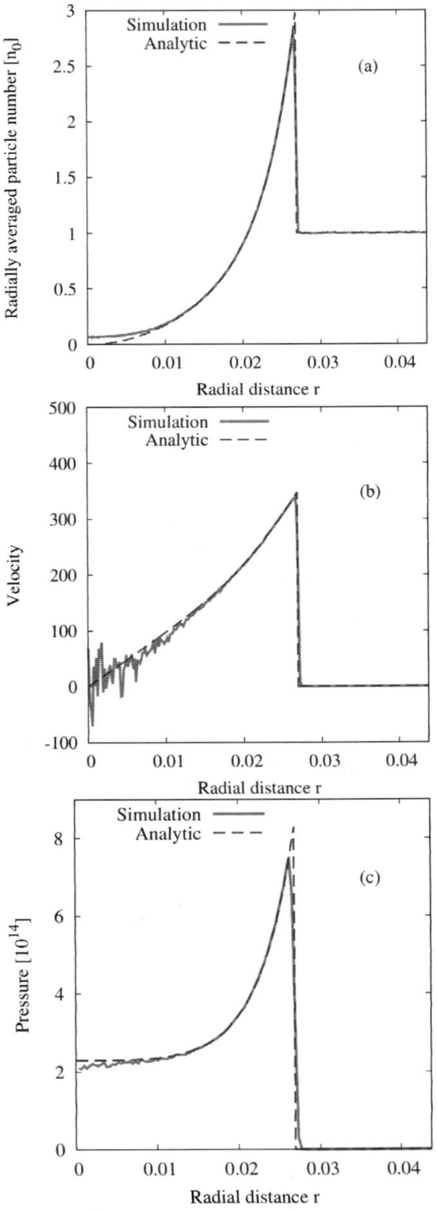

Figure 1. Radial profiles of the particle density (a), velocity (b), and pressure (c) in the Sedov test

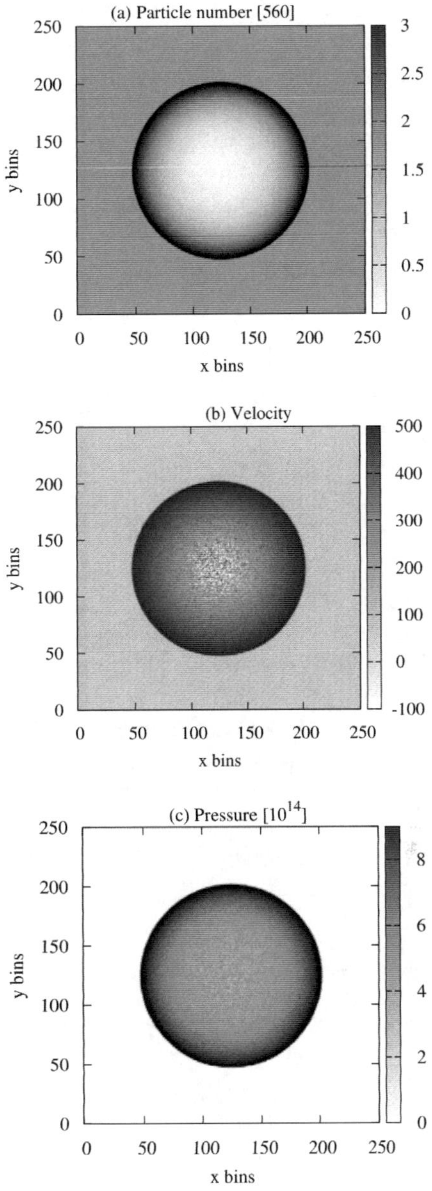

Figure 2. Number of particles (a), velocity (b), and pressure (c) per bin in the Sedov test

setting $v_{\text{in}} = 37180$ for particles with radial distance $r \leq r_{\text{blast}}$. For this study we chose an adaptive timestep with $\Delta t = \Delta x / v_{max}$, whereas v_{max} is the maximum absolute particle velocity at the current timestep. With a number of particles in the blast region of 1840 the resulting explosion energy is $E_{\text{blast}} \sim 1.27176 \times 10^{12}$. Figures 1(a)-(c) show the the density, radial velocity, and pressure profiles at timestep 420 with $t = 2.57877 \times 10^{-5}$, averaged over r together with the analytic solution. The latter is obtained by the publicly available *sedov3.f* code [25]. Figures 2(a)-(c) show the particle numbers, bulk velocity, and pressure per bin in the simulation space at the same timestep. We find that the density of matter at the shock front is slightly reduced. However, as previously explained, this is an expected result of simulation due to the finite spatial resolution. The particle number in the center of the simulations has higher values than the analytic solution which can be attributed to the finite minimal number of test-particles in one cell. The small value of the latter on the other hand leads to large fluctuations in radial velocity at the center of the simulation as seen in Figs. 1(b) and 2(b) as well as in the pressure profile shown in Fig. 1(c). The pressure is obtained from the stress tensor (see [21] for more details) via a radial average of the pressure per bin, shown in Fig. 2(c). Overall, we see a very good agreement between the analytic solution and the kinetic simulations.

5. Summary

In this work we present a transport code which combines the computational advantages of the Direct Simulation Monte Carlo technique with spatial resolution of the Point-of-Closest-Approach method to solve the collision integral of the Boltzmann equation. This enables us to simulate systems with a large number of test-particles in the hydrodynamic regime with a high spatial accuracy. To ensure that the hydrodynamic regime can be successfully reproduced we perform shock wave studies whereas the present work is focused on the Sedov shock wave test. To facilitate the comparison of our simulation to analytic solutions of shock wave propagation, we model particle interactions via elastic scatterings with the number of degrees of freedom given by the number of dimensions in the simulation. The current work verifies that our code is able to handle matter in the hydrodynamic regime and reproduce the Sedov shock with high accuracy. To our knowledge, there are no particle methods that addressed the Sedov shock test and our work represents the first of such studies. In the future we aim to apply our code to study the dynamics of core-collapse supernovae and matter in inertial confinement fusion capsules.

References

[1] Landau L and Lifshitz E 1975 *Fluid Mechanics* 1st ed (*Course of Theoretical Physics* vol 6) (Pergamon Press)
[2] Gingold R A and Monaghan J J 1977 *Monthly Notices of the Royal Astronomical Society* **181** 375–389
[3] Chapman S and Cowling T G 1970 *The mathematical theory of non-uniform gases* (Cambridge: University Press, 1970, 3rd ed., p. 47)
[4] Wong C Y 1982 *Physical Review C* **25** 1460–1475
[5] Boyd I D, Chen G and Candler G V 1995 *Physics of Fluids* **7** 210–219
[6] Yap Y W and Sader J E 2012 *Physics of Fluids* **24** 032004
[7] Bauer W, Bertsch G F, Cassing W and Mosel U 1986 *Physical Review C* **34** 2127–2133
[8] Aichelin J and Stöcker H 1986 *Physics Letters B* **176** 14–19
[9] Bouras I, Molnár E, Niemi H, Xu Z, El A, Fochler O, Greiner C and Rischke D H 2010 *Physical Review C* **82** 024910
[10] Kortemeyer G, Bauer W, Haglin K, Murray J and Pratt S 1995 *Physical Review C* **52** 2714–2724
[11] Casanova M, Larroche O and Matte J P 1991 *Physical Review Letter* **67**(16) 2143–2146
[12] Hernquist L and Ostriker J P 1992 *Astrophys. Journal* **386** 375–397
[13] Liebendörfer M, Rampp M, Janka H and Mezzacappa A 2005 *Astrophysical Journal* **620** 840–860
[14] Liebendörfer M, Whitehouse S C and Fischer T 2009 *Astrophysical Journal* **698** 1174–1190
[15] Abdikamalov E, Burrows A, Ott C D, Löffler F, O'Connor E, Dolence J C and Schnetter E 2012 *Astrophysial Journal* **755** 111
[16] Alexander F J, Garcia A L and Alder B J 1995 *Physical Review Letters* **74** 5212–5215
[17] Bird G A 1970 *Physics of Fluids* **13** 2676–2681
[18] Bertsch G F and Das Gupta S 1988 *Physics Reports* **160** 189–233
[19] Cugnon J, Mizutani T and Vandermeulen J 1981 *Nuclear Physics A* **352** 505–534
[20] Stöcker H and Greiner W 1986 *Physics Reports* **137** 277–392
[21] Sagert I, Colbry D, Strother T, Pickett R and Bauer W 2012 *ArXiv e-prints* (*Preprint* `1210.8084`)
[22] Taylor G 1950 *Royal Society of London Proceedings Series A* **201** 159–174
[23] Sedov L 1959 *Journal of Applied Mathematics and Mechanics* **23** 568–571
[24] Rosswog S and Price D 2007 *Monthly Notices of the Royal Astronomical Society* **379** 915–931
[25] Timmes F X URL `http://cococubed.asu.edu/code pages/sedov.shtml`

AUTHOR INDEX